# ASHRAE GreenGuide
## The Design, Construction, and Operation of Sustainable Buildings

This publication was developed under the auspices of the GreenGuide Subcommittee of TC 1.10, Energy Resources, and TC 2.8, Building Environmental Impacts and Sustainability. TC 1.10 and TG 2 BIE, Buildings' Impacts on the Environment, were merged in October 2002 to form TC 2.8.

## LIST OF CONTRIBUTORS

**John M. Swift, Jr.**
*Cannon Design, Boston, MA*

**Thomas Lawrence**
*University of Georgia, Athens, GA*

**H. Jay Enck**
*Commissioning & Green Building Solutions, Buford, GA*

**Malcolm Lewis**
*CTG Energetics, Irvine, CA*

**David L. Grumman,**
*Editor of the first edition of the* GreenGuide,
*Grumman/Butkus Associates, Evanston, IL*

**Neil Moiseev**
*Shen Milsom & Wilke, Inc., New York, NY*

**John Lane**
*Evapco, Inc., Taneytown, MD*

**John Andrepont**
*The Cool Solutions Company, Lisle, IL*

**Paul Torcellini and Michael Deru**
*National Renewable Energy Laboratory, Golden, CO*

**Wladyslaw Jan Kowalski**
*Pennsylvania State University, University Park, PA*

**Steven Rosen**
*Autodesk, Boston, MA*

**Jerry Ackerman**
*Clearwater Systems Corporation, Essex, CT*

**Jordan L. Heiman**
*St. Louis, MO*

**Mark Mendell, Michael Forth, and James Bones**
*Cannon Design, Boston, MA*

**Len Damiano**
*Green Building Controls Subcommittee Chair,
ASHRAE TC 1.4, and Ebtron*

**Kimberly Barker**
*ASHRAE TC 1.4, and Siemens*

**Bill Becker**
*Chicago ITT and Urban Wind Design, Chicago, IL*

**Bion Howard**
*Building Environmental Science and Technology,
Hilton Head, SC*

**Mark Hertel**
*ASHRAE TC 6.7, and SunEarth, Inc.*

**Constantinos A. Balaras**
*Institute for Environmental Research and
Sustainable Development,
National Observatory of Athens (NOA)*

**E. Mitchell Swann**
*MDC Systems Corp., LLC*

**Ainul Abedin**
*Past President, ASHRAE Pakistan Chapter*

**Brian A. Rock**
*School of Architecture and Urban Design,
The University of Kansas, Lawrence, KS*

**Amy Butterfield**
*Georgia Institute of Technology, Atlanta, GA*

**Michael Gallivan**
*Turner Construction Co., Inc.*

**Mark Hydemann and Glenn Friedman**
*Taylor Engineering, Alameda, CA*

**Ron Perkins**
*Supersymmetry USA, Navasota, TX*

**Vikas Patnaik and Mick Schwedler**
*Trane, Co., LaCrosse, WI*

**Hal Levin**
*Building Ecology Research Group, Santa Cruz, CA*

**Krishnan Gowri**
*Pacific Northwest Laboratory, Richland, WA*

**Gail S. Brager**
*University of California, Berkely, CA*

**Dean Borges**
*University of Nevada, Reno, NV*

**Paul McGregor**
*McGregor & Associates, Lake Cove, Australia*

**Brad Jones**
*Sebesta Blomberg, Boston, MA*

**David Bearg**
*Life Energy Associates, Newton, MA*

**Kevin Cross**
*Honeywell, Ft. Collins, CO*

**Karl Stum**
*Summit Building Engineering, LLC, Vancouver, WA*

**Guy S. Frankenfield**
*Natgun Corporation*

**Dean S. Borges**
*University of Nevada, Reno, NV*

**Eddie Leonardi**
*School of Mechanical and Manufacturing Engineering,
The University of New South Wales, Sydney, Australia*

# ASHRAE GreenGuide
## The Design, Construction, and Operation of Sustainable Buildings

AMERICAN SOCIETY OF HEATING, REFRIGERATING
AND AIR-CONDITIONING ENGINEERS, INC.

ELSEVIER

AMSTERDAM • BOSTON • HEIDELBERG • LONDON
NEW YORK • OXFORD • PARIS • SAN DIEGO
SAN FRANCISCO • SINGAPORE • SYDNEY • TOKYO
Butterworth-Heinemann is an imprint of Elsevier

ISBN 1-933742-07-0
ISBN 978-1-933742-07-6

©2006 American Society of
Heating, Refrigerating and
Air-Conditioning Engineers, Inc.
1791 Tullie Circle, NE
Atlanta, GA 30329
www.ashrae.org

Butterworth-Heinemann is
an imprint of Elsevier
30 Corporate Drive, Suite 400,
Burlington, MA 01803, USA

Cover design by Tracy Becker.

Library of Congress Cataloging-in-Publication Data

ASHRAE greenguide : the design, construction, and operation of sustainable buildings. -- 2nd ed.
    p. cm.
    Summary: "The ASHRAE GreenGuide was developed primarily to provide guidance to designers of HVAC&R systems in how to participate effectively on design teams charged with producing green buildings"--Provided by publisher.
    ISBN 13: 978-1-933742-07-6 (hardcover)
    1. Sustainable buildings--Design and construction. 2. Sustainable architecture. 3. Buildings--Environmental engineering. I. American Society of Heating, Refrigerating and Air-Conditioning Engineers. II. Title: ASHRAE green guide.

TH880.A83 2006
720'.47--dc22
                    2006029914

# ASHRAE STAFF

### SPECIAL PUBLICATIONS

**Mildred Geshwiler**
*Editor*

**Christina Helms**
*Associate Editor*

**Cindy Sheffield Michaels**
*Assistant Editor*

**Michell Phillips**
*Administrative Assistant*

### PUBLISHING SERVICES

**David Soltis**
*Manager*

**Tracy Becker**
*Graphic Applications Specialist*

**Jayne Jackson**
*Publication Traffic Administrator*

### PUBLISHER

**W. Stephen Comstock**

## Tomorrow's Child

Without a name, an unseen face,
And knowing not the time or place,
Tomorrow's Child, though yet unborn,
I saw you first last Tuesday morn.
A wise friend introduced us two,
And through his shining point of view
I saw a day, which you would see,
A day for you, and not for me.
Knowing you has changed my thinking,
Never having had an inkling
That perhaps the things I do
Might someday threaten you.
Tomorrow's Child, my daughter-son,
I'm afraid I've just begun
To think of you and of your good,
Though always having known I should.
Begin I will to weigh the cost
Of what I squander, what is lost,
If ever I forget that you
Will someday come to live here too.

by Glenn Thomas, ©1996

Reprinted from *Mid-Course Correction: Toward a Sustainable Enterprise: The Interface Model* by Ray Anderson. Chelsea Green Publishing Company, 1999.

# CONTENTS

# SECTION 2: THE DESIGN PROCESS

# GREENTIPS

## BUILDING-TYPE GREENTIPS

## SIDEBARS

# FOREWORD

by William Coad

---

*Mechanical engineering* has been defined as "the applied science of energy conversion." ASHRAE is the preeminent technical society, representing engineers practicing in the fields of heating, refrigeration, and air conditioning, the technology that utilizes approximately one-third of the global nonrenewable energy consumed annually.

ASHRAE membership has actively pursued more effective means of utilizing these precious nonrenewable resources for many decades from the standpoints of source availability, efficiency of utilization, and technology of substituting with renewable sources. One significant publication in *ASHRAE Transactions* is a paper authored in 1951 by G.W. Gleason, Dean of Engineering at Oregon State University, titled "Energy—Choose it Wisely Today for Safety Tomorrow." The flip side of the energy coin is the environment and, again, ASHRAE has historically dealt with the impact that the practice of the HVAC&R sciences have had upon both the indoor and the global environment.

However, the engineering community, to a great extent, serves the needs and desires of accepted economic norms and the consuming public, a large majority of whom have not embraced the energy/environmental ethic. As a result, much of the technology in energy effectiveness and environmental sensitivity that ASHRAE members have developed over this past century has had limited impact upon society.

In 1975, when ASHRAE published Standard 90-75, that standard served as our initial outreach effort to develop an awareness of the energy ethic and to extend our capabilities throughout society as a whole. Since that time updated revisions of Standard 90 have moved the science ahead. In 1993, the chapter on "Energy Resources" was added to the *ASHRAE Handbook—Fundamentals*. In 2002 ASHRAE entered into a Partnering Agreement with the US Green Building Council, and it is intended that the second edition of this design guide will continue to assist the organization in their efforts at promoting sustainable design, as well as the many other organizations that have advocated for high-performance building design.

The consuming public and other representative groups of building professionals continue to become more and more aware of the societal need to provide buildings that are more energy resource effective and environmentally compatible. This publication, authored and edited by ASHRAE volunteers, is intended to complement those efforts.

ASHRAE will continue to advance its leadership through initiatives such as "The Sustainability Roadmap." Information on this effort can be found on ASHRAE's Engineering for Sustainability Web site, www.engineeringfor sustainability.org/.

# PREFACE TO THE SECOND EDITION
by John Swift

When the first edition of the *ASHRAE GreenGuide* was developed, it was intended that the Guide would be a continuous work in progress. This second edition of the Guide fulfills that intent and represents ASHRAE's continued commitment to leadership in the areas of high-performance building design and operation.

The second edition of the *ASHRAE GreenGuide* includes a new chapter on LEED Guidance for Mechanical Engineers and a new chapter on building systems' impact on the local environment, both indoor and outdoor. There are 20 new ASHRAE GreenTips, including a new version of the GreenTip that focuses on specific building types. Some of the chapters from the first edition have been reorganized in an attempt to more accurately mirror the path that an actual project would take from pre-design to post-occupancy. Content has been added and edited in all of the chapters, with significant updates in the subject areas of building automation systems, renewable energy options, CHP and GSHP systems, and construction issues. Dual units have been provided, as well as more international HVAC engineering representation in the editorial process. Graphs, photographs, renderings, and diagrams have been added where necessary to provide a more complete overview of specific subject matter. In addition, references have been added and updated in order to make the second edition current as of its date of publication.

The *ASHRAE GreenGuide* is primarily for HVAC&R designers, but it will also be a useful reference for architects, owners, building managers, operators, contractors, and others in the building industry who want to understand some of the technical issues regarding high-performance design from an integrated, buildings systems perspective. Considerable emphasis is placed on teamwork and close coordination between parties.

## HOW TO USE THIS *ASHRAE GREENGUIDE*

The Guide is not intended to be "the last word" on the technical aspects of green design nor, for that matter, a design guide proper. Throughout the Guide, numerous techniques, processes, measures, or special systems are described succinctly in a modified outline or bullet form, always in the same format. These are called ASHRAE GreenTips. Each GreenTip concludes with a listing of other sources—books, magazine articles, research papers, organizations, Web sites—that may be referenced for greater detail.

This document is intended to be used more as a *reference* than as something one would read in sequence from beginning to end. The table of contents is the best place for any reader to get an overall view of what is covered in this document. All readers should take the time to read Chapter 1, "Green/Sustainable High-Performance Design," which provides some essential definitions and meanings of key terms. Chapter 2, "Background and Fundamentals," might well be skipped by the more experienced designer-readers. This chapter covers the background of the green design movement and what other organizations have done, and it reviews some engineering fundamentals that govern the technical aspects of green design. Chapter 3 provides an introduction to the commissioning process, a critical component that needs to be addressed *from the beginning* on all truly successful high-performance building projects, and Chapter 4 covers "Architectural Design Impacts." Chapter 5, "The Design Process—Early Stages," is essential reading for all who are interested in how the green design process works. Chapter 6 provides an overview of the LEED certification process for mechanical engineers.

The nitty-gritty engineering aspects start in Chapter 7 and run through Chapter 16. This is where the reader will find virtually all the practical suggestions for possible incorporation in a green design, the ASHRAE GreenTips. Chapters 17–18 cover what happens after the project's design is done—that is, during construction and after. There are some sound advice and helpful tips in that section, and even though it covers a post-design time frame, *reading that section should not be put off until construction begins*.

The last section of the book, the Afterword, is the Preface to the First Edition written by David Grumman, editor of the first edition. The hard work and leadership that David Grumman, Sheila Hayter, Jordan Heiman, and many other ASHRAE volunteers provided produced a solid foundation on which to build this and subsequent editions of the *ASHRAE GreenGuide*. Their commitment continues with the new group of contributors to this second edition.

# ACKNOWLEDGMENTS

The following individuals provided written materials and editorial content and formed the Senior Editorial Group of the ASHRAE TC 2.8 *GreenGuide* Subcommittee for the second edition:

**John M. Swift, Jr.**
*Cannon Design, Boston, MA*

**Thomas Lawrence**
*University of Georgia, Athens, GA*

**H. Jay Enck**
*Commissioning & Green Building Solutions, Buford, GA*

**Malcolm Lewis**
*CTG Energetics, Irvine, CA*

The following individuals contributed written materials and/or editorial comments on various topics for the second edition of the *ASHRAE GreenGuide*. All or portions of these contributions have been incorporated, with editing.

**David L. Grumman**
*Editor of the first edition of the* GreenGuide,
*Grumman/Butkus Associates, Evanston, IL*

**Neil Moiseev**
*Shen Milsom & Wilke, Inc., New York, NY*

**John Lane**
*Evapco, Inc., Taneytown, MD*

**John Andrepont**
*The Cool Solutions Company, Lisle, IL*

**Paul Torcellini and Michael Deru**
*National Renewable Energy Laboratory, Golden, CO*

**Wladyslaw Jan Kowalski**
*Pennsylvania State University, University Park, PA*

**Steven Rosen**
*Autodesk, Boston, MA*

**Jerry Ackerman**
*Clearwater Systems Corporation, Essex, CT*

**Jordan L. Heiman**
*St. Louis, MO*

**Mark Mendell, Michael Forth, and James Bones**
*Cannon Design, Boston, MA*

**Len Damiano**
*Green Building Controls Subcommittee Chair, ASHRAE TC 1.4, and Ebtron*

**Kimberly Barker**
*ASHRAE TC 1.4, and Siemens*

**Bill Becker**
*Chicago ITT and Urban Wind Design, Chicago, IL*

**Bion Howard**
*Building Environmental Science and Technology, Hilton Head, SC*

**Mark Hertel**
*ASHRAE TC 6.7, and SunEarth, Inc.*

**Constantinos A. Balaras**
*Institute for Environmental Research & Sustainable Development,*
*National Observatory of Athens (NOA)*

**E. Mitchell Swann**
*MDC Systems Corp., LLC*

**Ainul Abedin**
*Past President, ASHRAE Pakistan Chapter*

**Brian A. Rock**
*School of Architecture and Urban Design, The University of Kansas, Lawrence, KS*

**Amy Butterfield**
*Georgia Institute of Technology, Atlanta, GA*

**Michael Gallivan**
*Turner Construction Co., Inc.*

**Mark Hydemann and Glenn Friedman**
*Taylor Engineering, Alameda, CA*

**Ron Perkins**
*Supersymmetry USA, Navasota, TX*

**Vikas Patnaik and Mick Schwedler**
*Trane, Co., LaCrosse, WI*

**Hal Levin**
*Building Ecology Research Group, Santa Cruz, CA*

**Krishnan Gowri**
*Pacific Northwest Laboratory, Richland, WA*

**Gail S. Brager**
*University of California, Berkely, CA*

**Dean Borges**
*University of Nevada, Reno, NV*

**Paul McGregor**
*McGregor & Associates, Lake Cove, Australia*

**Brad Jones**
*Sebesta Blomberg, Boston, MA*

**David Bearg**
*Life Energy Associates, Newton, MA*

**Kevin Cross**
*Honeywell, Ft. Collins, CO*

**Karl Stum**
*Summit Building Engineering, LLC, Vancouver, WA*

**Guy S. Frankenfield**
*Natgun Corporation*

**Dean S. Borges**
*University of Nevada, Reno, NV*

The editorial staff of Special Publications at ASHRAE provided significant contributions. Christina Helms, Cindy Michaels, Micki Geshwiler, and Steve Comstock provided great support in this effort. In addition, the incoming president of ASHRAE, Terry Townsend, championed the effort and provided motivation to all ASHRAE members to contribute.

The *GreenGuide* Subcommittee of ASHRAE Technical Committee TC 1.10, Energy Resources, was responsible for creating the first edition of this Guide. (Just prior to its completion, TC 1.10 merged with Task Group BIE, Buildings' Impact on the Environment, to form TC 2.8, Building Environmental Impact and Sustainability.) Members of that subcommittee were David L. Grumman, Fellow ASHRAE,

Chair; Jordan L. Heiman, Fellow ASHRAE; and Sheila Hayter, Chair of TC 1.10. Sheila Hayter created the subcommittee and initiated the initial discussions and meetings. Jordan Heiman was responsible for identifying authors, carrying on most communications with them, and creating the bibliography. David Grumman was responsible for creating the topic format, assembling the various chapters, and editing the document.

Prior to approval by TC 2.8, the document was reviewed by a three-person panel consisting of Theodore Pannkoke; William Coad, Fellow ASHRAE, Presidential Member; and Thomas Cappellin—all members of TC 2.8.

The idea for this publication was initiated by 1999–2000 ASHRAE President Jim Wolf and carried forward by then President-Elect (and, subsequently, President) William J. Coad.

All work performed—by the authors, editors, developing subcommittee, review panel, and TC participants—was voluntary.

# Section 1: Basics

# 1

# GREEN/SUSTAINABLE
# HIGH-PERFORMANCE DESIGN

## INTRODUCTION

In recent years, much information has been put forth about the impact of the built environment (e.g., buildings) on the natural environment. This information has been both written and spoken, and there have been conferences and seminars on the subject and organizations have sprung up devoted specifically to this issue. Not only have the messages contained in this outpouring of information attempted simply to explain what this issue is, but they have variously promoted the concept of "green" design, exhorted to action, strived to motivate, warned of consequences from ignoring it, and instructed how to do it.

While this vast amount of promotion has often been helpful, much has been either largely irrelevant or simply not useful to the practicing designer of HVAC&R systems and equipment for buildings (i.e., to the ASHRAE member involved on a day-to-day basis in the mechanical/electrical building system design process). Based on input received from grassroots ASHRAE members, a need was felt for guidance on the green-building concept *specifically directed toward such practitioners*. A desire was also expressed that it contains information of direct practical use. This Guide is an attempt to meet that need.

*Green* is one of those words that can have more than a half-dozen meanings, depending on circumstances. One of these is the greenery of nature (grass, trees, and leaves). It is this reference to nature—symbolic, if you will—that is the meaning this term denotes in this publication. While not all things in nature are green, we believe that the term *green* serves as a fitting verbal symbol of the concept and practices this Guide strives to promote. While *green* is a fitting symbol, it does not completely encompass the full meaning of sustainability, which is maintaining ecological balance.

The difference between a *green* and *sustainable* design is the degree to which the design helps to maintain this ecological balance. Some characteristics of green design have no impact in terms of maintaining ecological balance, including indoor environmental quality (IEQ), an important element of green design. Many green design characteristics, such as reduced energy usage and pollution, do have positive long-term effect. This Guide contains green and sustainable design elements but is not intended to cover the full breadth of sustainability.

A design that is green/sustainable is a design that minimizes the negative human impacts on the natural surroundings, materials, resources, and processes that prevail in nature. It is not necessarily a concept that denies the need for any human impact, for human existence is part of nature too. Rather, it endorses the belief that human-kind can exist, multiply, build, and prosper in accord with nature and the earth's natural processes without inflicting irreversible damage to those processes and the long-term habitability of the planet.

Our definition of *green buildings* inevitably extends beyond the concerns of HVAC&R designers alone, since the very concept places an emphasis on integrated design of mechanical, electrical, architectural, and other systems.

Specifically, the view of this chapter's authors is that *a green/sustainable building design is one that achieves high performance, over the full life cycle, in the following areas*:

- Minimizing natural resource consumption through more efficient utilization of nonrenewable natural resources, land, water, and construction materials, including utilization of renewable energy resources to achieve net zero energy consumption.
- Minimizing emissions that negatively impact our indoor environment and the atmosphere of our planet, especially those related to indoor air quality (IAQ), greenhouse gases, global warming, particulates, or acid rain.
- Minimizing discharge of solid waste and liquid effluents, including demolition and occupant waste, sewer, and stormwater, and the associated infrastructure required to accommodate removal.
- Minimal negative impacts on site ecosystems.
- Maximum quality of indoor environment, including air quality, thermal regime, illumination, acoustics/noise, and visual aspects to provide comfortable human physiological and psychological perceptions.

It should be noted that the above five bullets are compatible with the definitions of other organizations.

Ultimately, even if a project does not have stated green/sustainable goals, the overall approaches, processes, and concepts presented in this Guide provide a design philosophy that can be useful for any project. Using the principles of this Guide, an owner or a member of his or her team can document the objectives and criteria to include in a project, forming the foundation for a collaborative integrated project delivery approach, which can lower design, construction, and operational costs, resulting in a lower total cost for the life of the project.

## RELATIONSHIP TO SUSTAINABILITY

The related term *sustainable design* is very commonly used, almost to the point of losing any consistent meaning. While there have been some rather varied and complex definitions put forth (see sidebar titled "Some Definitions and Views of Sustainability from Other Sources"), we prefer a simple one (very similar to the third one in the sidebar). *Sustainability* **is "providing for the needs of the present without detracting from the ability to fulfill the needs of the future."**

While this is a simple and good general definition when applied to planet Earth as a whole, it is difficult to apply it in a meaningful way, without being arbitrary, to an individual earthly component (building, automobile, industrial plant, oil field) on the planet. Ultimately, sustainability of the planet depends on the *collective contribution to sustainability* of the various elements of the planet; but who is to say just how much each element should contribute?

The above discussion suggests that the concepts of "green design" and "sustainable design" have no absolutes—that is, they cannot be defined in black-and-white terms. These terms are more useful when thought of as a mindset—a goal to be sought, a process to follow. This Guide is a means of inducing designers of the built environment to strategies that can be utilized in developing a "green/sustainable design" and set forth some practical techniques to help practitioners achieve the goal of green design and, thus, make a significant contribution to Earth's sustainability.

## "GOOD" DESIGN

*Good design* might be said to be the process that results in a well-designed building.

Does *good* design intrinsically mean that green design has been achieved as well? More significantly, does green design automatically incorporate the characteristics of good design? It is important to clarify this question for users of this Guide because many definitions of green design *do* assume that it includes at least some, if not all, of the characteristics of good design (Grondzik 2001).

The broad characteristics of good building design, encompassing both the engineering and non-engineering disciplines, might be briefly defined as meeting the defined objectives and criteria of the owner. As owners expand their criteria to include green and sustainable objectives, good designs can transition to green and sustainable designs.

Thus, the authors choose to make the distinction between the characteristics of good design and green design—but at the same time **strongly advocate that buildings should strive to achieve both**. In summary, green design does not necessarily incorporate many important characteristics of good design; but good design, on the other hand, does not yet include those green design characteristics that this Guide strives to promote—though many would say it should.

## COMMITMENT TO GREEN/SUSTAINABLE HIGH-PERFORMANCE PROJECTS

Green projects require more than a project team with good intentions; they require commitment from the owner, early documentation of sustainable/green goals documented by the Owner's Project Requirement (OPR) document, and the designer's documented basis of design. The most successful projects incorporating green design are ones with dedicated, proactive owners who are willing to examine the entire spectrum of ownership—from design to construction to long-term operation of their facilities. These owners understand that green buildings require more planning, better execution, and better operational procedures, requiring a firm commitment to changing how building projects are designed, constructed, operated, and maintained to achieve a lower total cost of ownership and lower long-term environmental impacts.

Implementing green/sustainable practices does raise the initial design "soft" costs associated with a project. Additional design services, commissioning, and certain green features may add as much as 0.7% to 2% (varies based on project size) of total project cost to the front end. Implementing the commissioning process early in the predesign phase of a project adds to the initial budget item but, in most cases, actually reduces costs in excess of the additional budget cost before the end of design with additional savings before the end of construction. Experienced practitioners of green design have also found that this investment, more often than not, has led to lower total project costs before construction is completed, resulting in a net savings.

In addition, significant savings and improved productivity of the building occupants can be realized for the life of the building, lowering the total cost of ownership. To achieve life-long benefits also requires operating procedures for monitoring performance, making adjustments (continuing commissioning) when needed, and appropriate maintenance.

## WHAT DRIVES GREEN PROJECTS

Green building advocates can cite plenty of reasons why buildings *should* be designed "greenly." (See "Justifications for Green Design" on page 14.) The fact that these reasons exist does not make it happen, nor does the existence of designers—or design firms—with green design experience. *The main driver of green building design is the motivation of the owner*—the one who initiates the creation of a project, the one who pays for it (or who carries the burden of its financing), the one who has (or has identified) the need to be met by the project in question. If the owner does not believe that green design is needed, thinks it is unimportant, or thinks it is of secondary importance to other needs, then it will not happen.

Some owners of new and existing buildings have already discovered the benefits of commissioning without any intent to "green" their new or existing buildings. Owners with no commissioning experience often believe that their design and construction teams will provide a quality project without the commissioning process. What these owners often receive is a project where most of the materials and equipment were installed and have the appearance of operating as intended, only to discover after the end of the warranty period that the systems do not perform as expected, leaving the owner with the impacts of poor building performance due to higher operation and maintenance costs and reduced occupant satisfaction and productivity, all of which significantly increase the total cost of ownership and environmental impacts.

The commissioning process plays a key role in assisting a project delivery team and owner to reduce project risk, lower design and construction costs, and reduce the total cost of ownership. Beginning in the predesign stage, documenting the objectives, criteria, and basis of design places the delivery team on a solid foundation for success, providing clear direction and a benchmark for success against which the design, construction, and operation efforts will be judged. This is particularly important in "green" projects when clear objectives and criteria are needed to alert the design team of the owner's commitment to sustainability and a green building.

In the very early stages of a building's development—perhaps during the designer interview process or before a designer has even been engaged—an owner may become informed on the latest trends in building design. This may occur through an owner doing research, conferring with others in the field, or discussing the merits of green design with the designer/design firm the owner intends to hire.

This initial interaction between the owner and the design professional is where the design firm with green design experience can be very effective in turning a project not initially so destined into one that's a candidate for green design. When an owner engages a designer, it is because the owner has faith in the professional ability of that designer and is inclined to listen to that designer's ideas on what the building's design direction and themes should be. As the designer works with the owner to meet their defined objectives and criteria of the project, the designer has the opportunity to identify approaches that can meet those objectives and criteria in a green/sustainable way. Thus, *designers should regard the very early contacts with a potential owner as a golden opportunity to steer the project in a green direction for the benefit of the owner and the planet.*

## INGREDIENTS OF A SUCCESSFUL GREEN PROJECT ENDEAVOR

The following ingredients are essential in delivering a successful green design:

- **Commitment** from the entire project team, starting with the owner.
- **Establishing Owner's Project Requirements (OPRs), including green design goals,** early in the design process.
- Integration of team ideas.
- **Effective execution** throughout the project's phases—from predesign through the end of its useful service life.

### Establishing Green Design Goals Early

Establishing goals early in the project planning stages is key to developing a successful green design and minimizing costs. It is easy to say that goals need to be established, but many designers and owners struggle with what green design is and what green/sustainable goals should be established. Some typical questions are: What does it cost to design and construct a green project? Where do you get the best return for the investment? How far should the team go to accomplish a green design?

Today there are many guides a team can use with ideas on which green/sustainable principles should be considered. Chapter 5, "The Design Process—Early Stages," presents several rating systems and references on environmental performance improvement. The essence of these documents is guidance on how to reduce the impact the building will have on the environment. While the approaches and goals contained in each differ, all suggest common principles that designers may find helpful to apply to their particular projects.

### Integration of Team Ideas

No green project will be successful if the various project stakeholders are not included in the process. These stakeholders include the owner, their operations staff, commissioning authority (CxA), design disciplines, contractors, and users. These stakeholders, if known, should work in close coordination, beginning in the earliest stages. Use the commissioning process, as discussed in Chapter 3, to obtain input from the various stakeholders and develop the OPR document that defines the owner's objectives and criteria, including the green/sustainable project goals that set the foundation for what the team is tasked with delivering but also benchmark for success. Based on the owner's stated objectives and criteria, the integrated team works together with a clear direction and focus to deliver the project. No longer can the mechanical, electrical, and plumbing engineers and landscape architects become involved only after the building's form and space arrangements are set (i.e., end of schematic design); that is far too late for the necessary cross-pollenization of ideas between engineering and architectural disciplines that must occur for green design to be effective.

### Effective Execution

Still, all the good intentions an owner and project team may have during the early stages are meaningless without follow-through, not only during the construction process, but thereafter over the entire life of the facility by those responsible for operation and maintenance. Thus, ideally, success requires a committed owner, use of the commissioning process, the owner's defined objectives and criteria, including green/sustainable project goals established in predesign, knowledgeable design practitioners working in concert to develop a green/sustainable design, competent contractors who buy into the green concept, and, finally, operators armed with the necessary tools and who are properly trained and dedicated to keeping the facility operating at peak performance for its life.

## THE ENGINEERING/ENERGY CONSERVATION ETHIC

Since the 1973 oil embargo, the HVAC&R industry has continued to improve the efficiency of air-conditioning systems and equipment, promulgated energy conservation standards, developed energy-efficient designs, experimented with a wide variety of design approaches, strived for good IAQ, and shared the lessons learned with industry colleagues. As in the past, efforts must be continued to find new and better solutions to improve energy efficiency, further reduce our dependence on nonrenewable energy sources, and increase the comfort of people in the buildings they occupy.

Most designers have guided owners through life-cycle analyses of various options, identified approaches to improve building efficiency, and developed strategies to meet the stated goals of owners. They have also had owners reject their ideas because of too-long payback periods due to project budget constraints. Despite those setbacks, progress toward green/sustainable design is becoming more prevalent, moving to become an industry standard practice. It is thus incumbent on our industry to recognize the impact its work has on the environment, which goes beyond matters of first cost, recurring cost, and even life-cycle cost. The ethic of the industry requires its practitioners to strive to identify these environmental costs and assign values to them, values that represent the total cost to society rather than just conventional measurements of capital.

The building industry must strive to inform both itself and its clients about the value of natural capital, and it must begin to base decisions on other than just the traditional metrics of financial, manufactured, and human capital. It took nature a very long time—3.8 billion years—to generate the earth's natural capital. The amazing fact is that, in the last 200 years, humans have consumed most of this natural capital while, at the same time, Earth's population has grown exponentially (Hawken et al. 1999). As a result, countries are positioning to secure and protect their economies and the resources needed to support their growth, resulting in escalating project costs. Society, led by the building industry, must develop more efficient use of our resources, including recycling of demolition materials, in order to meet the needs of the future.

## INCENTIVES

For both individuals and firms in the design and construction profession, many incentives exist to develop green/sustainable projects. As with any aspect of business practice that adds value to a project, fees and client expectations must be carefully managed. While a client may balk at added fees charged for the commissioning, additional coordination, and studies necessary to meet green/sustainable goals for systems that are part of a traditional project scope, some clients welcome additional services of the right type. First, the appetite of the project client and full project team for a green/sustainable project must be gauged; then the commensurate level of commissioning, design, construction, and operational services can be provided. The OPRs, defined early and documented in writing, facilitate such understandings, providing a more defined initial fee proposal and acceptance.

Green project capabilities can positively impact building professionals' careers and the firms that employ them. Firms can enhance these capabilities by providing leadership on green issues, building individual competencies, providing ongoing support for professional development in relevant areas, rewarding accomplishments, marketing or promoting green success stories, and building their clientele's interest in green/sustainable design.

When properly delivered, green design/construction/operation capabilities can enhance service to clients, build repeat business, obtain public relations and marketing value, and increase demand for these services, especially among the architects or owners that represent a substantial proportion of many firms' billings. In addition, many businesses can better retain employees and raise their satisfaction level. Finally, green/sustainable project competency can reduce risks in practice: knowledge of green issues is necessary to manage risk when participating in aggressively green projects.

Some projects may lend themselves to the incorporation of green concepts without incurring additional costs beyond the business's own investment in the knowledge of, and experience with, green projects. Other sustainable design projects may require additional services for a greater number of design considerations not customarily included in ordinary design fees, such as extended energy analysis, daylighting penetration, and solar load analysis, or extended review (of materials and components, environmental impacts such as embodied energy, transportation, construction, composition, and IAQ performance). Other categories of additional services may include project-specific research or training of the engineering team where required and client education and communication.

If the project team engineer can demonstrate how these green/sustainable attributes can benefit the owner so as to include them into the green/sustainable goals defined in the OPRs, then an opportunity exists for providing additional services that can benefit the project in terms of specific positive results. With such goals accepted by the owner, these services can be proposed, and hopefully accepted, at the outset of the project. Owners then receive the desired level of service from each person needed to be successful in delivering a green/sustainable project. Allowing the project to benefit, as has been demonstrated on many projects, where case studies are available, green/sustainable design has resulted in lower construction and long-term operating and ownership costs. The owner who understands these savings is more likely to accept the short-term costs of additional services in pursuit of the lower project costs and long-term benefits of optimized life-cycle cost. If, due to a designer's enhanced engineering capabilities, the need to include one or more ancillary consultants on a design team is eliminated or reduced, then the short-term benefit to the client may be quite clear. Even when a sustainability or environmental consultant is added to the project team, there remain opportunities to extend engineering scope within the limits of the competency and experience of the engineers on the design team.

Many forms of incentives can be conferred on a project through judicious selection and pursuit of sustainable design goals. Today's HVAC&R professionals can add value to their offerings and distinguish themselves in the marketplace by leveraging external incentives to reduce clients' project costs. Sustainable engineering design also provides opportunities to develop internal incentives to reinforce green goals; such incentives can help the design team make green choices throughout the design and construction process. These two broad categories of incentives, external and internal, can each encompass both direct and indirect financial benefits and nonfinancial advantages.

The familiar range of energy incentives offered by many utility-administered, state-mandated, demand-side management programs represent one obvious form of external financial incentive—the rebate check available to owners or utility ratepayers who pursue energy conservation measures in buildings and mechanical/electrical systems. Of course, the availability of these and other financial incentives varies from place to place and project to project. But many other external financial incentives can reduce project costs where applicable, such as:

- sustainable design tax credits (now offered by New York)
- marketable emissions credits
- tax rebates
- brownfield funds
- historic preservation funds
- community redevelopment funds
- economic development funds
- charitable foundation funds

The European Union's support programs have, in the past and currently under the Intelligent Energy-Europe program (EIE), provided additional funding sources for projects that support the promotion of energy efficiency and use of alternative energy sources in buildings. For example, the eco-buildings projects aim at a new approach for the design, construction, and operation of new and/or refurbished buildings, which is based on the best combination of the double approach: to reduce substantially, and, if possible, to avoid, the demand for heating, cooling, and lighting and to supply the necessary heating, cooling, and lighting in the most efficient way and based as much as possible on renewable energy sources and polygeneration. Information on eco-building demonstration projects throughout Europe, supported by the European Commission (DG TREN), is available at www.sara-project.net.

Incentives internal to a project itself can be created through contractual arrangements. The simplest example is an added fee for added scope, as discussed above.

At the other end of the spectrum, some proactive firms offer design review and commissioning services on a performance basis. Construction project insurance offerings have included "rebates" if no IAQ claims are made a predetermined number of years after construction is completed. If both client and firm are willing, there are many ways to create internal incentives for green engineering design on a project-specific basis. (See further discussion on incentive fees in Chapter 16.)

Many individuals and groups have done valuable work in exploring various incentives for green design. The *ASHRAE Journal* and many other industry periodicals carry case studies with green engineering elements. Assembling and familiarizing oneself with the many successfully completed green projects provides useful tools for selling and delivering green engineering services.

## BUILDING TEAM SPIRIT

The essence of building team spirit is acknowledging that everyone on the team is important to the success of a project, from the visionaries on the design team to the laborers at the construction site. Each contributor must have pride in his or her work and needs to feel the efforts put forth are valued and appreciated. Sustainable development requires buy-in by all parties, from the owner initiating the project, to the CADD operator in the design office, to the installing subcontractors, to the maintenance person keeping the facility functioning—each plays an indispensable role. Ultimately the owner is the most important member of the team. It is important to understand that within the owner's organization there are often different departments or staff that focus on user activities, design and construction, or operation and maintenance and do not necessarily communicate or are invited to communicate their objectives and criteria to the design and construction team. Success of the project and the long-term benefits of established green/sustainable goals can be easily lost if all of the owner's staff is not included. For example, operators with insufficient training and understanding of the project's intended operation can inadvertently negate the benefits through improper operation.

### International Perspective

**REGULATIONS AND COMMENTARY**

Society has recognized that previous industrial and developmental actions caused long-term damage to our environment, resulting in loss of food sources and plant and animal species and changes to the earth's climate. As a result of learning from past mistakes and studying the environment, the international community identified certain actions that threaten our ecosystem's biodiversity, and, consequently, it developed several governmental regulations designed to protect our environment. Thus, in this sense, the green design initiative began with the implementation of building regulations. An example is the regulated phasing out of fully halogenated chlorofluorocarbons (CFCs) and partially halogenated refrigerant hydro-chlorofluorocarbons (HCFCs).

In Europe, the Directive on the Energy Performance of Buildings (EPBD) (European Commission, Energy Performance of Buildings, Directive 2002/91/EC of the European Parliament and of the Council; Official Journal of the European Communities, Brussels 2002) has been in effect since January 4, 2006, throughout the European Union (EU). All EU member states are obligated to bring into force national laws, regulations, and administrative provisions for setting minimum requirements on the energy performance of new and existing buildings that are subject to major renovations and for energy performance certification of buildings. Additional requirements include regular inspection of building systems and installations, an assessment of the existing facilities, and provision of advice on possible improvements and alternative solutions. The objective is to properly design new buildings and renovate existing buildings in a manner that will use the minimum nonrenewable energy, produce minimum air pollution as a result of the building operating systems, and minimize construction waste, all with acceptable investment and operating costs, while improving the indoor environment for comfort, health, and safety.

An energy performance certificate (EPC) will be issued when buildings are constructed, sold, or rented out. The EPC will document the energy performance of the building, expressed as a numeric indicator that allows benchmarking. The certificate will also include recommendations for cost-effective improvement of the energy performance, and it will be valid for up to ten years.

According to EPBD, minimum energy performance requirements are set for new buildings and for major renovation of large existing buildings in each EU member state. Energy performance will have to be upgraded in order to meet minimum requirements that are technically, functionally, and economically feasible. In the case of large new buildings, alternative energy supply systems should be considered, for example, decentralized energy supply systems based on renewable energy, combined heat and power, district or block heating or cooling, heat pumps, etc. The inspection of boilers and air-conditioning units on a regular base is compulsory. To further promote the efforts, public buildings will have to display the EPC visible to the public.

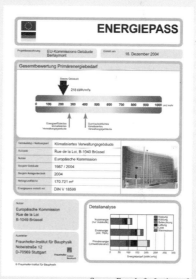

Source: Fraunhofer Institute of Building Physics, Germany.

**The energy performance certificate awarded to the European Commission's HQs "Berlaymont" building, Brussels, Belgium.**

## Justifications for Green Design

### DOING THE RIGHT THING

The motivations and reasons for implementing green buildings are diverse but can be condensed into essentially wanting to do the right thing to protect the earth's resources. For some, a wake-up call occurred in 1973 with the oil embargo—and with it a realization that there may be a need to manage our planet's finite resources.

### REGULATIONS

Society has recognized that previous industrial and developmental actions caused long-term damage to our environment, resulting in loss of food sources and plant and animal species and changes to the earth's climate. As a result of learning from past mistakes and studying the environment, the international community identified certain actions that threaten our ecosystem's biodiversity—and, consequently, it developed several governmental regulations designed to protect our environment. Thus, in this sense, the green design initiative began with the implementation of building regulations. An example is the regulated phasing out of CFCs.

### LOWERING OWNERSHIP COSTS

A third driver for green design is lowering the total cost of ownership in terms of construction costs, resource management and energy efficiency, and operational costs, including marketing and public relations. Examples include providing a better set of construction documents that reduce or eliminate change orders, controlling site stormwater for use in irrigation, incorporating energy-efficiency measures in HVAC design, developing maintenance strategies to ensure continued high-level building performance and higher occupant satisfaction, and reducing marketing and administrative costs.

Case studies on commissioning show that construction and operating costs can be reduced from 1 to 70 times the initial cost of commissioning. A recent study by an international engineering firm indicated that treating stormwater on-site cost one-third as much as having the state or local government treat stormwater at a central facility, significantly lowering the burden to the tax base. A 123,000 ft² (11,400 m²) higher education building constructed in 1997 in Atlanta, Georgia—a building that already had many sustainable principles applied during its design—provides an example of how commissioning, measurement, and verification play a critical role in ensuring that the sustainable attributes designed into the building are actually realized a $1.00 per square foot savings when recommissioned, lowering the total cost of ownership. Recommissioning identified several seemingly inconspicuous operational practices that were causing higher-than-needed consumption in the following areas: chilled-water cooling, by 40%; steam, by 59%; and electricity, by 15%. Implementing continuing measurement and verification (M&V) ensures that the building will continue to perform as designed, again lowering the total cost of ownership.

More esoteric is the cost of unhappy occupants, which includes administrative costs, marketing costs due to more frequent tenant turnover, or increased business cost due to absenteeism or reduced productivity, as well as the impact on marketplace image. Marketplace image is a significant driving force in promoting green/sustainable design. Green/sustainable projects provide owners an opportunity to distinguish themselves in public as well as promote their business or project in order to obtain the desired result. Promotion of green/sustainable attributes of a project can help with public relations and help overcome community resistance to a new project.

## INCREASED PRODUCTIVITY

Another driver for green design is the recognition of increased productivity from a building that is comfortable and enjoyable and provides healthy indoor conditions.

Comfortable occupants are less distracted, able to focus better on their tasks/activities, and appreciate the physiological benefits good green design provides. A case study conducted by Pacific Northwest Laboratory points out many interesting observations about human response to daylight harvesting, outside views, and thermal comfort. The study, which compares worker productivity in two buildings owned by the same manufacturer, illustrates both the positive and negative impact the application of green design principles can have on human productivity. (Heerwagen, Judith H., Pacific Northwest Laboratory. *Assessing the Human and Organizational Impacts of Green Buildings.*)

The first building is an older, smaller industrial facility divided into offices and a manufacturing area. It has high ribbon windows around the perimeter walls in both office and manufacturing areas, providing only limited daylight harvesting. It has an employee lounge, a small outdoor seating area with picnic tables, and conference rooms.

The newer facility is 50% larger with energy-efficient features, such as large-scale use of daylight harvesting, energy-efficient florescent lamps, daylight harvesting controls, DDC HVAC controls, environmentally sensitive building materials, and a fitness center at each end of the manufacturing area.

The study focuses on individual quality-of-work issues and the manufacturer's own production performance parameters. The study provides a mixed review of green design and gives insight on what conditions need to be avoided. While occupants perceived satisfaction and comfort stemming from daylighting and outside views, they also expressed complaints regarding glare and lack of thermal comfort. The green building studied did not appear to have controlled the quality of daylight through proper glazing selection, which may be the cause of the complaints about glare and thermal comfort.

Subsequent chapters discuss specific design parameters relative to building envelope design, including daylight harvesting, energy efficiency, and thermal comfort.

See also Chapter 16 for further data on the relationship between human labor costs and other costs of operating a business and for some further examples.

## FILLING A DESIGN NEED

There are increasing numbers of building owners and developers asking for green design services. As a result, there is considerable business for design professionals who can master the principles of green design and provide leadership in this arena.

Some publications that demonstrate the drivers of green design include *Economic Renewal Guide* (Kinsley 1997), *Natural Capitalism* (Hawken, Lovins, and Lovins 1999), and the *Earth From Above* books (Arthus-Bertrand 2002).

## SOME DEFINITIONS AND VIEWS OF SUSTAINABILITY FROM OTHER SOURCES

- "Everyone talks about sustainability, but no one knows what it is." *(Dr. Karl-Henrik Robert, founder of the organization The Natural Step)*
- "Humanity must rediscover its ancient ability to recognize and live within the cycles of the natural world." *(The Natural Step for Business)*
- Development is sustainable "if it meets the needs of the present without compromising the ability of future generations to meet their own needs." *(Brundtland Commission of the United Nations)*
- To be sustainable, "a society needs to meet three conditions: Its rates of use of renewable resources should not exceed their rates of regeneration; its rates of use of non-renewable resources should not exceed the rate at which sustainable renewable substitutes are developed; and its rates of pollution emissions should not exceed the assimilative capacity of the environment." *(Herman Daly)*
- "Sustainability is a state or process that can be maintained indefinitely. The principles of sustainability integrate three closely intertwined elements – the environment, the economy, and the social system – into a system that can be maintained in a healthy state indefinitely." *(Design Ecology Project)*
- "In this disorganized, fast-paced world, we have reached a critical point. Now is the time to re-think the way we work, to balance our most important assets." *(Paola Antonelli, Curator, Department of Architecture and Design, New York City Museum of Modern Art)*

# 2

# Background and Fundamentals

The use of green engineering concepts has evolved quite rapidly in recent years and is today a legitimate and spreading movement in the HVAC&R and related engineering professions. Much of this recent work has been driven by the emergence of green architecture, also commonly referred to as *sustainable* or *environmentally conscious* architecture; this, in turn, is being encouraged by increased client demand for more sustainable buildings.

The emergence of green building engineering is best understood in the context of the movement in architecture toward sustainable buildings and communities. Detailed reviews of this movement appear elsewhere and fall outside the scope of this document. A brief review of the history and background of the green design movement is provided, followed by a discussion of its applicability. Several leading methodologies for performing and evaluating green building design efforts are reviewed.

## SUSTAINABILITY IN ARCHITECTURE

Prior to the industrial revolution, building efforts were often directed throughout design and construction by a single architect—the so-called "master builder" model. The master builder alone bore full responsibility for the design and construction of the building, including any "engineering" required. This model lent itself to a building designed as one system, with the means of providing heat, light, water, and other building services often closely integrated into the architectural elements. Sustainability, semantically if not conceptually, predates these eras, and some modern unsustainable practices had yet to arise. Sustainability in itself was not the goal of yesteryear's master builders. Yet some of the resulting structures appear to have achieved an admirable combination of great longevity and sustainability in construction, operation, and maintenance. It would be interesting to compare the ecological footprint (a concept discussed later in this book) of, say, Roman structures from two millennia ago heated by radiant floors to a 20th century structure of comparable size, site, and use.

In the 19th century, as ever more complicated technologies and the scientific method developed, the discipline of engineering emerged separate from architecture. This change was not arbitrary or willful but rather was due to the increasing complexity of design tools and construction technologies and a burgeoning range of available materials and techniques. This complexity continued to grow throughout the 20th century and continues today. With the architect transformed from master builder to lead design consultant, most HVAC&R engineering practices performed work predominantly as a subconsultant to the architect, whose firm, in turn, was retained by the client. Hand-in-hand with these trends emerged the 20th century doctrine of "buildings-over-nature," an approach still widely demanded by clients and supplied by architectural and engineering firms.

Under this approach—buildings designed under the architect as prime consultant following the "buildings-over-nature" paradigm—the architect conceives the shell and interior design concepts first. Only then does the architect turn to structural engineers, then HVAC&R engineers, then electrical engineers, etc. (Not coincidentally, this hierarchy and sequence of engineering involvement mirrors the relative expense of the subsystems being designed.)

With notable exceptions, this sequence has reinforced the trend toward "buildings-over-nature": relying on the brute force of sizable HVAC&R systems that are resource-intensive to build and maintain—and energy-intensive to operate—to achieve conditions acceptable for human occupancy. In this approach to the design process, many opportunities to integrate architectural elements with engineered systems are missed—often because it's too late. Even with an integrated design team to "bridge back" over the gaps in the traditional design process, a sustainable building with optimally engineered subsystems will not result if not done by professionals with appropriate knowledge and insight.

## When Green Design Is Applicable

One leading trend in architecture, especially in the design of smaller buildings, is to "invite nature in" as an alternative to walling it off with a shell and then providing sufficiently powerful mechanical/electrical systems to perpetuate this isolation. This situation presents a significant opportunity for engineers today. Architects and clients who take this approach require fresh and complementary engineering approaches, not tradition-bound engineering that incorporates extra capacity to "overcome" the natural forces a design team may have "invited" into a building. Natural ventilation and hybrid mechanical/natural ventilation, radiant heating, and radiant cooling are examples of the tools with which today's engineers are increasingly required to acquire fluency. One example of a building designed with this in mind is the GAP, Inc., in San Bruno, California. (See Chapter 8 for some GreenTips relating to alternative ventilation techniques and see Watson and Chapman [2002] for radiant heating/cooling design guidance.)

Fortunately, many information sources are available, and many projects have been accomplished and documented to inform engineers. Further, new tools for understanding and defending engineering decisions in such projects are merging, including a revised ASHRAE thermal comfort standard, Standard 55 (latest approved edition), that provides an adaptive design method more applicable to buildings that interact more freely with the outdoor environment.

Another more widely demanded approach to green HVAC&R engineering presents another significant opportunity for engineers. This approach applies to projects ranging from flagship green building projects to more conventional ones where the client has only some "appetite" for green. The demand for environmentally conscious engineering is evidenced by the expansion of engineering groups, either within or outside architectural practices, that have built a reputation for a green approach to building design.

Work done by others on such projects can inform engineers' building designs. In addition, many informative resources are available to help engineers better understand the principles, techniques, and details of green building design and to raise environmental consciousness in their engineering practices. By learning and acquiring appropriate resources, obtaining project-based experience, and finding like-minded professionals, engineers can reorient their thinking to deliver better services to their clients.

Green HVAC engineering can be provided for its own sake, independent of any client or architect demand. The appetite for environmentally conscious engineering must be carefully gauged and opportunities to educate the design team carefully seized. In this way, engineers can bring greater value to their projects and distinguish themselves from competing individuals and firms.

## Embodied Energy and Life-Cycle Assessment

Building materials used in the construction and operation of buildings have energy embodied in them due to the manufacturing, transportation, and installation processes of converting raw materials to final products. The material selection process should also consider the environmental impact of demolition and disposal after the service life of the products. Life-cycle assessment (LCA) databases and tools are used to calculate and compare the embodied energy of common building materials and products. Designers should give preference to resource-efficient materials and reduce waste by recycling and reusing whenever possible.

A variety of options exists for the HVAC designer to consider if conducting an LCA analysis. The International Organization for Standardization (ISO) 14000 series of standards on environmental management serves as a method to govern development of these tools. LCA tools are available from both private commercial as well as governmental or public domain sources. The *BEES* (Building for Environmental and Economic Sustainability) tool was developed by the National Institute for Standards and Technology (NIST) in the United States with support from the

US Environmental Protection Agency (EPA). The *TRACI* (Tools for the Reduction and Assessment of Chemical and Other Environmental Impacts) from the EPA focuses primarily on chemical releases and raw materials usage in products. A few tools produced by commercial firms, such as ENVEST2, SimaPro, and GaBi in Europe and the Athena Environmental Impact Estimator in Canada, are available (ASHRAE's commercialization policy prevents detailed comparison or discussion of these).

## Green Building Rating Systems/Environmental Performance Improvement Programs

There are two general types of programs that exist to encourage green building design. One type might be termed a rating system per se and the other a guide or program to encourage and assist designers in achieving green building design. (This book is in the latter category.)

**Green Building Rating Systems.** There are various rating systems developed by reputable organizations in both the United States and internationally that attempt to provide a rating to indicate how well a building meets prescribed requirements and to determine whether a building design is green or not or the degree of "greenness" it has. They all provide useful tools to identify and prioritize key environmental issues. These tools incorporate a coordinated method for accomplishing, validating, and benchmarking sustainably designed projects.

As with any generalized method, each has its own limitations and may not apply directly to every project's regional and other specific aspects. For example, as Walter Grondzik notes in his paper addressing how extensively the rating systems address *all* elements of IEQ (see Chapter 1), some of those elements are overlooked (Grondzik 2001).

It should not be implied that this Guide advocates exclusion of these elements. By advocating both *green* as well as *good* design in any building design endeavor, this is avoided. Yet at the same time this Guide does not intend to *discourage* use of the mentioned rating systems or other programs by ASHRAE members or by the design teams in which its members participate. Thus, while this Guide does not endorse or recommend use of any one rating system or program, it is also not opposed to the use of one if it is perceived that such will be of help in achieving a green design. Each method may play a valuable role in increasing an engineer's ability to help deliver sustainably designed projects.

The leading *rating* method in the United States is the LEED program, created by the US Green Building Council (USGBC). LEED stands for *leadership in energy and environmental design*. The USGBC started offering this system in 1998 and describes it as a voluntary, consensus-based, and market-driven building rating system. It evaluates environmental performance from a "whole building" perspective over a building's life cycle, providing a numerical standard for what constitutes a "green building." Their goal is to raise awareness of the benefits of building green, which is transforming the marketplace. LEED has been applied to numerous projects over a range of project certification levels, and its use has grown rapidly over the past several years. Rating systems are now in place for new construction (LEED-NC), existing buildings (LEED-EB), core and shell construction (LEED-CS), and commercial interiors (LEED-CI). For residential projects, LEED programs are also in development for new homes (LEED-H) and neighborhood developments (LEED-ND). LEED also qualifies individuals as LEED consultants, although it does not require such consultants on projects seeking LEED ratings. The LEED program and registered building projects are being established in other countries, such as India, Australia, Canada, and China.

Another rating method that was originally developed in Canada and is being introduced in the United States is the Green Globes program. Green Globes is an online auditing tool that overlaps much of the same concepts as LEED. While similar in the intent of helping a building owner or designer develop a sustainable design, Green Globes is primarily a self-assessment tool (although third-party assessment is an option) and also provides recommendations for the project team to follow for improving the sustainability of the design. In the UK, Green Globes is known as the Global Environmental Method (GEM) program.

The Building Research Establishment Environmental Assessment Method (BREEAM®) rating program was established in the UK in 1990. This is a voluntary, consensus-based, market-oriented assessment program. With one mandatory and two optional assessment areas, BREEAM® encourages and benchmarks sustainably designed office buildings. The mandatory assessment area is the potential environmental impact of the building; the two optional areas are design process and operation/maintenance. Several other countries and regions have developed or are developing related spin-offs inspired by BREEAM®.

**Environmental Performance Improvement Programs.** Several other programs exist or have been conducted in the past decade with the purpose of guiding or encouraging green-building design but without assigning a specific rating. They are:

1.  *GBC, the Green Building Challenge*

Canada began a process in 1998 called the *Green Building Challenge*, involving about 20 countries developing and testing a second-generation environmental performance assessment system. That system is designed to reflect the very different priorities, technologies, building traditions, and even cultural values that exist in various regions and countries.

The GBC offers criteria for the following general categories: context factors, transportation, resource consumption, environmental loadings, IEQ, service quality, economics, and management.

2.  *C-2000, Integrated Design Process*

Natural Resources Canada initiated the C-2000 program in 1993 as a small demonstration of achieving very high levels of building performance. It was assumed that some level of financial incentive would be required at various stages of the building process to make the program a success. Its technical requirements cover energy performance, environmental impacts, indoor environment, functionality, and a range of other parameters.

Experience with the program in the first group of projects to which it was applied yielded the fact that the program's guidance in producing integrated design was the main reason high-performance levels were achieved. The program was then called the *Integrated Design Process*, and it now focuses its project interventions on design advice in the very early stages.

3.  *CBIP, Commercial Buildings Incentive Program*

An offshoot of the C-2000 Program, CBIP was an effort to create a larger national program to move the Canadian building industry toward energy efficiency. Its financial incentives provide incremental costs for the design process. Because it attempts to address a larger audience, it is simpler than the C-2000 program in that it does not require customized support and that it concentrates on energy performance.

4.  *Other Resources*

Other guides and methods include:

- *The Whole Building Design Guide*
- *Green Building Advisor*
- California Collaborative for High Performance Schools (CHPS)
- *Minnesota Sustainable Design Guide*
- *New York High Performance Building Guidelines*

Work referred to by architects includes:
- *AIA Environmental Resource Guide*
- *The Hannover Principles*
- *GreenSpec*
- *The Natural Step*
- ISO 14000, www.iso.org/iso/en/iso9000-14000/index.html
- BEES 3.0, www.bfrl.nist.gov/oae/software/bees.html
- TRACI, epa.gov/ORD/NRMRL/std/sab/iam_traci.htm

These and additional works by such entities as the National Renewable Energy Laboratory (NREL) and the Rocky Mountain Institute (RMI) are listed in the references or at the end of this chapter.

## FUNDAMENTALS OF RELATED ENGINEERING TOPICS

Understanding the basic tenets that define the engineer's profession is imperative for thoughtful design. While this Guide is not intended to serve as an engineering textbook, it is helpful to review key fundamentals of engineering that influence the design of sustainable buildings from the perspective of the HVAC engineer. These include the first and second laws of thermodynamics, heat transfer, and fluid systems. This will provide the reader with insights into the opportunities available for energy conservation as well as other green-building design opportunities.

The fundamental engineering concepts are first presented in this section. (Applications of these principals to HVAC&R systems are discussed in the following section.) The purpose of this section is to identify key relationships and dependencies using the language of the engineer: the fundamental laws and formulae. It is assumed that the reader is a practicing engineer with some experience rather than a neophyte, so this section is not meant to be a primer on the science behind HVAC&R. However, before delving into the greener aspects of the topic it is good to remind the reader of the basics.

Webster defines *formula* as "a general fact, rule or principle expressed in mathematical symbols." In HVAC, the facts, rules, or principles we are most concerned with are the laws of thermodynamics. It has been said that if you can write an equation for a problem, you have the solution. But we must not confuse a "plug-and-chug" approach using formulae with engineering: the successful engineer must be able to both write the formula and understand its underlying principles.

A formula is composed simply of constants and variables. Constants are defined (given) and cannot be influenced. Therefore, we can only resolve a problem (influence an outcome) by manipulating its variables. When this simple premise is understood, key relationships and dependencies manifest themselves and the solution becomes obvious.

### Thermodynamic Laws

The laws of thermodynamics are at the core of the analysis and design of HVAC&R systems. This section briefly summarizes the first and second laws and their implications in green design.

The first law in its basic form is

$$Q - (W_{flow} + W_{shaft}) = \Delta U + \Delta E_{potential} + \Delta E_{kinetic} \,.$$

For a system in steady state and substituting in for the internal, potential, and kinetic energy terms leads to the following:

$$\dot{Q} - \dot{W} = \dot{m}[(u_2 - u_1) + (p_2 v_2 - p_1 v_1) + (V_2^2 - V_1^2)/2 + g(z_2 - z_1)]$$

where

$\dot{Q}$      = heat transferred to or from the system; the dotted symbol refers to the rate of heat being transferred

$E$      = energy contained in the system (potential or kinetic)

$\dot{W}$      = work produced or required by the system; the dotted symbol refers to the rate of work being done

$u$      = internal energy of the fluid (i.e., water, steam, air, refrigerant) per unit mass

$m$      = mass of fluid

$pv$      = product of the pressure and specific volume of the fluid

$V$      = velocity of the fluid in the system

$h$      = enthalpy of the fluid per unit mass, expressed as $(u + pv)$

$z$      = height or potential energy of the fluid

1 and 2 = subscripts denoting *before* and *after* states of the parameter

(*Note: A dot over a symbol means "the rate of transfer of."*)

The internal energy ($u$) and flow energy ($pv$) terms can be combined into the fluid enthalpy, given by

$$h = u + pv.$$

The second law is represented by several equations involving the change in entropy of the fluid, but for the purposes of making decisions on energy and green design, studying the Carnot cycle, as represented on temperature-entropy coordinates, is particularly useful.

One common application of the first-law equation to a building HVAC system is the combustion processes generating heat to raise the temperature of a fluid for providing heat to a building. When looking at the heating means, be it a boiler, hot water generator, or warm air furnace, the terms for work ($W$), changes in kinetic energy $(V_2^2 - V_1^2)/2$, and potential energy $(z_2 - z_1)$ are small in comparison to enthalpy difference, so the first law becomes

$$\dot{Q} \cong \dot{m}(h_2 - h_1) \,.$$

## Green Design Implications of Thermodynamic Laws

There are two types of energy: stored (potential) energy and the energy of motion, called *kinetic energy*. Regardless of its form, however, the first law of thermodynamics always applies. For a closed system, in essence, it says:

*Energy cannot be created or destroyed.*

A closed system is one in which energy and materials do not flow across the system boundary. The first law is why energy efficiency and green design are a necessity. If we could create energy, there would be no reason to conserve it. We must be aware that we are largely dependent on sources of energy that are in finite supply. Therefore, it is logical to use less energy of this type as a rule and to move toward renewable, more efficient energy sources in general.

If *energy* is the ability to do work, then what happens when we tap that potential? The result is threefold: work, heat, and entropy. Work is the transfer of energy by mechanical means, such as a fan or pump. Heat refers to a transfer of energy from one object to another because of a temperature difference. And entropy, simply stated, is an indicator of the state of disorder of a system.

The second law of thermodynamics helps us to appreciate the relevance of sustainable design even more:

*All processes irreversibly increase the entropy of a system and its environment.*

If you understand that the Earth is our system, then you realize that the limited amount of usable energy we have been granted (first law) will eventually and irreversibly be converted into unusable energy (second law), which brings us full circle. Of course, the earth is not a completely closed system in that energy is entering (via solar radiation) and leaving (such as earth radiating energy out into space). Regardless, our dependence on energy in a useful form, and the immutable laws of nature, set the tone for proper (green) design: *use energy judiciously and effectively.*

## Fundamentals of Heat Transfer

Heat travels in three ways: conduction, convection, and radiation. Note the following general correlations:

Conduction ≈ Heat transfer by molecular motion within a material or between materials in direct contact
Convection ≈ Energy exchange via contact between a fluid in motion and a solid
Radiation ≈ No contact required; heat transfer by electromagnetic waves

In real-world situations, heat transfer occurs via all three modes at the same time. Depending on the problem type, one or two of these modes will generally dominate the rate of heat transfer at any given moment. But to keep things simple, we will discuss each mode of heat transfer separately.

**Conduction.** Consider heat transfer through a portion of the building shell (wall, window, door, floor, or roof). The process can be expressed as follows:

$$Q = UA\Delta T$$

where $Q$ is the amount of heat transferred, $A$ is the exposed surface area, and the temperature delta ($\Delta T$) is the difference between the two boundaries of the wall (outdoor air and indoor air).

The rate at which heat is transferred by conduction is controlled by the overall heat transfer coefficient $U$:

$$U = 1/\Sigma R$$

where $\Sigma R$ is the overall thermal resistance for the material layers of the system in question. The overall thermal resistance will typically include terms for convective heat transfer resistances in effect on both the inside and outside surfaces (see "Convection" discussion below).

In the equation for conduction through a portion of the building shell, there are no fixed variables other than the fact that the outdoor air temperature is dictated by the project's location. The design team does have the ability to minimize $Q$ by minimizing the U-factor, area, and temperature difference. Therefore, steps that can be taken include:

- Limit conductance by providing an envelope with a low overall U-factor. This can be accomplished through the use of insulating materials, air spaces in glass and walls, thermal breaks in construction, etc.
- Create an indoor environment wherein the space dry bulb can be higher or lower than traditional design norms in the summer and winter, respectively. This can be done using dehumidification and/or humidification as long as you stay within acceptable ranges (see ASHRAE Standard 55).
- Limit the surface area exposed to large temperature differentials by increasing shading, both on the building and occurring naturally through the use of landscaping. Considering low-profile, semi-buried buildings is another option.

**Convection.** There are numerous formulas describing energy transfer through convection. The *ASHRAE Handbook—Fundamentals* gives at least 12 factors used in determining convective heat transfer coefficients, and it lists no fewer than 25 equations for calculating heat transfer through forced convection. We will limit this discussion to the comparison of natural versus forced convection.

Natural convection is often called free convection and is primarily due to differences in density and the action of gravity. To see convection in action, observe a LAVA® lamp. The "lava" gains heat from the lightbulb and rises; as it cools, it falls again. Replace the lightbulb with a hot-water-filled finned tube and swap the "lava" for air, and you get a fair idea of how convection works in a heating application. The lesson from this fairly obvious example is that natural convection is a simple law of nature that can be used to the designer's benefit in a number of ways. Forced convection occurs when the fluid (air, water, etc.) movement is done via an external mover, such as a fan or pump.

Recommendations for HVAC design related to convective heat transfer include the following:

- Consider the use of displacement ventilation when appropriate. In a displacement ventilation design, cool air is supplied at a low level and returned at the higher levels of a room, and the design relies on the natural increase in buoyancy as air is warmed by heat generation sources such as people or equipment. This has the benefit of better IAQ since the upward flow of air naturally "lifts" pollutants out of the occupied space.

• Locate returns or exhaust directly over heat-generating equipment such as copiers and refrigerators, removing heat from the space at the source and thus lowering the imposed space sensible cooling load.

• Provide baseboard heat to warm the exterior envelope surfaces to reduce radiant heat loss from occupants to walls and windows. (Note: While this may increase comfort, it can increase heat loss through the surface by action of the natural convective flow, reducing the surface heat transfer coefficient.)

• Alternatively, use perimeter radiant ceiling panels to counteract perimeter heating loss.

• Apply passive ventilation or hybrid ventilation, provided extreme outside conditions do not preclude its successful application.

By supplying low and returning high, you reduce the need for mixing that accompanies traditional overhead supply and return systems. By removing sensible load at the source, you reduce the sensible load in the space and, in turn, the supply air required to handle it. By allowing ventilation air to enter and exit a building naturally, the need for forced ventilation through air-handling systems and ductwork is eliminated. All of these steps decrease the amount of work required to address loads within a space, which leads to our next section.

**Radiation.** Heat transfer via radiation presents a unique challenge and opportunity for the designer. We have all stood next to a cold window and felt chilled even though the ambient temperature was at a comfortable level. The same holds true for sunny days when one can get too warm even though the thermostat says all is well. The simplified form of the equation describing radiant heat transfer is

$$Q = \varepsilon\sigma A(T_1^4 - T_2^4),$$

where $\varepsilon$ is emissivity, $\sigma$ is the Stefan-Boltzmann constant, $A$ is surface area, and the temperature ($T$) terms are the absolute temperature difference between the radiant object (subscript 1, with emissivity of $\varepsilon$) and its surroundings (subscript 2, a blackbody).

Emissivity is a property that reflects the ability of that material to emit thermal radiation energy relative to the maximum theoretically possible at the material's temperature. Emissivity is a function of both the material itself and surface conditions. A dull black surface such as charcoal has an emissivity close to 1 (that of a blackbody), while shiny metallic surfaces have lower values, more in the range of 0.1 to 0.4. A related property for thermal radiation is the material absorptivity, which reflects that material's ability to absorb incoming thermal radiation. A material with absorptivity of 0.8 will absorb 80% of the incoming thermal radiation. In general, one can consider the material's absorptivity and emissivity to be the same value.

Surfaces with higher emissivity will absorb and emit more thermal energy. But notice the dramatic difference changing the temperature difference can make; the rate at which an object radiates or absorbs heat is proportional to the difference in the fourth powers of the absolute temperatures involved.

When the designer is faced with the challenge of minimizing the heat transferred by radiant means, the following steps can be taken for situations where cooling loads dominate:

- Explore the possibility of eliminating or drastically reducing the area ($A$) directly exposed to the radiant source through shading or other means. For most building applications, the radiant source is the sun, which can be treated as an object emitting energy at 5800 K, or 10,000°F.
- Recommend the use of "cool roof" technologies that balance the emissivity and absorptivity of the surface to minimize the net solar heat gain to the roof.
- With glazing, the designer should evaluate the trade-off of using a low-emissivity material with other selective (reflective) coatings.
- Avoid dark colors on the building exterior, which typically have a higher emissivity and absorb more heat.
- Limit east and west exposures, especially those with a large amount of glass.
- Offset the radiant load. For example, in a large atrium with a large glazing exposure and/or exterior walls, offsetting the radiant gains from the envelope with radiant cooling in the floor will produce a net effect that is significantly more comfortable for the occupant.

For situations where heating loads are significant and you are looking to maximize the heat gained, $Q$, as with solar collection or passive heating, do just the opposite of the above; increase exposure, maximize surface area, and use dark colors and high-e materials (ASHRAE 1996).

When heating or cooling with radiant panels, remember the power of the temperature difference to maximize thermal efficiency. You may be able to accommodate the architect's aesthetic sense by minimizing the need for excessive radiant surface area of high-e materials. For example, refer to Chapter 6, Figure 1, "Radiation Heat Transfer from Heated Ceiling, Floor, or Wall Panel," in *ASHRAE Handbook—Systems and Equipment*. Note that, assuming a 70°F room temperature and other factors being equal, raising the effective panel temperature from 100°F to 200°F would raise the radiant heat transfer by a factor of 5.7! In turn, significantly less panel surface area would be required to accomplish the same amount of heating. Standing under a higher temperature panel may result in a warmer head relative to the other body areas and, consequently, may cause discomfort.

### Fundamentals of Fluid Flow

The analysis of fluid flow and systems is also a fundamental concept for HVAC designers. For an incompressible, steady-flowing fluid, the Bernoulli equation governs. This equation is based on the conservation of energy principle and states that between Points 1 and 2 within a system, the following relationship holds:

$$0 = \frac{P_2 - P_1}{\rho g} + \frac{V_2^2 - V_1^2}{2g} + (z_2 - z_1)$$

where $P$ is the fluid pressure, $V$ is the fluid velocity, and $z$ is the elevation at Points 1 and 2.

When a fluid passes through a fan or pump, additional energy is input to the system in the form of an increase in pressure and perhaps fluid velocity. The power required to move a fluid involves, in essence, a modification of the Bernoulli equation with the left-hand side not being zero but reflecting the additional energy input to the fluid.

## APPLICATIONS TO HVAC&R SYSTEMS AND PROCESSES

### Power Generation

For steam boilers, with the water undergoing phase change, the enthalpy difference is much larger than it is for either the water in hot water generators or the air in warm air furnaces, where the enthalpy change is proportional to the temperature rise of the fluid.

Combustion of the fuel/air mixture results in a high temperature of the flue gases being discharged to the chimney or vent. *Green design requires examining the recovery of the energy in the flue gases for possible reuse—as in preheating outside air, heating hot water for domestic use, or serving as a heat source for other building heating purposes such as snow-melting systems.*

When looking at the generation of power for on-site power production (as, for example, with cogeneration systems) or at vapor-compression refrigeration systems in building air conditioning (where the work produced or required is a major energy concern), looking at the first and second laws together is necessary.

The theoretically most efficient power generation system is given by the Carnot cycle. This can be represented as a rectangle on temperature-entropy coordinates (see Figure 2-1). The area of the rectangle represents the total energy (both work and thermal) involved in the case of a power plant generating power or of vaporcompression refrigeration air-conditioning or heat-pump systems. For example, in Figure 2-1 the shaded area represents the heat load transported by a vapor-compression refrigeration cycle ($Q_{load}$) and the cross-hatched area the amount of work output necessary by the compressor to move the heat ($W$). The total of the two represents the heat rejected to the environment at the condenser ($Q_{rejected}$).

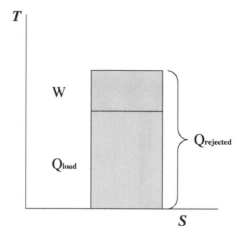

**Figure 2-1 Carnot cycle for refrigeration.**

For power generation, the upper temperature, representing the high temperature in the Carnot cycle, $T_h$, should be as high as possible to maximize work output. Similarly, the lower temperature in the Carnot cycle, $T_c$, should be as low as possible. While real power generation systems do not operate precisely on the Carnot cycle, the principle is still valid.

Similarly, for refrigeration, where one wants to minimize the work input (power required), $T_h$ should be as low as possible and $T_c$ as high as possible.

### Power Production

Many buildings or building groups use on-site electric power generation plants that, when combined with some sort of heat recovery, are termed *cogeneration plants*.

A common modern generation plant may employ a gas turbine, which requires compressing air before combustion of fuel in a combustion chamber; this, in turn, yields high-temperature combustion gases at high pressure, which expand through the turbine, generating power. The work required to compress a fluid is a function of the fluid density and compressibility, with low-density fluids (such as air) requiring more work to achieve a given pressure increase than higher-density fluids (such as liquids like water). This results in relatively low-pressure gases entering the turbine and high-temperature exhaust from the turbine at the lower atmospheric pressure obtained after expansion.

Modern combined-cycle power plants use high-temperature turbine exhaust as the heating medium for heat recovery steam generators. These steam generators operate at much higher pressures than gas turbines since water first must be pumped to the high pressures entering the steam generators, and this high-density fluid requires much less energy to achieve the high pressure than the air in the gas turbine cycle. The steam expands through a steam turbine to low condensing temperatures, $T_c$, for heat rejection in a condenser. For on-site power generation without combined cycles, the gas turbine exhaust is used as a heat source for other building energy needs, achieving the lower $T_c$ in that manner.

### Air Conditioning

In a refrigeration plant, where the end-purpose is the cooling of air for building cooling purposes, the chiller or direct expansion refrigerant should have the cooling fluid at as high a temperature ($T_c$) as possible to minimize work input. Of course, this temperature is limited by the temperature to which the air must be cooled. The closer $T_c$ is to that air temperature, however, the more heat transfer surface area will be required in the heat exchanger used to cool the air, in both the cooling coils in direct expansion systems or in chilled-water coils.

The high temperature ($T_h$) at which heat is rejected is limited by the temperature of the heat sink. This would be the ambient air in an air-cooled condenser or by the approach to the air wet-bulb temperature with water-cooled condensers used in conjunction with cooling towers. Again, the closer the refrigerant $T_h$ is to the coolant temperature, the larger and more expensive will be the heat exchanger (condenser) required.

In real cycles, the work produced by the turbines (gas or steam) is the enthalpy difference across the turbine multiplied by the mass flow rate. The net power generated is the difference between this enthalpy difference and that required for the compression of combustion air plus the work to pump the water in the steam part of the cycle.

For the refrigeration cycle using expansion valves, the work input is the enthalpy difference across the compressor.

### Hydraulic Machine (Pumps, Fans) Similarity Analysis

The performance of a pump or fan at one operating point can be used to predict the performance of the same (or similar) equipment at a different operating condition. This is done using similarity relations. Since the energy required in pumping water needed in chilled-water systems—and that needed to power the fans used in air-distribution systems—can be quite substantial, a closer look at these relations can be useful in analyzing possibilities for saving energy. Similarity relations, at the point of maximum efficiency, yield

$$\left.\frac{Pv}{N^3 D^5}\right|_{\text{Maximum Efficiency}} = \text{Constant}$$

and

$$\left.\frac{Q}{ND^3}\right|_{\text{Maximum Efficiency}} = \text{Constant},$$

where

$Q$ = volumetric flow rate,

$N$ = impeller rotative speed,

$D$ = impeller diameter,

$P$ = power input to the pump or fan, and

$v$ = specific volume of the fluid.

When considering performance of the same pump or fan at different operating conditions, the impeller diameter is the same. The first similarity relation listed then reduces to

$$\left.\frac{Pv}{N^3}\right|_{\text{Maximum Efficiency}} = \text{Constant}.$$

The power input needed to run the pump or fan is therefore proportional to $N^3$— or to the *cube of the rotating speed*. This gives rise to the use of variable-speed drives, where part loads on the system result in decreased flow requirements. The reduction in flow is accomplished by lowering the pump or fan rotating speed, $N$. Input power to the fan or pump will therefore decrease as the cube of $N$.

The use of variable-speed (or variable-frequency) drives does involve additional controls and engineering design analysis of operating conditions to ensure that the system will perform. This will add some to the initial cost of the system. However, the cost of equipment and controls needed to implement a variable-speed drive has been decreasing and has reached the point that even relatively small power equipment (1–2 hp) can economically be controlled with a variable-speed system.

### Forced Convection and Mass Transfer

In HVAC, mass transfer by forced convection is used to facilitate a thermodynamic transfer of energy. Though all types of fluids and gases are moved in HVAC applications, the discussion here is limited to water and air. Note in the following the key relationship between mass flow rates, the fluid's thermal properties, and the load. Additional technical information on fluid transport is contained in Chapter 8, "Space Thermal/Comfort Delivery Systems."

**Water.** The amount of energy required to address a sensible load $Q$ within a system is found to be proportional to the fluid mass flow rate $m$ into the system and to the temperature difference $\Delta T$ between the system and the fluid. This can be expressed as

$$Q_s = mcv\Delta T,$$

where $c$ is the specific heat at constant pressure $p$ of the medium. For temperature changes that are not too great, $c_p$ can be considered a constant. The sensible heat formula for water in IP is

$$Q_s = (\text{gpm})(500)\Delta T.$$

The key relationships are clear. If $Q$ is a given, then the only variables are flow and temperature difference: the greater the temperature difference ($\Delta T$), the less flow that will be required, and vice versa. If the load $Q$ is variable, then one can react by varying flow, $\Delta T$, or both. Recommended strategies include:

- Use high $\Delta T$ designs whenever feasible, barring any negative efficiency impacts that may result from a lowered chiller COP. The greater the temperature difference, the less flow that will be required.
- Maintain a constant $\Delta T$ and vary the flow in response to load. The value of this strategy becomes clearer when considered with the fan/pump affinity laws discussed later in this chapter.

**Moist Air.** When working with air, we are rarely dealing with a dry gas. Instead, air is usually a mixture of water and air. In turn, air-side energy calculations not only involve the temperature difference (sensible heat) but also must factor in the changes of state (latent heat). The sensible load in standard air is similar to that discussed above for water and can be quantified (in I-P)[1] as follows:

$$Q_s = (\text{cfm})(1.08)\Delta T$$

The amount of energy required to address a latent load $Q$ within a system is found to be proportional to the fluid mass flow rate $m$ into the system and to the humidity ratio difference $\Delta W$ between the air in the system and the air entering the system, which can be expressed as

$$Q_L = ml\Delta W,$$

where $l$ is the latent heat of vaporization for water. For temperature changes that are not too great, $l$ can be considered a constant. The latent heat equations (in I-P) for standard air are:

$$Q_L = (\text{cfm})(0.68)\Delta W_{\text{grains/(lb)}}$$

$$Q_L = (\text{cfm})(4840)\Delta W_{\text{lb/lb}}$$

The key relationships again are obvious. If $Q$ is a given, then the only variables are flow and moisture content of the air delivered. The greater the humidity ratio difference (i.e., the dryer the supply air), the less flow that will be required.

---

1. These are standard rule-of-thumb relations with constants set only for I-P units. Similar relationships exist for calculations in SI units.

Last, addressing total heat in a space leads us to the following:

$$Q_T = Q_s + Q_L = m\Delta h$$

where $\Delta h$ is the difference in enthalpy between the air/water vapor in the space and the supply air/water vapor. For standard air, the process (in I-P) is

$$Q_T = (cfm)(4.45)\Delta h .$$

In all of the air formulas, the relationship remains. The only actions at our disposal are varying flow rates and altering thermal properties. To minimize the flow rate and energy required in air systems, consider the following possible steps:

- Move less air by supplying colder air and increasing the temperature difference when possible. The designer will need to account for any negative chiller efficiency impacts due to lowered COP. High induction diffusers or fan terminal units can be used to temper the supply air at the point of distribution to avoid drafts and discomfort.
- Maintain a constant $\Delta T$ and vary the airflow in response to load.
- Use energy recovery to pretreat outdoor air to minimize mixed air differentials and, in turn, reduce the load associated with the dictated ventilation quantity.
- Consider passive preheating strategies, such as buried duct or solar walls, when introducing outdoor air in the winter, again to minimize the mixed air differentials and the associated load.
- Create an indoor environment wherein the space dry bulb can be higher than traditional design norms in the summer. Lowering relative humidity levels by means of passive dehumidification via return air bypass can allow a higher space temperature, which equates to a higher $\Delta T$ in reference to the supply air.
- Create an indoor environment wherein the space dry bulb can be lower than traditional design norms in the winter (see ASHRAE Standard 55). Utilize radiant task heating for comfort in lower ambient air temperatures (Chapman et al. 2001).
- Use hybrid radiant/convective systems where conditions, artifacts, or special needs of individuals conflict with comfort of viewers, visitors, or personnel.
- Capture sensible load at the source and remove it from the space. In turn, less supply air is then required to handle the space load.
- Use stratification strategies such as displacement ventilation and remove the ceiling load from the space comfort equation. Note that the central equipment still sees this load, but we do not have to move as much air in the space.

## Work and Power

We know that it takes energy to change the thermal properties of our heating and cooling medium, i.e., temperature, enthalpy. When possible, we want to take advantage of natural forces such as free convection and thermal radiation heat transfer. For the remainder of the load, we will need to transfer energy using forced convection and mass transfer. To move all of that mass, we need to do work; work is quantified in terms of input power, and we want to use as little power as possible.

## Efficiencies

If moving a fluid, the efficiency $E$ of the device and the motor/drive combination, combined with the specific gravity (sg) of the fluid, will determine the amount of power required. For a pump, the equation for motor horsepower (mhp) is

$$\text{mhp} = \frac{(\text{gpm})(\Delta P \text{ ft} - \text{hd})(\text{sg})}{(3960)E_{pump}E_{motordrive}} \quad \text{or} \quad \text{mkW} = \frac{(\text{L/s})(\Delta P \text{ kPa})(\text{sg})}{(1000)E_{pump}E_{motordrive}}$$

and, for a fan,

$$\text{mhp} = \frac{(\text{cfm})(\Delta P \text{ in. w.g.})(\text{sg})}{(6356)E_{fan}E_{motordrive}} \quad \text{or} \quad \text{mkW} = \frac{(\text{L/s})(\Delta P \text{ kPa})(\text{sg})}{(1000)E_{fan}E_{motordrive}}$$

In most instances the specific gravity is fixed because the fluid is determined by the design (i.e., air at sea level, water, water with glycol, etc.). So the key actions that can be taken are to minimize flow and pressure drop and to maximize equipment and drive efficiencies. We have discussed flow already, but regarding pressure drop and efficiency, consider these steps:

- To the extent possible, minimize pressure drop by increasing conduit size (duct, pipe) and limiting the number of fittings. Obviously there is a breakeven point where increased size becomes unworkable or not cost-effective, but that is a matter of judgment and/or technical evaluation.
- Keep fluid specific gravity in check. For example, adding glycol not only decreases thermodynamic performance, but it also increases the amount of pumping energy required to move the fluid.
- Select equipment based on judicial review of the efficiency curves. Maximize efficiency, but avoid unsteady locations or fringe selections on the operating curves.
- Specify *premium* efficiency motors, not just high-efficiency motors. The difference is usually worth the added cost, which can be confirmed through a life-cycle cost analysis.
- Consider different pump speeds and types to maximize efficiency and performance (i.e., end suction-vertical in-line, 1750–3600 rpm).
- Evaluate the right fan for the right need, such as forward-curved vs. backward-inclined, prop versus centrifugal, plenum vs. housed.

## Pump or Fan Affinity Laws

The affinity laws are often called the fan laws or the pump laws depending on which equipment you are using. The fact is that, regardless of the medium—be it air, water, or maple syrup—the affinity laws present a series of powerful relationships that can make or break a design.

$$\frac{rpm_2}{rpm_1} = \frac{Flow_2}{Flow_1}$$

$$\frac{\Delta P_2}{\Delta P_1} = \left(\frac{Flow_2}{Flow_1}\right)^2$$

$$\frac{mhp_2}{mhp_1} = \left(\frac{Flow_2}{Flow_1}\right)^3$$

The relationships are simple. Flow and speed are directly proportional. But the static pressure in a system varies by the *square* of the flow, and the required motor horsepower varies by the *cube* of the flow! What does this mean in terms of energy required? See the following example.

Consider a typical office building at 10,000 ft$^2$ (929 m$^2$) with a varying load. Peak load is 200 ft$^2$/ton (5.28 m$^2$/kW), which translates to 600,000 Btu/h (176 kW). The original design utilized a constant-flow, variable-temperature multizone air-handling system. Design airflow is based on a 55°F (12.8°C) supply temperature, 75°F (23.9°C) zone temperature for a 20°F (11.1 K) $\Delta T$. Fans were selected for 28,000 cfm (13,215 L/s) at 4 in. total static pressure (1.0 kPa), with a fan efficiency of 75% and motor efficiency equal to 95%.

$$mhp = \frac{(cfm)(\Delta P \text{ in. w.g.})(sg)}{(6356)\varepsilon_{Fan}\varepsilon_{Motor/Drive}} = \frac{(28,000)(4 \text{ in. w.g.})(1)}{(6356)(0.75)(0.95)} \approx 25 \text{ hp}$$

The owner is considering replacing the aging system with a constant-temperature, variable-volume system. The average load diversity within the building is 75%, which means that the fan at any given time will have to deliver, on average, no more than 21,000 cfm (9,910 L/s). Based on fan energy alone, what kind of energy savings can we expect?

$$\frac{mhp_2}{mhp_1} = \left(\frac{Flow_2}{Flow_1}\right)^3 = \frac{mhp_2}{25 \text{ hp}} = \left(\frac{2100}{28000}\right)^3$$

$$mhp_2 = 10.5 \text{ hp}$$

$$\Delta mhp = 25 - 10.5 = 14.5 \text{ hp}$$

$$1 \text{ hp} = 2545 \text{ Btu/h}$$

$$(14.5 \text{ hp})(2545 \text{ Btu/h} - \text{hp}) = 36,903 \text{ Btu/h}$$

Note that a 25% decrease in flow equates to a 58% decrease in required power, and these principles apply to variable-flow water systems as well. Clearly there is an incentive to minimize flow and pressure drop whenever possible.

## SUMMARY OF ENGINEERING APPLICATION FUNDAMENTALS

- Understanding a formula's key relationships and dependencies will often make green-oriented measures obvious.
- Because we have a limited supply of energy and cannot create it, a good sustainable design will work to conserve energy.
- When designing building envelopes, heat transfer by conduction and solar radiation should be optimized for the energy transfer process (heating, cooling) involved.
- When designing heating and cooling systems, the use of heat transfer and ventilation through natural means, such as natural convection or radiation, should be considered from the beginning of the design process and utilized fully when possible.
- When forced convection and mass transfer are necessary, minimize flow and pressure drop, maximize the effect of thermal characteristic (temperature, enthalpy, or humidity ratio) differences, and specify equipment with the highest efficiencies.

## REFERENCES AND RESOURCES

### Published

AIA. 1996. *Environmental Resource Guide*. Edited by Joseph Demkin. New York: John Wiley & Sons.

Brand, S. 1995. *How Buildings Learn: What Happens After They're Built*. New York: Viking Penguin USA.

McDonough, W. 1992. The Hannover Principles: Design for Sustainability. Presentation, Earth Summit, Brazil.

### Online

Advanced Buildings Benchmark™ (E-Benchmark), www.poweryourdesign.com

American Institute of Architects, www.aia.org

BREEAM®, www.breeam.org

BuildingGreen (for purchase), www.greenbuildingadvisor.com

California Collaborative for High Performance Schools (CHPS), www.chps.net

Center of Excellence for Sustainable Development (CESD), Smart Communities Network, www.smartcommunities.ncat.org

Energy Efficiency and Renewable Energy Network (EREN) and High Performance Buildings Research Initiative, www.eere.energy.gov and www.eere.energy.gov/buildings/highperformance

Green Building Challenge, www.greenbuilding.ca

Green Globes, www.greenglobes.com

Lawrence Berkeley National Laboratories, Environmental Energy Technologies Division, http://eetd.lbl.gov/

LEED Green Building Rating System®, USGBC, www.usgbc.org

*Minnesota Sustainable Design Guide*, www.sustainabledesignguide.umn.edu

National Renewable Energy Laboratory, Center for Buildings and Thermal Systems, www.nrel.gov/buildings_thermal/

*New York City High Performance Building Design Guidelines* (1999), www.nyc.gov/html/ddc/html/ddcgreen/documents/guidelines.pdf

Oikos: Green Building Source, www.oikos.com

Rocky Mountain Institute, www.rmi.org

Sustainable Buildings Industry Council, www.sbicouncil.org

Sustainable Communities Network (SCN), Sustainable Building Resource Directory (focused on Mid-Atlantic region of the US), www.sbrd.org

*Whole Building Design Guide*, www.wbdg.org

# Section 2:
# The Design Process

# 3

# COMMISSIONING

The commissioning process is a quality-oriented process for achieving, verifying, and documenting that the performance of facilities, systems, and assemblies meet defined objectives and criteria. *ASHRAE Guideline 0, The Commissioning Process,* contains the fundamental objectives of commissioning and provides limited guidance on conducting the process. The Owner's Project Requirements (OPRs) are the foundation for both the commissioning process and for defining the objectives and criteria that will guide the project delivery team. Commissioning is not only an accepted part of good project delivery but an essential, and in some cases required, part of green building design and construction. *ASHRAE Guideline 1, The HVAC Commissioning Process* (latest approved version), should be followed as minimum practice. The US Green Building Council's LEED manual also provides commissioning guidance. There is also a brief section on commissioning in Chapter 5, "The Design Process—The Early Stages," of this Guide that specifically addresses commissioning activities in design.

An important part of green design is verification that the goals defined by the owner and integrated by the design and construction team are actually achieved as intended, from the first day of occupancy. This verification involves all stakeholders, from designers to construction contractors to operating staff to occupants. Commissioning is not an exercise in blame; it is, rather, a collaborative effort to identify and reduce potential design, construction, and operational problems by resolving them early in the process at the least cost to everyone.

The 1993 National Conference on Building Commissioning provided the definition: "*Commissioning is a systematic process of assuring that a building performs in accordance with the design intent and the owner's operational needs.*" This process provides many benefits to the owner, the design and construction teams, building occupants, and building operators. What owner—particularly a long-term owner—doesn't want reduced risk, fewer change orders (and the resulting cost avoidance), improved energy efficiency, lower operating costs, satisfied tenants/occupants, and a building that operates as intended from day one of occupancy?

What contractor or designer doesn't dream of a project with few or no problems or callbacks and their resulting additional costs? Although numerous commissioning service models exist, it is this author's opinion that commissioning must be performed by a third-party provider or the owner's own commissioning team and begin early in predesign to receive maximum benefit from the commissioning process. Starting early improves designer and contractor quality control processes, identifies and helps resolve problems during design, when corrective action is the least expensive, and during construction, when the contractor has the materials and resources on site for efficient corrective action minimizing post-occupancy repairs.

Commissioning can be broken down into five phases—predesign, design, construction, acceptance, warranty/continuous commissioning (or recommissioning). Distinct commissioning activities occur during each of the five phases. The commissioning process begins in the predesign phase with development of the OPRs. During design, the CxA provides checklists to designers to assist them in their design quality control process and alert the designers on the specifics the CxA will be focusing on during the commissioning design reviews. During construction, the CxA provides construction checklists to the contractors to assist them in their quality control process and, as in the design reviews, verifies that the contractor's quality control process is working. These efforts significantly improve the chances that the systems being commissioned will have minor modifications during performance testing and reduce the delivery team's efforts. The acceptance phase verifies through testing that the systems perform as intended and helps resolve issues prior to occupancy. During the warranty phase, the CxA monitors system performance and verifies that training provided was understood by operators. The CxA assists operators in better understanding their systems, adjusting the systems for maximum performance, which helps prevent inappropriate modifications by the operators due to lack of understanding. Commissioning authorities can also help integrate measurement and verification (M&V) plans and procedures that can be used to identify when a building begins operating outside of allowable tolerances, signaling the owner that corrective action is needed to maintain performance. How and to what extent an owner incorporates commissioning generally depends on the owner's understanding of commissioning. Generally, owners start with construction phase commissioning and soon see how much more they would have benefited by starting in the predesign phase. Other key factors are how long the owner holds the property; the owner's staff capabilities and funding methodology for design, construction, and operation; the project schedule; and ownership experience. This chapter will cover the selection and role of the commissioning provider, a discussion of various commissioning models, the choice of building systems for commissioning, and the long-term benefits provided by verification of project goals.

One of the most beneficial attributes of sustainable development principles contained in the US Green Building Council's (USGBC) Leadership in Environmental and Energy Design (LEED) rating system is the inclusion of commissioning.

If a building's green features do not perform, there is little benefit in having incorporated them in the design. Commissioning provides verification that the building systems operate as intended.

## CxA'S ROLE AND OWNER'S PROJECT REQUIREMENTS

It is the CxA's role to lead the collaborative team effort required to balance competing interests in the owner's favor. To accomplish this task, a benchmark is needed. This benchmark is a document called the Owner's Project Requirements (OPRs).

The OPR document is a written document that details the functional requirements of a project and the expectations of how it will be used and operated. This includes project and design goals, measurable performance criteria, budgets, schedules, success criteria, and supporting information (specific information that should be included can be found in ASHRAE Guideline 0).

In the predesign phase of a project, an owner may express orally (or in a formal written document) the basic requirements of the project. This is called *predesign programming data*. This information may typically include justification for the project, program analysis/requirements, intended building use, basic construction materials and methods, proposed systems, project schedule, and general information (such as attaining LEED certification).

A key element of a CxA's role in the preferred predesign phase model is to develop the OPR document. Much of the predesign programming data could be part of the OPR document. However, since there is such a large amount of information generated in the programming phase of a project, only the concepts most important to the owner—to be tracked through design and construction—should be included. Information extraneous to the actual design, such as justification for the facility, permitting details, history or policy issues, etc., should not become part of the OPR document.

The OPR document forms the basis from which the commissioning provider verifies that the developed project meets the needs and requirements of the owner. An effective commissioning process depends on a clear, concise, and comprehensive OPR document with benchmarks for each of the objectives and criteria. This written document details the functional requirements of the facility and the expectations of how it will be used and operated. The OPR document includes project and design goals, measurable performance criteria, budgets, owner directives, schedules, and supporting information. It also includes information necessary for all disciplines to properly plan, design, construct, operate, and maintain systems and assemblies.

If no formal program exists, the OPR document can be used to assist with identifying the criteria the design team is tasked with meeting. However, the main purpose of the OPRs is to document the owner's objectives and criteria. The designer's basis-of-design documents the assumptions the designers made to meet the OPRs, and a summary of this information is provided to the operators of the project after the design and construction team have long left the project. As such, the OPR document provides the benchmark against which the design, construction, and project operating performance can be measured.

The OPR document is a living document that is updated by the commissioning provider throughout the life of the project, recording the concepts, calculations, decisions, and product selections used to meet the OPRs, as well as to satisfy applicable regulatory requirements, standards, and guidelines.

Further discussion of the CxA's role during specific phases is included in the next section.

## COMMISSIONING PHASES

The role of the CxA varies according to the phase of the project when commissioning starts. Because of the nature of construction, the further along construction is, the more difficult and expensive changes become. Historically, owners and contractors set up a contingency fund intended to cover the unpredictable cost of changes in a project. If the design does not meet the owner's needs, that owner may be forced to accept the project as is because changes to meet what an owner may really need would be too costly at that point. If the CxA is engaged as late as the construction phase, there is some—but very limited—opportunity to address potential design problems. The longer an owner waits to engage a CxA, the less influence the CxA has on resolving problems cost effectively. Using a comprehensive commissioning process that starts during predesign is the best approach, and the role the CxA plays during construction (as outlined in this chapter) is based on the assumption of utilizing a best-practices approach.

### Predesign

The contrast between the quality and quantity of information provided by owners is often related to their development experience. Institutional owners who have developed many buildings and who have held those properties for extended periods of time have often developed over the years the information that design teams need in order to understand an owner's basic needs. As previously stated, one of the greatest values of involving a CxA in the predesign phase is to develop a comprehensive OPR document that serves as the project benchmark, guiding all project team members. This improves project efficiency because the requirements are clearly defined before the design process begins, reducing/eliminating the need for redesign.

### Design Phase

During the design phase of a project, the CxA develops a design-phase commissioning plan, design checklists, and specifications that incorporate commissioning into the project. The commissioning plan provides a narrative of the design-phase commissioning process and establishes roles and responsibilities and a schedule that includes commissioning activities. The design checklists reflect objectives and criteria the designers should check during their quality control process and communicate to the designers the specific focus of the CxA's design reviews. Typically three reviews occur in most design phases. The specifications identify the roles and responsibilities of the project team, the systems that will be commissioned, and the criteria for acceptance of the commissioned systems.

There are two different schools of thought about what the CxA should focus on during a design review. The first is to look only for commissionability of the systems; that is, to only identify design elements that prevent functional testing. An example would be ductwork design that would not allow accurate flow measurements because its configuration causes turbulence. Design reviews that only assess commissionability provide valuable information and address how a system would be tested to verify commissioned system performance.

The second type of design-phase commissioning is more comprehensive; it combines quality review to assess the design, a constructability review to minimize potential change orders, a verification of the designer's quality control process to identify possibly confusing or conflicting information in the contract documents, a value engineering review that does not compromise the serviceability or performance of a facility, and possible alternatives the design team should consider. Third-party comprehensive design-phase commissioning provides great benefit to the project team and owner. (See further discussion in the "Commissioning Models" section later in this chapter.)

The CxA's comprehensive design reviews assess the design against the OPR document and evaluate whether the design meets the owner's objectives and criteria. This review provides a "second pair of eyes" for the owner and designers; it reduces project risk, helps identify and resolve confusing information, evaluates constructability, identifies value engineering opportunities, and evaluates the designer's quality control process. Design-phase commissioning promotes communication, identifies disconnects, questions design elements that appear incorrect, and shares experience to produce a better set of contract documents and a better building.

Several engineering trade magazines, several long-term owners, and insurance companies who provide errors and omissions (E&O) coverage to the design community have all voiced their concerns about the quality of construction documents and have charted how E&O premiums are affected as a result of judgments or settlements. Design-phase commissioning reduces the risks of change orders, accompanying construction delays, and E&O claims, and it helps clarify construction documents. Design-phase commissioning, if correctly implemented, is a seamless process that provides benefits to the entire project team.

One of the roles of the CxA is to assemble a review team experienced in the type of facility being designed. Generally, the CxA has a team of reviewers with specific background and experience to review the disciplines selected for design-phase commissioning. This process often requires the most senior individuals as part of the design-phase commissioning provider team. (See "Selection of a CxA" later in this chapter.) One design review process developed by one company over the years is as follows:

- Written comments from the reviewers are provided to the design team and owner.
- Comments are reviewed, and the design team responds back to the commissioning review team with written responses.
- Meetings are scheduled between the review team and the designers to adjudicate comments as necessary, allowing the owner to understand the issues and having an opportunity to provide direction as needed.
- Design concerns, comments, and actions taken are recorded in the design review document. Changes are made as agreed and the commissioning review team verifies the change and closes the issue as appropriate.

Using a best-practices approach, the design-phase commissioning process could occur four different times during the project related to phase completion: at 100% of schematic design, 100% of design development, 95% complete construction documents, and 100% of construction documents. (Combining the review of the first two phases could be done on smaller projects.) An advantage of four reviews, however, is that the design is evaluated based on the OPR document goals *before* the design development phase starts.

Changes to optimize building performance, daylighting considerations, system selection, and stacking/massing synergies can best be addressed during schematic design review. Review during design development allows the team to identify potential problems and constructability perspectives early enough to resolve many issues before the construction documents phase starts.

The quantity of design concerns typically increases most in the 95% construction document phase because more detail is provided about each building system and component. The concerns identified at that point typically revolve around details—finishes, coordination conflicts, etc.—and resolving these concerns provides clearer direction to the contractors, resulting in better cost and schedule predictions.

Depending on the schedule of the project, it is not uncommon that concerns identified at the 95% construction documents phase often go unaddressed by a design team. This is especially true in fast track projects when designers, responding to owner and contractor demands for documents, struggle to finish and deliver their work product. This is why the 100% review is so important.

The financial benefits of design-phase commissioning are immense. On a project for the State of Louisiana, design-phase commissioning was responsible for $3.4 million in first-cost savings and $5 million in contingency savings on a $100 million project.

### Construction Phase

The CxA's role during the construction phase is to review the 100% complete construction documents and submittals; develop and/or accept/integrate contractor construction checklists; identify and track issues to resolution; develop, direct, and verify functional performance tests; observe construction of commissioned systems; review the O&M manuals provided under the contract; and provide a systems manual. The purpose of these activities is to verify that the OPRs have been met, commissioned systems are serviceable, commissioned systems perform as intended from the beginning, and operational personnel receive the training and documentation necessary for maintaining building performance.

The CxA's review of 100% construction documents is to verify that the concerns addressed during design have been resolved, reducing the risk of contractors building flaws into the project. Sometimes agreement between the designers and the commissioning design review team is not reached during the design phase. An example of this would be a disagreement over building pressurization control: the designer may feel that the design provides adequate control, and the CxA may disagree. The CxA must verify whether or not the building is correctly pressurized through performance tests. If the designer is correct, the CxA closes the issue after verification. If pressurization is an issue, then the team has the opportunity to correct the concern before more serious and costly ramifications can occur. Systems failing to perform are identified and the project team works to resolve the issue while the entire team is still engaged in the project.

The CxA reviews submittals to look for potential performance problems. An example of this would be a contractor's ductwork shop drawings, showing high-pressure loss fittings that would increase energy consumption. The CxA review activity does not, nor is it meant to, take the place of the designer's review, which should reveal whether the contractors are following the designer's intent. Several purposes can be combined by the CxA in his or her review of the submittals, depending on the role the owner defines for the CxA. For instance, if the CxA is also assisting the team with LEED certification, the CxA can verify that the sustainable development goals identified in the OPRs are being met.

At the start of construction, a CxA may choose to be part of pre-bid conferences between the general contractor or construction manager to provide an oral description of commissioning activities, to describe the general roles and responsibilities the contractor will be asked to fulfill in the commissioning process, and to answer questions. Clear communication with the contractors during pre-bid has proven important in preventing high bids due to a "fear factor" from contractors unfamiliar with the commissioning process.

In the early stages of construction, the CxA develops a commissioning plan that defines the commissioning process, the roles and responsibilities of the project team, lines of communication, systems being commissioned, and a schedule of commissioning activities. The CxA conducts an initial commissioning scoping meeting where the commissioning plan is reviewed by the project team and, based on this information, a final commissioning plan is developed and implemented.

Throughout construction the CxA observes the work to identify conditions that would impair preventive maintenance or repair, hinder operation of the system as intended, and/or compromise useful service life and to verify other sustainability goals such as IAQ management during construction. The CxA develops activities to help perfect installation procedures at the start of the specific construction activity and coordinates activities to help ensure that the contractor's quality control process is working through verification of construction checklists, start-up procedures, and testing and balancing. When contractors have completed their construction checklist for a specific system being commissioned, they are stating that their systems will perform as intended; the CxA verifies this by directing and witnessing functional performance testing.

### Acceptance Phase

The functional testing phase of the commissioning process is often referred to as the *acceptance phase*. With designer and contractor input, the CxA develops system tests (functional tests) to check that the systems perform as intended under a variety of conditions. Contractors under the CxA's direction execute the test procedures while the CxA records results to verify performance. The tests should verify performance at the component level through inter- and intrasystem levels. Another practice is to also have the contractors simulate failure conditions to verify alerts and alarms as well as system reaction and interaction with associated systems. Problems identified are resolved while contractors and materials are still on site and the designers are engaged.

### Warranty Phase

The first year of a project is critical to finding and resolving issues that arise, and the CxA plays an important role in helping ensure that a facility performs at its optimum. The warranty period is also the period when the contractors and manufacturers are responsible for the materials and systems installed and the only time the owner has to identify warranty repairs without additional construction costs. As such, the CxA has specific responsibilities during this critical period to assist the owner and operational staff in identifying problems and assisting them with resolution at the least interruption to the occupants.

During the first couple of months, the CxA verifies that systems are performing as intended through monitoring of system operation. Many systems cannot be fully tested until the building is occupied. There might be a small percentage of system components that pass functional testing but, under actual load, fail to perform as intended. These components must be identified in the warranty period and replaced or repaired as necessary. The CxA's role in conjunction with the operational staff is to search out problems that only become evident under actual load. To accomplish this, the commissioning provider performs several specific tasks.

The commissioning provider identifies system points to trend, verifying efficient system operation; installs independent data loggers to measure parameters beyond the capabilities of the building automation system (BAS); and monitors utility consumption. Using the trend system data from the selected BAS input points, the CxA analyzes the information, looking for operational sequences that consume natural resources unnecessarily and conditions that could compromise occupant satisfaction within the working environment. In addition, the CxA also looks for conditions that could result in building failure, such as high humidity in interstitial spaces in the building's interior, hot spots in the electrical distribution system, or analysis of electrical system harmonics where power quality is essential to the owner. These functions can only be tested after occupancy.

Additionally, some systems, such as the heating and cooling equipment, can only be fully tested when the season allows testing under design load conditions. The CxA works closely with the operational staff to identify and help resolve issues that become apparent in the warranty period and verifies that the operational staff fully understands and meets their warranty responsibilities. In addition, the CxA should provide to the operational staff the specific functional test procedures developed for their use in maintaining building performance for the life of the facility. By having the CxA work with the operational staff during the warranty period, the operators gain valuable insight into how the building should operate and what to look for to ensure continued performance. This helps to overcome a typical industry problem in owners' experience: the bypassing of system components and controls because of the operational staff's lack of understanding.

## Continuous Commissioning Phase

The warranty period is only one year in the life of a building that may remain in service for 50 to 100 years or more. Most of the time, remodeling and change of building use are the main reasons for change in building system operation and component performance. For owners to get maximum performance from their facilities, they must know when systems fall outside of allowable performance tolerances. This is best done through a measurement and verification (M&V) process, where operating parameters are tracked and compared to a benchmark. (Where the term "M&V" is used subsequently, it means that actual measurement and verification is to be done as part of the performance evaluation process being discussed.)

The *International Performance Measurement and Performance Protocol* (*IPMPP*), published by the US Department of Energy (DOE) through its Office of Energy Efficiency and Renewable Energy, provides several methods to establish operational benchmarks for energy consumption, as does *ASHRAE Guideline 14-2002, Measurement of Energy and Demand Savings*. With a green design, there are other parameters that must also be measured, including water consumption, waste genera-tion, recycling, pesticide use, etc. The operational tracking of these parameters reduces total cost of ownership, impact on the environment, and building occupants' quality of life. An owner can obtain guidance on integrating sustainable operation practices by adopting the USGBC LEED EB program at this stage of the building's life.

See also Chapter 18, "Operation/Maintenance/Performance Evaluation."

## SELECTION OF A CxA

There are several organizations that provide certification for CxAs. Two of these organizations, the Building Commissioning Association (BCA) and The University of Wisconsin College of Engineering—Department of Engineering Professional Development (UW), provide a comprehensive CxA certification program, requiring several days of classroom instruction with an exam and verified commissioning experience with letters of recommendation from their peers and project owners that demonstrate their ability to implement the commissioning process on both a small and large project. Other organizations generally conduct a one-day training class followed by an exam to receive a CxA certification. As with finding a doctor, lawyer, contractor, or design professional, the key element is that the commissioning provider should have experience in the types of systems an owner wants commis-sioned. In other words, an owner must match the experience with the job. A good CxA generally has a broad range of knowledge: hands-on experience in operations and maintenance, design, construction, and investigation of building/system fail-ures. CxAs must also be detail-oriented, good communicators, and able to provide a collaborative approach that engages the project team.

## SELECTION OF SYSTEMS TO COMMISSION

Commissioning of all systems using the whole-building approach has proven to be beneficial. However, due to budget constraints, owners may want to look at commissioning systems that will yield the greatest benefit to them. Long-term owners have an advantage and can apply their experience of where they have encountered problems historically and elect to commission only those systems. Others who do not have that depth of experience may wish to talk with long-term owners or insurance providers to gain perspective.

Commissioning was originally developed on the West Coast, where energy effi-ciency was the prime driver. The commissioning process has expanded beyond the original commissioning of HVAC systems to include building envelope, electrical, plumbing, security, etc., in addition to the HVAC systems. This expansion from HVAC is often referred to as *whole-building commissioning*.

There are many factors that define which building systems should be commis-sioned, but there are no published standards yet to help guide owners through such selection. It often depends on the associated risk of *not* commissioning. E&O insurance providers publish graphs of claims against design professionals by discipline. Interestingly enough, 80% of the claims against architects are for moisture intrusion. A 1984 World Health Organization committee report suggested that up to 30% of new and remodeled buildings worldwide may be the subject of excessive complaints related to IAQ. The reasons for sick buildings include inadequate ventilation, chemical pollutants from both indoor and outdoor sources, and biological contaminates. The problem of mold and mildew is being called the asbestos of the 2000s.

Commissioning provides several benefits, two of which are risk reduction and lower total cost of ownership. Based on a specific climate such as Phoenix, Arizona, the risk of *not* commissioning the building envelope is much less than in Atlanta, Georgia. Based on functional requirements, not commissioning the security systems in a conventional office building may have minimum risk compared to a federal courthouse. So what should be commissioned?

The best time for determining what systems should be commissioned is during the development of the OPRs. Generally, there are three main system categories that should be commissioned: building envelope, mechanical systems, and electrical systems. Subsystems that could also be included are irrigation and/or process water systems. Depending on the functional requirements of a facility and the complexity of systems, additional systems that may be commissioned include security, voice/data, selected elements of fire and life safety, and daylighting controls.

The US Green Building Council's LEED rating system recognizes the benefits of commissioning and its importance to green building design, construction, and operation. While the LEED reference guide does not specifically identify which building systems must be commissioned, it does point out that energy-efficient use of natural resources, IEQ, and productivity are important goals of sustainable/green projects. As such, the LEED reference guide does imply that building systems that affect energy consumption, water usage, and IEQ should be commissioned.

Each element of green design needs verification to ensure that the design, construction, and operation of the green, high-performance facility meet the expectations of the team and realize the financial return envisioned by the owner.

## COMMISSIONING MODELS

Independent third-party commissioning is the preferred commissioning approach because it significantly reduces the potential for conflict of interest. It also allows for integration of commissioning professionals specialized to meet a project's specific needs, such as building envelope, security, mechanical conveyances, labs, etc.

Commissioning as part of the project design professional's responsibility does not typically result in an unbiased presentation of issues. Design teams do not intentionally provide bad designs; they are working with schedules, budgets, and multiple players in the process. Even if the CxA is a different individual from any design team member but is still within an organization or sister company with the same upper leadership, it is common practice not to bring attention to negative issues. Commissioning conducted under this model has a high probability of minimizing design issues during design and passing responsibility on to the construction team during construction, resulting in "after-the-fact" solutions and corresponding costs and delays. The earlier an issue is identified and resolved, the least cost there is to the project.

Commissioning as part of the general contractor's responsibility has many of the same problems as the model using the design professional. Contractors, by the very nature of the construction business, are focused on schedule and budget. This focus is not always in favor of the owner. Most contractors are quality-minded and do their best to identify problems and assist with resolutions (though often to their detriment because they inadvertently take responsibility for the design in doing so). If a constructability issue arises that will adversely affect the schedule or budget, the contractor may choose to "fix" the issue and hope that it does not create a warranty callback. The main problem is that many of the issues are discovered too late in the process, again resulting in change orders, construction delays, and additional costs.

Mechanical contractors have the same problems as general contractors in providing commissioning services. They work with a schedule and budget and review their own work or that of a supervisor or co-worker. The mechanical contractor's knowledge and experience are usually limited to the mechanical field; thus, they lack the background to commission other systems.

To truly be an owner's advocate, the CxA must owe allegiance to no one but the owner. A third-party CxA will verify that the goals defined by the owner and integrated by the design and construction team are actually achieved as intended, from the first day of occupancy. If the CxA is separate from the design professional or contractor, he/she will provide unbiased reporting of issues to the team and guide them toward timely solutions without finger pointing, delays, and liability.

## A COMMISSIONING CHECKLIST

Finally, from a somewhat different perspective and as a guide, following is a checklist on commissioning used by one design firm.

### One Design Firm's Commissioning Checklist

❑ Begin the commissioning process during the design phase; carry out a full commissioning process from lighting to energy systems to occupancy sensors, etc.

❑ Verify and ensure that fundamental building elements and systems are designed, installed, and calibrated to operate as intended.

❑ Engage a commissioning authority (CxA) that is independent of both the design and construction team.

❑ Develop Owner's Project Requirements (OPRs) and review designer's basis of design to verify requirements have been met.

❑ Incorporate commissioning requirements into project contract documents.

❑ Develop and utilize a commissioning plan.

❑ Verify installation, functional performance, training, and operation and maintenance documentation.

❑ Complete a commissioning report.

❑ Additional commissioning:

–Conduct a focused review of the design prior to the construction documents phase.

–Conduct a focused review of the construction documents near completion of the construction document development effort that is prior to the issuing of documents for construction.

–Conduct reviews of contractor submittals that are relevant to systems being commissioned.

–Provide information required for recommissioning systems in a single document to the owner.

–Have a contract in place to review with operational staff current building operation and condition of outstanding issues relative to original or seasonal commissioning and to provide assistance resolving issues within the one-year warranty period.

❑ Encourage long-term energy management strategies.

-Provide for the ongoing accountability and optimization of building energy and water consumption performance over time.

-Design and specify equipment to be installed in base building systems to allow for comparison, management, and optimization of actual vs. estimated energy and water performance.

-Employ measurement and verification (M&V) functions where applicable.

-Tie contractor final payments to documented M&V system performance and include in the commissioning report.

-Provide for an ongoing M&V system maintenance and operating plan in building operations and maintenance manuals.

❑ Provide for the ongoing accountability and optimization of building energy and water consumption performance over time.

❑ Operate the building ventilation system at maximum fresh air for at least several days (and ideally several weeks) after final finish materials have been installed before occupancy.

❑ Provide for the ongoing accountability of waste streams, including hazardous pollutants.

❑ Use environmentally safe cleaning materials.

❑ Train operation and maintenance workers.

# 4

# ARCHITECTURAL DESIGN IMPACTS

One of an architect's primary functions as part of the design team is to create an environment. This environment has both a psychological and a physiological effect on the occupants, which, in turn, impacts human productivity, building operational efficiency, and effectiveness of natural resource use.

Site location, building orientation and geometry, building envelope, arrangement of spaces, and local climatic characteristics are all elements the design team must address, and the result will have a distinct impact on both the occupants' environment and the efficiency of the building. Buildings that effectively combine their surroundings with the application of green design principles generally provide a project with greater psychological and physiological benefits for the occupants. Buildings that also implement sustainability reduce carbon emissions, use fewer resources for construction and operation, are constructed for longer useful service life, and require less to maintain.

As described in Chapter 3, developing the OPRs is an essential precursor to identifying a project's objectives and criteria, including sustainable/green design goals. The OPR document creates the benchmarks used in assessing the various options a design team develops to meet the project's objectives and criteria during design. This chapter is intended to help designers understand the impacts some architectural decisions have and how these decisions affect sustainable/green project goals.

## SITE LOCATION

Consideration of the implications of site selection is essential to minimize the negative environmental impacts that may accompany a project, from construction activities to those that will occupy the facility. Prudent site selection can lower first cost, operating and maintenance costs, environmental cost, and people cost. Sustainable/green design should consider the larger cost of projects that encroach on animal habitats, prime farmland, or public parks. Other considerations are transportation of materials and labor to construct the project; loss of land that supports biodiversity; the highways, roads, and bridges required to provide access to the facility; the infrastructure needed to support operation of the facility; and the proximity of the facility to residential and other services for building occupants to reduce natural resources needed for transportation.

While design engineers may have little say about the above considerations, it is wise for the architect to involve the engineers early in the site selection process, when possible. Matters such as building form and orientation, nearby pollution sources, ambient air quality, groundwater levels, site drainage, availability of or access to various energy sources (including renewables), and other not-so-obvious characteristics can have implications for a successful design in the later stages.

Further guidance is offered by various rating systems, one of which is the US Green Building Council's LEED rating systems for new construction, commercial interiors, core and shell, and existing buildings, which have specific information on site selection and other site-related concerns.

## SITE ORIENTATION

Building orientation affects many aspects of green design, ranging from energy performance to visual stimulation of the building occupant. Considerations of solar orientation; prevailing winds; availability of natural light; shading created by natural vegetation, topography, or adjacent structures; and views—all impact the designer's choice of how to orient the building on a site. Site orientation can also affect landscaping choices and irrigation water consumption. The benefits, drawbacks, and trade-offs should be weighed when choosing the orientation, and the engineer members of the design team can be particularly helpful here.

Buildings that minimize east and west exposures, especially where a lot of glass is used, are generally more energy efficient because of the huge solar heat gains associated with east- and west-facing elevations during cooling months. If a goal of the owner is to use natural breezes to help meet cooling requirements, then the building needs to be oriented with operable windows and the dominant elevations perpendicular to the prevailing breezes to capture windward/leeward effects and better draw outdoor air through the building. (In some instances this may conflict with minimizing east and west elevations to limit solar heat gain.)

Here is where computer simulations, performed by an experienced energy analyst and yielding fast and factual results, can assist the design team by evaluating nuances in building orientation and the effect various stacking and massing options have on building performance.

## BUILDING FORM/GEOMETRY

A building's form (stacking, massing, and overall geometry) has a significant impact on a building's functionality, energy efficiency, and occupant performance. One of the most important considerations in green design is the effect form has on natural lighting.

Glazing size, orientation, and an occupant's distance from glazing, in addition to glazing characteristics, determine the quality and quantity of natural light reaching a building's interior, as well as occupant views of the surrounding environment. The most desirable natural light comes from the north; it has the least solar heat gain associated with it and is composed of diffused light, which does not cause glare.

The distance natural light will travel into the interior of a building is dependent on window and ceiling height. The quantity of light is dependent on the glazing area. The quality of light is determined by orientation and glazing characteristics. The usefulness of natural light to meet task lighting requirements is a function of light quality on task surfaces. All of these factors affect the form a building takes to meet the requirement for natural daylight harvesting.

Several sources of information are available to assist designers with daylight harvesting strategies. First, Chapter 13, "Lighting Systems," in this Guide offers some basic considerations on daylight harvesting, concentrating on applicability, pros and cons, and cost. Two others, directed more toward architectural design aspects, are

- "Tips for Daylighting with Window," available at eande.lbl.gov/BTP/pub/designguide/download.htm, and
- "Daylighting Design" by Benjamin Evans, in *Time-Saver Standards for Architectural Design Data*, McGraw-Hill, Inc., 1997.

Daylight harvesting is only one of many green factors that may influence building geometry. Buildings designed for natural ventilation could be configured in a form to best capture prevailing breezes and direct them for most beneficial use. Stepping a building back as it rises in height could allow solar access to an adjacent property. The roof of a stepped building could also be vegetated, reducing the quantity and rate of stormwater to be treated.

## BUILDING ENVELOPE

The building envelop performs the primary function of keeping the weather out (and, when feasible, letting its good aspects in), and its design is a key factor that defines how well a building and its occupants perform. (Examples of how the factors discussed in this chapter affect occupant performance are contained in Judith Heerwagen's productivity studies performed at a manufacturing facility designed to maximize the use of natural light [Heerwagen 2001].) Construction materials and techniques of the building envelope dictate the useful service life of the building, IAQ, HVAC sizing, structural design, and maintenance costs, all of which have significant impact on the environment and the total cost of ownership.

### Daylight Harvesting and Energy

Access to outdoor views and natural light have positive psychological and physiological effects on building occupants, but, as noted in Heerwagen's study, too much light and glare can have negative psychological and physiological impacts (Heerwagen 2001). Analysis of the building envelope utilizing daylight harvesting simulation programs can help a design team optimize building geometry, define glazing characteristics based on glazing orientation, and provide essential information needed in performing an energy analysis of the facility. (See Figures 4-1 and 4-2 for daylight harvesting examples.)

Daylight harvesting programs only provide one side of what a design team must consider when creating a building envelope. Honing it to minimize heating and cooling energy consumption requires energy modeling of the building, which is where the engineering side of the design team comes in during this phase of design. The books in the ASHRAE *Advanced Energy Design Guide* series also have tips and offer guidance on building envelope design without conducting energy modeling.

Several software programs exist that allow the results of the daylight harvesting model to be entered into the energy model (see Figure 5-5 in the next chapter). This combination of programs allows evaluation of different glazing characteristics, HVAC system types, and life-cycle costs of various combinations to determine which best meets project goals. Programs such as *Radiance* and *e-quest* are available over the Internet at no cost; these are terrific tools in developing a green design. Other software suites sold by vendors (such as the *IES VE* suite) offer integrated modeling concepts that help evaluate and optimize design solutions.

## Moisture Intrusion

Although a primary function of the building envelope is protecting the building interior and its occupants from inclement weather, an astonishing fact is that 80% of insurance claims against architects are related to moisture intrusion through the building envelope. Further, moisture intrusion is a leading cause of sick building syndrome. Water can enter through the building envelope by three methods: direct rainwater intrusion, water vapor transmission, and negative pressurization (unwanted infiltration).

Design teams often use "belt-and-suspenders" approaches to try to avoid direct rainwater intrusion but then fail to test the design and installation to ensure that the design intent is met. Chapter 3 discusses in detail how commissioning helps ensure that the building performs as intended and verifies that this aspect of green design intent is met.

Often overlooked in design is water vapor transmission into and across the building envelope. Appropriate members of the design team should examine each proposed building envelope assembly type and conduct a vapor transmission analysis for each. Calculation methods for evaluating vapor transmission and determining the likelihood of moisture collecting within the building envelope can be found in the *ASHRAE Handbook—Fundamentals*. IAQ problems and building failure

Figure 4-1   The Hauptmann Woodward Research Laboratory Building in Buffalo, NY, optimizes daylight harvesting with distinctive architectural geometry and building form.

**Figure 4-2 Example of north-facing assembly space that optimizes daylight harvesting while minimizing solar gain.**

resulting from moisture collecting within the building envelope has occurred in most areas of the United States and Canada.

While negative pressurization of a building in an arid climate generally has little air quality impact, IAQ problems *can* result when it occurs in a hot and humid—and sometimes even a moderate —climate. The resulting infiltration of humid air, in addition to being an added air-conditioning cost, can result in condensation in unexpected—and sometimes unseen—places. The ensuing problems (such as mold, mildew, spore production, etc.) can be so severe as to result in building evacuation and extensive remedial costs, sometimes even exceeding the original cost of the building. (Having to build a building twice is not sustainable!) Design teams need to be very conscious of building pressurization and ensure that the building envelope is appropriately pressurized for the climate and intended building use. Here, in particular, coordination of HVAC design with building envelope design is critical to achieving good IEQ and in controlling energy consumption.

## ARRANGEMENT/GROUPING OF SPACES

Although the owner's program, functional needs, daylight harvesting constraints, aesthetics, and many non-engineering green factors go into an architect's determination of how spaces are grouped and arranged in a building, what results can

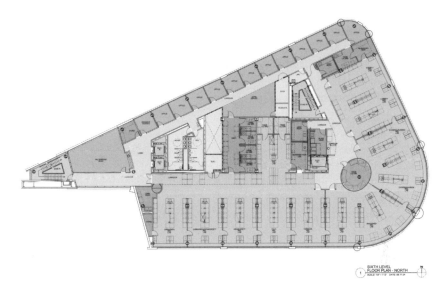

**Figure 4-3** **Good design practice for a typical floor layout optimizes scheduling capabilities and daylight harvesting while promoting interaction.**

also impact how efficiently the HVAC system performs. (See Figure 4-3 for an example floor layout.)

Avoiding unnecessary energy use by providing the capability to shut down or scale back the operation of systems serving building areas not being used is a basic green design principle (use only what is needed), and doing this depends in part on how spaces are arranged or grouped. If a department or group of occupants is known to work on a different schedule than most others, having that area served by a separate air-handling system, for instance, would avoid the need to run one or more large air-handling systems to accommodate the needs of that one group.

While this is only one factor of many that an architect must consider, there is no reason why the HVAC engineer should not ensure that the architect is aware of this factor where it may be applicable. It is also important that the designers help educate the building operators specifically in how the building is intended to be operated from the design perspective and provide strategies that can be employed to reduce utility consumption.

## CLIMATIC IMPACTS

Climatic factors are those conditions, features, or influences external to the building that can have an impact on the building. Some are natural and some are man-made. The key characteristics are: ambient temperature and humidity patterns, ambient air quality, potential pollution sources, solar availability and intensity, wind patterns, soil conditions, freshwater availability and quality, and site drainage.

It is important to understand microclimate issues within a region. For example, there are different parts of San Francisco that have significantly different climates with respect to temperature, humidity, and sunshine.

The climatic characteristics of a site obviously have an impact on how the building performs, especially its energy performance and impact on its surroundings. The design team should be aware of such key characteristics, with each member examining them from the standpoint of his or her own expertise: How will each affect my portion of the design? How will the climate affect the overall design? Can the local climate be utilized or accommodated in a way to further the goal of sustainable/green design for this building, including minimizing carbon emmissions?

Sources for climatic data include the following:

- Regional Climate Data Web site, www.wrcc.dri.edu/rcc.html
- National Oceanic and Atmosphere Administration Web site, www.noaa.gov/climate.html
- *2005 ASHRAE Handbook—Fundamentals*

## INTERDEPENDENCY

The site, resource availability, each element of the design, IEQ, and operability are all interdependently related. Focusing on any one element more than the others can compromise the effectiveness, performance, and cost. Defining the OPRs in the predesign phase of a project documenting the owner's objectives and criteria, including sustainable/green goals, sets a solid foundation for integrated design and improves delivery efficiency as well as the balancing of sustainable principles with how the building will function and the owner's needs.

## BUILDING-TYPE GREENTIPS

A new type of GreenTip is introduced in this second edition of the *ASHRAE GreenGuide*. These tips are called *Building-Type GreenTips*. The intent is to give the design engineer a very general idea of concepts to start with that are specific to certain building types. References are included for more detailed study and analysis of design options. As the Guide is updated in the future, the intent is to provide a greater variety of this type of GreenTip.

## ASHRAE Building-Type GreenTip #1: Performing Arts Spaces

### GENERAL DESCRIPTION

Performing arts spaces include dance studios, black box theaters, recital halls, rehearsal halls, practice rooms, performance halls with stage and fixed seating, control rooms, back-house spaces, and support areas.

### HIGH-PERFORMANCE STRATEGIES

#### Acoustics

1. Clearly understand different criteria for noise criteria levels to be achieved in different type of spaces.
2. Consider 2 and 4 in. liners for large ducts serving spaces with noise criteria levels of 25 and lower.
3. Locate equipment as far away from low noise criteria spaces as practical.
4. Work closely with the acoustic consultant, structural engineer, architect, and construction manager to integrate strategies that eliminate the distribution of vibration and equipment noise from the HVAC systems to the performance spaces.
5. Design duct distribution to eliminate noise transfer between acoustically sensitive spaces. This can be done by using duct liner, additional elbows to isolate sound travel, sound attenuators, etc.
6. Do not route piping systems through or above spaces that are acoustically sensitive.

#### Energy Considerations

1. Demand-control ventilation for high-occupancy spaces.
2. Heat recovery for spaces served by air-handling units (AHUs) with 100% outdoor air capability or over 50% outdoor air component.
3. Consider strategies that allow the significant heat gain from the theatrical lighting equipment to stratify rather than handling all of the equipment heat gain within the "conditioned space" zones in the building.
4. Because of the significant variation in the cooling load throughout the day, incorporating a thermal energy storage (TES) system into the central plant design will reduce the size of the chiller plant equipment, saving capital costs, along with energy and operational costs.

#### Occupant Comfort

1. Consider underfloor supply air/displacement air strategies for large halls with fixed seating.

2. Consider stage air distribution separately from seating air distribution.
3. Consider $CO_2$ sensors in all spaces that have infrequent, dense occupancy.
4. Consider humidification control for all spaces where musical instruments and vocalists will practice, store equipment, and perform.

## KEY ELEMENTS OF COST

1. If properly integrated, an underfloor distribution system should not add significant capital costs to the project.
2. Heat recovery strategies should be assessed using life-cycle analyses. All components of the strategy must be taken into account, including the negative aspects, such as adding fan static pressure and, therefore, using more fan energy when heat wheel or heat pipe strategies are considered.

## SOURCES OF FURTHER INFORMATION

Bauman, F.S., and A. Daly. 2003. *Underfloor Air Distribution Design Guide.* Atlanta: American Society of Heating, Refrigerating and Air-Conditioning Engineers, Inc.

## ASHRAE Building-Type GreenTip #2: Health Care Facilities

### GENERAL DESCRIPTION

Health care facilities are infrastructure-intensive and include many different types of spaces. The HVAC systems for these different types of spaces must be designed to address the specific needs of the spaces being served. The first considerations should always be safety and infection control. In addition, optimizing energy efficiency and positively affecting the patient experience should also be important design team goals.

### HIGH-PERFORMANCE STRATEGIES

#### Safety and Infection Control

1. Consider HEPA filtration for all air-handling equipment serving the facility.
2. Consider air distribution strategies in operating rooms and trauma rooms that zone the spaces from most clean to least clean. Start with the most clean zone being the operation/thermal plume location at the patient, the zone around the doctors, the zone around the room, and then the zone outside the room.
3. Pressurize rooms consistent with AIA and/or ASHRAE guidelines.
4. Provide air exchange rates in excess of AIA guidelines in operating rooms, intensive care units (ICUs), isolation rooms, trauma rooms, and patient rooms.
5. Redundancy of equipment should be designed for fail-safe operation and optimal full- and part-load energy-efficient operation.
6. Intake/exhaust location strategies should be modeled to ensure no reintroduction of exhaust into the building.

#### Energy Considerations

1. Heat recovery for spaces served by AHUs with 100% outdoor air capability.
2. Utilize variable-air-volume (VAV) systems in noncritical spaces working in conjunction with lighting occupancy sensors.

#### Occupant Comfort

1. Acoustics of systems and spaces must be designed with patient comfort in mind.
2. Daylight and views should be provided while minimizing the HVAC load impact of these benefits.
3. Provide individual temperature control of patient rooms with the capability of adjustment by patient.

4. Building pressurization relationships/odor issues should be carefully mapped and addressed in the design and operation of the building.

## KEY ELEMENTS OF COST

1. HEPA filtration costs are significant in both first cost and operating cost. The engineer should work closely with the infection control specialists at the health care facility to determine cost/benefit assessment of the filtration strategies.
2. Heat recovery strategies should be assessed using life-cycle analyses. All components of the strategy must be taken into account, including the negative aspects, such as adding fan static pressure and, therefore, using more fan energy when heat wheel or heat pipe strategies are considered.

## SOURCES OF FURTHER INFORMATION

AIA. 2006. *Guidelines for Design and Construction of Health Care Facilities.* Washington, DC: American Institute of Architects.

ASHE. *Green Guide for Health Care.* American Society of Healthcare Engineering. Available from www.gghc.org.

ASHRAE. 2003. *HVAC Design Manual for Hospitals and Clinics.* Atlanta: American Society of Heating, Refrigerating and Air-Conditioning Engineers, Inc.

The Center for Health Design, www.healthdesign.org.

NFPA. 2005. *NFPA 99, Standard for Health Care Facilities.* Quincy, MA: National Fire Protection Agency.

## ASHRAE Building-Type GreenTip #3: Laboratory Facilities

### GENERAL DESCRIPTION

Laboratory facilities are infrastructure-intensive and include many different types of spaces. The HVAC systems for these different types of spaces must be designed to address the specific needs of the spaces being served. The first considerations should always be safety and system redundancy to ensure the sustainability of laboratory studies. Life-cycle cost analyses for different system options is critical in developing the right balance between first costs and operating costs.

### HIGH-PERFORMANCE STRATEGIES

#### Safety

1. Fume hood design and associated air distribution and controls must be designed to protect the users and the validity of the laboratory work.
2. Pressurize rooms consistent with the *ASHRAE Laboratory Design Guide* and any other code-required standards. Utilize building pressurization mapping to develop air distribution, exchange rate, and control strategies.
3. Optimize air exchange rates to ensure occupant safety while minimizing energy usage.
4. Chemical, biological, and nuclear storage and handling exhaust and ventilation systems must be designed to protect against indoor pollution, outdoor pollution, and fire hazards.
5. Intake/exhaust location strategies should be modeled to ensure that lab exhaust air is not reintroduced back into the building's air-handling system.

#### Redundancy

1. Consider a centralized lab exhaust system with a redundant ($n + 1$) exhaust fan setup.
2. Redundant central chilled-water, steam or hydronic heating, air-handling, and humidification systems should be designed for fail-safe operation and to optimize full-load and part-load efficiency of all equipment.

#### Energy Considerations

1. Heat recovery for spaces served by AHUs with 100% outdoor air capability or over 50% outdoor air component.
2. Utilize VAV systems to minimize air exchange rates during unoccupied hours.

3. Consider low-flow fume hoods with constant volume controls where this concept can be properly applied.

## Occupant Comfort

1. Air systems should be designed to allow for a collaborative working environment. Acoustic criteria should be adhered to in order to maintain acceptable levels of noise control.
2. Daylight and views should be considered where lab work will not be adversely affected.

## KEY ELEMENTS OF COST

1. Heat recovery strategies should be assessed using life-cycle analyses. All components of the strategy must be taken into account, including the negative aspects, such as adding fan static pressure and, therefore, using more fan energy when heat wheel or heat pipe strategies are considered.
2. Low-flow fume hoods should be evaluated considering the impact of reducing the sizes of air-handling, heating, cooling, and humidification systems.

## SOURCES OF FURTHER INFORMATION

Labs 21 Environmental Performance Criteria, www.labs21century.gov.

McIntosh, I.B.D., C.B. Dorgan, and C.E. Dorgan. 2002. *ASHRAE Laboratory Design Guide*. Atlanta: American Society of Heating, Refrigerating and Air-Conditioning Engineers, Inc.

NFPA. 2004. *NFPA 45, Standard on Fire Protection for Laboratories using Chemicals*. Quincy, MA: National Fire Protection Association.

## ASHRAE Building-Type GreenTip #4: Student Residence Halls

### GENERAL DESCRIPTION

Student residence halls are made up primarily of living spaces (bedrooms, living rooms, kitchen areas, common spaces, study spaces, etc.). Most of these buildings also have central laundry facilities, assembly/main lobby areas, and central meeting/study rooms. Some of these spaces also include classrooms, central kitchen and dining facilities, etc. The strategies outlined below can also be applied to hotels and multi-unit residential complexes, including downtown luxury condominium developments.

### HIGH-PERFORMANCE STRATEGIES

#### Energy Considerations

1. Heat recovery for spaces served by AHUs with 100% outdoor air capability serving living units (exhaust from toilet rooms/supply air to occupied spaces).
2. Utilize VAV systems or induction systems for public/common spaces.
3. Natural ventilation and hybrid natural ventilation strategies. (See Figure 4-4 for an example of natural ventilation use.)
4. Utilize electronically commutated motors (ECMs) for fan-coil units.
5. Utilize GSHP where feasible.

#### Occupant Comfort

1. Systems should be designed to appropriately control noise in occupied spaces.
2. Daylight and views should be optimized while minimizing load impact on the building.
3. Consider providing occupant control in all bedrooms

### KEY ELEMENTS OF COST

1. While there is a premium to be paid in first costs for ECMs, many utility companies have energy rebate programs that make this concept acceptable, even on projects with tight budgets.
2. Heat recovery strategies should be assessed using life-cycle analyses. All components of the strategy must be taken into account, including the negative aspects, such as adding fan-static pressure and, therefore, using more fan energy when heat wheel or heat pipe strategies are considered.
3. Hybrid natural ventilation strategies could be utilized using operable windows, properly designed vents using the venturi effect to optimize natural airflow through the building, and shutdown of mechanical ventilation and cooling

systems during ambient temperature ranges between 60°F and 80°F. This will save significant operating costs. The costs of the operable windows and vents will need to be weighed against the energy savings.

## SOURCES OF FURTHER INFORMATION

ASHRAE. 2005. *2005 ASHRAE Handbook—Fundamentals*. Chapter 27, pp. 25.10–25.12. Atlanta: American Society of Heating, Refrigerating and Air-Conditioning Engineers, Inc.

BRESCU, BRE. 1999. *Natural Ventilation for Offices Guide and CD-ROM*. ÓBRE on behalf of the NatVent Consortium, Garston, Watford, UK, March.

Svensson C., and S.A. Aggerholm. 1998. Design tool for natural ventilation. *Proceedings of the ASHRAE IAQ '98 Conference, New Orleans, October 24–27*.

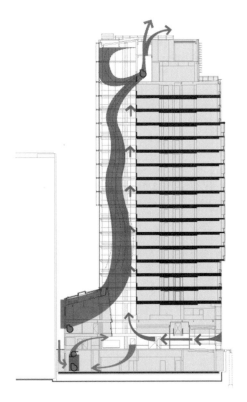

**Figure 4.4 Suffolk University 10 Sommer Street Residence Hall (Boston, MA)—natural ventilation in atrium optimizes views while minimizing solar heat gain.**

# ASHRAE Building-Type GreenTip #5:
## Athletic and Recreation Facilities

### GENERAL DESCRIPTION

Athletic and recreational spaces include pools, gymnasiums, cardio rooms, weight-training rooms, multipurpose rooms, courts, offices, and other support spaces, etc.

### HIGH-PERFORMANCE STRATEGIES

#### Energy Considerations

1. Demand control ventilation for high-occupancy spaces.
2. Heat recovery for spaces served by AHUs with 100% outdoor air capability or over 50% outdoor air component. (See Figure 4-5 for an example of heat recovery use.)
3. Consider strategies that allow the significant heat gain in high volume spaces to stratify rather than handling all of the heat gain within the "conditioned space" zones in the building.
4. Consider heat recovery/no mechanical cooling strategy for the pool area in moderate climates.

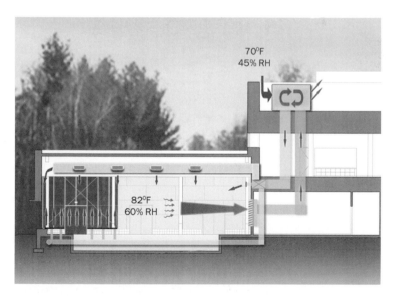

**Figure 4.5    University of Maine Pool HVAC system with heat recovery.**

5. Consider occupied/unoccupied mode for large locker room and toilet room areas to set back the air exchange rate in these spaces during unoccupied hours and save fan energy.
6. Consider heating pool water with waste heat from pool dehumidification system.

## Occupant Comfort

1. Consider $CO_2$ sensors in all spaces that have infrequent, dense occupancy.
2. Consider high-occupancy and low-occupancy modes for air-handling equipment in gymnasiums utilizing a manual switch and variable-frequency drives (VFDs).
3. Consider hybrid natural ventilation strategies in areas that do not have humidity control issues (i.e., pools, training rooms, etc.)

## KEY ELEMENTS OF COST

1. The pool strategy described above should reduce first costs and operating costs.
2. Heat recovery strategies should be assessed using life-cycle analyses. All components of the strategy must be taken into account, including the negative aspects, such as adding fan static pressure and, therefore, using more fan energy when heat wheel or heat pipe strategies are considered.
3. Demand control ventilation adds minimal first costs and often provides paybacks in one to two years.

# 5

# THE DESIGN PROCESS—EARLY STAGES

## OVERVIEW

The design process is the first crucial element in producing a green building. For design efficiency it is necessary to define the owner's objectives and criteria, including sustainable/green goals, before beginning the design to minimize the potential of increased design costs. Once designed, the building must be constructed, performance verified, and operated in a way that supports the green concept. If it is not designed with the intent to make it green, the desired results will never be achieved.

Figure 5-1 conceptually shows the impact of providing design input at succeeding stages of a project relative to the cost and effort required. The solid curve shows that it is much easier to have a major impact on the performance (potential energy

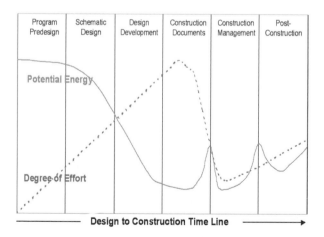

**Figure 5-1  Impact of early design input on building performance.**

savings, water efficiency, maintenance costs, etc.) of a building if you start at the very earliest stages of the design process; the available impacts diminish thereafter as you proceed through the subsequent design and construction phases. A corollary to this is that the cost of implementing changes to improve building performance rises at each successive stage of the project (cost is shown as the dotted curve of this graph).

Designers are often challenged and sometimes affronted by the idea of green design, for many feel they have been producing good designs for years. The experience of many is that they have been forced to design with low construction cost in mind, and when they offer opportunities to improve a building's design, they are often blocked by the owner due to budget constraints. The typical experience is that owners will not accept cost increases that do not show a return in potential savings in five years or less, and many demand eighteen months or less. Many owners, especially owners who own the project for life, allow longer return on investments and use life-cycle cost parameters to lower the total cost of ownership.

Achieving green or sustainable design goals requires a different approach than has been customarily applied. Engineers and other designers are asked to become advocates, not just objective designers. Some have expressed the view that significant reductions in energy usage and greenhouse gas emissions will never occur by simply tweaking current practice. In other words, simply installing high-efficiency systems or equipment will not reduce energy usage sufficiently. Sustainable design requires designers to take a holistic approach and go beyond designing for just the owner and building occupants; they need to look at the long-term environmental impacts the development of a building will create. This may make many uncomfortable because it seemingly asks them to go beyond their area of expertise.

Both first cost and operating costs can be reduced by applying sustainable/green principles. Correct orientation and correct selection of glazing can reduce HVAC equipment size and cost as can the use of recycled materials such as crushed concrete in place of virgin stone for soil stabilization and structural fill.

Using the commissioning process, the CxA can assist the design team with obtaining agreement on the objectives, the criteria, and the sustainable/green goals for the project through the development of the OPRs. The OPRs form the basis from which all design, construction, acceptance, and operational decisions are made (see Chapter 3 for more information). The OPR document provides the foundation of understanding for the designers to efficiently accomplish the task of designing a sustainable/green project within the owner's business model and the constraints and limitations of the project. This process allows improved design efficiency and better team integration because there are clear objectives and criteria established before design begins. While sustainable/green principles can lower first cost as well as operational costs, the soft costs of design do increase slightly due to the additional design and coordination efforts required of the team.

Starting in predesign and carrying through to post-occupancy is essential for the success of a sustainable/green design. It starts with examining every aspect of the process—from the owner's site selection to building configuration, from architectural elements to efficient construction and operation—and can only occur with an integrated approach. Defining the OPR document containing the project goals, even before site selection if possible, is the suggested starting point.

At the very start of the project, green design goals have to be discussed, correlated to the owner's objectives and criteria, agreed to, and in fact embraced by the extended project team. This is often done in a charette format or simply a session spent discussing the issues. As these goals are defined, they are included in the OPR document.

Goals for a project traditionally include the functional program, leasable or usable area, capital cost, schedule, project image, and similar issues. The charette simply puts environmental goals on a plane with the capital cost and other traditional goals.

One of the goals may be to achieve the environmental goals at the same or similar capital cost. (As with any goals, the environmental goals should be measurable and verifiable.) Another of the goals may be a specific green building rating system target, an energy target as well, or perhaps an energy target alone.

A typical set of goals for a green design project might be:

1. Achieve a level of energy use at least 50% lower than the DOE-compiled average levels for the same building type and region, both projected and in actual operation. (Actual energy numbers may be adjusted for actual-vs.-assumed climatic conditions and hours of usage.)

2. Achieve an actual peak aggregate electrical demand level not exceeding 4.5 $kW/ft^2$ (50 $kW/m^2$) of building gross area.

3. Provide at least 15% of the building's annual energy use (in operation) from renewable energy sources. (Such energy usage may be discounted from the aggregate energy use determined under Goal 1 above.)

4. Taking into account the determinations of Goals 1, 2, and 3 above, assess the impact of the lesser net energy use on raw energy resource use (including off-site) compared to that of a comparable but conventional building, including the changed environmental impacts from that resource use, and verify that the aggregate energy and environmental impacts are no greater.

5. Achieve a per capita (city) water usage 40% lower than the documented average for this building type and region.

6. Achieve an aggregate up-front capital cost for the project that does not exceed $x$ dollars per $ft^2$ ($m^2$) of building gross area, which has been deemed by the project team to be no higher than 102% of what a "conventional" building would cost.

7. Recycle (or arrange for the recycling of) at least 60% of the aggregate waste materials generated by the building.

8. By means of post-occupancy surveys of building users conducted periodically over a five-year period, achieve an aggregate satisfaction level of 85% or better. Survey shall solicit occupant satisfaction with the indoor environment as to the following dimensions: thermal comfort, air quality, acoustical quality, and visual/general comfort.

9. Obtain a gold-level USGBC LEED certification for the building.

(For an example of what one major firm has done, see the sidebar, "One Firm's Green Building Design Process Checklist," on page 98.)

## THE OWNER'S ROLE

Of all of the participants, it is the owner who is the most crucial in making a green building happen. With the owner's commitment, the design, construction, and operating teams will receive the motivation and empowerment needed to create a green design.

Key design team members can—and should—attempt to educate the owner on the long-term benefits of a sustainable/green design, particularly if the owner is unfamiliar with the concept. After all, experienced design team members are in the best position to sell the merits of green design. However, such a commitment on the part of the owner, to be effective, must be made early in the design process.

Specific roles that an owner can fill in making a sustainable/green design effort successful include the following:

- Expressing commitment and enthusiasm for the green endeavor
- Establishing a basic value system (i.e., what is important, what is not)
- Selecting a CxA
- Participating in selection of design team members
- Setting schedules and budgets
- Participating in the design process, especially the early stages
- Maintaining interest, commitment, and enthusiasm throughout the project.

Strictly speaking, the "owner" on a project could be a corporation or small business, hospital, university or college, office building developer, nonprofit organization, or even an individual. In any case, that owner will have a designated representative on the building project team, presumably one who is very familiar with the owner's views and philosophy and can speak for that owner with authority.

## THE DESIGN TEAM

### Setting It Up

One of the first tasks in a sustainable/green design project is forming the design team and the commissioning team. This team should include the design team leader (often the architect), the owner, the CxA, the design engineers, and operations staff. Much of the design team's successful functioning depends not just on having ideas about what should go into the project but on being able to analyze the ideas quickly and accurately for their impact. A good part of this analysis will fall to one of the engineering disciplines to accomplish.

A traditional project team includes the following members:

- Owner
- Project manager
- Architect
- HVAC&R engineer
- Plumbing/fire protection engineer
- Electrical engineer
- Lighting designer
- Structural engineer
- Landscaping/site specialist
- Civil engineer
- Code enforcement official

An expanded project team for a sustainable/green design with commissioning would also include:

- Energy analyst
- Environmental design consultant
- Commissioning authority
- Construction manager/contractor
- Cost estimator
- Building operator
- Building users/occupants

The above lists the *possible roles* that might need to be filled on a reasonably large design project. Some roles may not be applicable or even needed on certain types of projects (e.g., civil engineer or landscaping/site specialist), and other roles may not be feasible to have represented in the early stages of project (e.g., building operator, building users, code enforcement official). Further, the variety of roles does not mean that there needs to be an equal number of distinct individuals to fill them; one individual may fill several roles, e.g., the architect often serves as project manager, the HVAC&R engineer as plumbing engineer or energy analyst, the electrical engineer as lighting designer, and a contractor on the team as cost estimator. Likewise, depending on the type of project, there could be other specialists as well.

Certain nontraditional roles are particularly important in green design:

**Energy Analyst**—Although this role has existed for some time, it assumes a much more intense and timely function in sustainable/green design, as there is a need to quickly evaluate various ideas (and interactions between them) in terms of impact on energy. These can range from different building forms and architectural features to different mechanical and electrical systems. The person in this role must be intimately familiar with energy and daylight analysis modeling tools and able to provide feedback on ideas expressed reasonably quickly. In short, he or she is a much more integral part of a sustainable/green design team than in a traditional design effort. In this respect, for a sizable project, it might be difficult for a single person to fill this role plus another as well.

**Environmental Design Consultant (EDC)**—As owners begin to request sustainable/green buildings from the design professions, a new discipline has emerged: the EDC. The role of this person is to help teams recognize design synergies and opportunities to implement sustainable and green features without increasing construction costs. When the CxA is also the EDC, the owner and designers benefit from an improved integration of the design process across disciplines, with the intent of creating an outcome with much lower environmental impact and higher user satisfaction. Leading projects show that this can often be accomplished without adding cost. The EDC has input in areas such as site, water, waste, materials, IEQ, energy, durability, envelope design, renewable energy, and transportation, and the CxA documents these objectives and criteria into the OPR document. Although this Guide will focus primarily on those areas pertinent to the HVAC&R design professional, it is becoming evident that this profession must broaden its sphere of concern in order to contribute meaningfully to the creation of sustainable/green buildings.

The CxA already has a collaborative relationship with the designers and, as an EDC, works with the HVAC&R team and others throughout the process to meet the owner's objectives and criteria. For example, a design reviewer may raise questions for the team to consider, such as:

- Is the building orientation optimized for minimum energy use?
- Is the combined system of building envelope, including glazing choices and the HVAC system, optimized for minimum energy use and lowest life-cycle cost?
- Are the loads, occupancy, and design conditions properly described?
- Are the proposed analytical tools adequate to the task of computing life-cycle costs and guiding design decisions?
- Is the proposed mechanical approach going to deliver excellent air quality to occupants under all conditions?
- Is the proposed mechanical approach going to deliver thermal comfort to occupants under all conditions?
- Is the proposed mechanical approach going to be easy to maintain? Is there enough space for mechanical equipment and adequate access to service and perhaps to eventually replace it?

- Is the proposed mechanical approach going to give appropriate control of the system to users?
- Is the proposed mechanical approach going to consume a minimum amount of parasitic energy to run pumps and fans?
- Are there site or other conditions likely to impact the mechanical system in unusual ways?
- Have all the impacts of the building on the site and surroundings been identified and taken into account?
- Are the proposed systems properly sized for the loads?

While it may seem that the role of the EDC is very similar to that of the energy analyst, the roles differ in that the EDC is more of a question-asker or issue-raiser, similar to the CxA asking questions as well as suggesting strategies or solutions for the team to consider. The EDC's brief is broader and more comprehensive in scope; his or her role is to stand back somewhat from the project and ask the broader questions regarding the environmental impact of the project. The CxA verifies that the sustainable/green goals contained in the owner's stated objectives and criteria are being met.

**Commissioning Authority (CxA)**—(Please refer also to Chapter 3, "Commissioning.") The CxA has the very important role of documenting the OPRs as early as possible, starting in the predesign phase of the project. This function is beneficial to both the owner and project team in that it condenses the mass of information into a single, cogent document; it records the owner's objectives, criteria, and goals and benchmarks for gauging success in achieving the defined requirements. The OPR document forms the basis from which all design, construction, acceptance, and operational decisions are made. Changes to the OPR document for various reasons are directed by the owner and generally provide clarification of an objective or goal or trade-offs due to budget constraints. In response to an OPR document, designers develop design concepts, having the benefit of well-defined requirements, and document their assumptions, studies, and accompanying calculations into the basis of design. The designer creates the basis of design to clearly convey the assumptions made in developing a design solution that fulfills the intent and criteria in the OPR document. The OPR document records the various changes in design direction, why they occurred, and the assumptions made by the design team. The document is updated as changes occur throughout the project and tracks why these changes were necessary.

Successful cost-effective application of green design principles must start early and be defined in predesign; this allows the team to look for synergies that help control hard and soft costs by more accurately defining design direction. The CxA helps the design team define what the owner is communicating and can help draw out how the owner expects the building to function and perform.

The CxA incorporates the information into the OPR document, which is used during the project as a benchmark for judging how well the project team meets the project requirements; it also serves as a written reminder of the goals and decisions that resulted in the final deliverable to the owner. The final version of the OPR document should be refined to approximately 20 to 30 pages in length and delivered along with the commissioning report as a reminder of the designer's original charge and the assumptions that were made with the owner's knowledge and direction.

The OPR document also serves as a guide to the building operating staff on how the facility was intended to operate and the features designed into the project. Development guidelines for the OPRs are contained in *ASHRAE Guideline 0, The Commissioning Process*.

## The Team's Role

Green design requires owners to make decisions sooner, design documents to be more complete and comprehensive, the construction process better coordinated, and operators better trained in maintaining facilities. All of this will impact the viability and success of a green project endeavor. Contractors not familiar with this project model may sound the cost and schedule alarm due to their inexperience in the new procedure. First-time application of sustainable development principles can result in slightly higher first costs, but this phenomenon will reverse itself as teams gain experience and improve their learning curve. As the building industry becomes more familiar with applying these principles, lower costs of ownership will result.

In addition to the standard tasks associated with a design project, the design team is responsible for developing and implementing new concepts that will create a green project. For most, this will require learning on their own time, becoming familiar with new advances in software tools, green materials, and alternative systems. There is an abundance of information; advances are occurring daily in the development of green products and materials as well as processes. The speed of these changes requires designers to add continuously to their knowledge base.

The greatest challenge to accomplishing green design is creating a team organizational structure that provides:

- Criteria for assessing how green the project should be.
- Strong leadership through the green design process to integrate team members.
- Clearly defined objectives after careful examination of design alternatives, costs, and schedule impacts.
- Documentation of success.
- Strong leadership by experienced green building practitioners leading the team through the decision process; this can help overcome confusion about applying green principles.
- Definition of what tasks are required to accomplish green design.
- Identification of who is responsible for each of the tasks.
- Identification of when tasks must be completed so as not to impede the design process or affect the project schedule.

- Establishment of criteria for the selection of green design features considered for incorporation into a project.
- Assistance with integrating selected green design goals into the construction documents.
- Definition of the level of effort required for each of the green project goals.
- Help to enable contractors overcome psychological and physical constraints.
- Establishment of how to track, measure, and document the success of accomplished project goals.

The designers must also help inform their clients that there are costs for the depletion of resources to be consumed beyond the cost of extraction. The practice of looking only at simple payback when analyzing alternatives based on extraction cost has never been realistic because there is no way to replace many resources at *any* cost.

Currently, most design teams are eager to develop green designs when given the opportunity but lack the experience of actually integrating green design into their projects. The addition of an experienced EDC will shorten the learning curve by helping them integrate their extensive knowledge that, when combined, will result in a cost-effective practical green design that meets the owner's requirements. In addition, most teams struggle with what makes a design green, how to incorporate green design principles, and the logistics of incorporating these principles into the design. Green design creates a need for a broader involvement of disciplines and a wider range of experience to ensure that a wider range of input and participation gets factored into the decision-making process.

The project team—from initial concept through construction documents, construction, and building operations—must work as an integrated unit to successfully achieve the goals set by the owner's objectives and criteria, creating better project performance, a basic principle of green design. This model will require the project team to investigate new approaches and process more information than ever before as they strive to increase performance and lower the total cost of ownership. The decision-making process must change from the traditional hierarchical method, with an emphasis on lowest first cost, to an integrated method focused on life-cycle cost. To achieve this requires close collaboration of the project team combined with innovative thinking among all disciplines.

The design team's responsibility as part of the project team is to assist the owner with setting sustainability/green goals that often include:

- Life-cycle cost optimization of energy-consuming systems, materials, and maintenance
- Systems integration and maintainability
- Minimization of environmental impact
- Documenting basis of design
- Assisting with training of building O&M staff during commissioning

## Team Leadership

The integrated building design process requires more effort between the team members to explore the various opportunities to incorporate sustainable principles into the project. For example, the architect, mechanical engineers, and electrical engineers must interact closely to develop a high-performance building that will provide an improved work environment, to lower operating costs, and to minimize consumption of natural resources. Strong leadership is essential to meet these objectives.

Designers should provide input to the OPR document to help establish sustainable/green project criteria. Designers can suggest criteria for sustainable/green goals or take the easier route by using one of the established rating systems developed through a balanced consensus process and add goals not contained in the established rating system.

## THE ENGINEER'S ROLE

The HVAC&R engineer is a crucial player in the design of a "green building." In fact, it is virtually impossible (and certainly not cost-effective) to design a green building without major involvement of that discipline. The HVAC&R engineer must get outside the normal "box" in which he or she lives and become more involved in the "why" of a design, as well as the "how." This means moving beyond simply responding to questions asked by others. It is important to participate in the decision making regarding how project goals will be achieved.

Engineers help analyze the various options to be considered, create mathematical computer models that are used to judge alternatives, provide creative input, and assist with development of new techniques and solutions. The HVAC engineer can be invaluable in helping the architect with building orientation considerations, floor plate form and dimension, and deciding which type of glazing will provide the maximum quantity of natural light, while at the same time analyzing the heat transfer characteristics of the glazing options. The HVAC engineer can also help the architect select structural systems and exterior walls to utilize thermal mass features to reduce equipment needs. Working with the electrical engineer and architect, the HVAC&R engineer can offer ideas and various options, such as incorporating daylighting and lighting controls to reduce artificial light when natural light is available, which, in turn, can result in lower cooling requirements and lowered HVAC requirements to meet peak load. Smaller equipment size translates into reduced structural and electrical requirements, lower operating and maintenance costs, and lower construction costs, all of which lower the total cost of ownership.

The plumbing engineer, working with the structural and civil engineers, and the landscape architect can reduce the facility's potable water, sewer, and stormwater conveyance requirements. Some examples are waterless urinals or use of stormwater or graywater for irrigation of vegetation or for flushing toilets. Depending on the type of building, water from condensate can be used for graywater applications or for cooling tower makeup. The design engineer must weigh the benefits of water-cooled condensers versus air-cooled condensers and the water versus electrical energy consumed by each. The engineer must examine the site climate, determine what alternatives and strategies can best be applied, and develop life-cycle analyses to guide the owner through the decision process posed by the maze of complex issues surrounding green design.

## PROJECT DELIVERY METHODS AND CONTRACTOR SELECTION

Successful projects depend upon the entire team of players involved: architect, engineers, program managers, construction managers, owner's representatives, facilities personnel, building users, and contractors. It is assumed that all parties will be ethical, reliable, diligent, and experienced. There are a number of project delivery methods that could be used to deliver the design and construction of a project. The three major methods to be briefly discussed here are the construction manager approach, the design/bid-build (D/B-B) approach, and the design/build (D/B) approach. It is important that the project delivery method be chosen early in the project for the same reasons that it is important when considering green design options. The discussion of the three methods below is intended only to relate to the effect this decision will have on the success of the optimized design and operation of the building.

### Construction Manager

The construction manager method is the process undertaken by public and private owners in which a firm with extensive experience in construction management and general contracting is hired during the design phase of the project to assess project capital costs and constructibility issues. This is especially important when considering design alternatives that are being considered in an effort to deliver a high-performance building to the client. The initial design process often includes a project definition stage, or programming, in which the owner works with the design professionals, the CxA, and the construction manager to define the specific scope of the project. The design professional utilizes this information to prepare a set of bidding documents that the construction manager uses to obtain bids from qualified subcontractors. The lowest responsible price is usually selected and the contractor then constructs the project.

*Advantages:* Budget control, buy-in of green concepts by contractors.

*Disadvantages:* Perceived lack of competitive general contractor and subcontractor pricing, innovative systems could be shelved due to overly conservative first cost estimates.

## Design/Bid-Build

Design/bid-build construction is the traditional process undertaken by public and private owners. The initial design process often includes a project definition stage, or programming, in which the owner works with the design professionals to define the specific scope of the project. The design professional utilizes this information to prepare a set of bidding documents. The bid documents are then available for qualified contractors and subcontractors to prepare pricing. The lowest responsible price is usually selected and the contractor then constructs the project.

*Advantages:* Usually results in the lowest first costs at the outset of the construction of a project.

*Disadvantages:* No contractor buy-in to green process and concepts, prequalification of contractors is difficult to do well.

## Design/Build

Design/build (D/B) construction is typically a response to a request-for-proposal (RFP) developed by an owner. The RFP is usually a document that defines the general scope of the project and then solicits price proposals to accomplish this work. The work effort to prepare the specific design of the project is to be included in the D/B offering. The D/B team usually consists of an architect, engineers for the various disciplines involved, a general contractor, and the trade subcontractors. This entire team should be in place until the project is turned over to the owner.

As the D/B team develops the design, it must respond to the premises defined in the original scope of work.

*Advantages:* Can result in lowest first cost, agreement by design/construction team with regard to design and building operation concepts.

*Disadvantages:* Can result in uneven distribution of risk among team members, can result in loss of design team members as owner advisors.

## Factors in Choosing an Approach

In all of the above scenarios, the team or contractors should be prequalified to perform the work prior to a request for pricing. Prequalification involves certain information that allows the selection of potential constructors:

- Experience in similar work
- Record of past performance by responsible references
- Financial capability
- Workload

Experience, past performance, and financial stability are all important factors. A record of exemplary performance and fiscal capability may be more important than experience in similar work. In any case, it is valuable to preselect the teams or contractors from whom you request pricing. Successful projects occur due to careful planning and implementation.

## SUCCESSFUL APPROACHES TO DESIGN

### Sustainable Design with System Commissioning

Universities throughout the country, recognizing the long-term cost benefits of sustainable design with system commissioning, are successfully implementing changes in how their projects are designed, constructed, and operated. Emory University's Whitehead Biomedical Research Building, for example, utilizes enthalpy wheels that recover 83% of the energy exhausted by the general exhaust system, captures air-conditioning condensate to displace potable water otherwise used for cooling tower makeup, and captures all the rainwater from the roof for site irrigation, displacing over three million gallons of potable water usage per year.

The following two sections describe two approaches to the green building design process that have proved successful.

### Low Energy Design Process: NREL's Experience

This energy design process was used to design and construct the thermal test facility (TTF) at the National Renewable Energy Laboratory (NREL). The TTF is a 10,000 ft$^2$ (929 m$^2$) office and light laboratory building constructed in 1996 in Golden, Colorado. Actual performance data collected for more than one year show that the TTF costs 63% less to operate than an equivalent building that complies with ASHRAE Standard 90.1. The process used is summarized in the sidebar, "NREL's Nine-Step Process for Low-Energy Building Design," on page 97.

Further information may be obtained from the *ASHRAE Journal* article "Optimizing Building and HVAC Systems" (Hayter et al. 1999).

### Integrated Design Process: Canada's Experience

Recent building design experience in North America and Europe led to the recognition of key factors that are relevant to the achievement of very high levels of environmental performance. These include use of an integrated design process (IDP), which incorporates passive and bioclimatic approaches and also includes an iterative process. (For a description of a specific program developed to investigate the feasibility of designing high-performance buildings using this process, see the sidebar on Canada's C-2000 program on page 99 of this chapter.)

## What Does *Integrated Design* Mean?

One of the key attributes of a well-designed, cost-effective green building is that it is designed in an "integrated" fashion, wherein all systems and components work together to produce overall functionality and environmental performance. This has a major impact on the design process for HVAC-related systems, as conceptual development must begin with HVAC system integration into the building form and into the approaches being taken to meet other green building aspects. For example:

- HVAC systems that employ natural ventilation and underfloor air distribution, often used in green buildings, can have major impacts on building form.
- Other building energy innovations, such as daylighting, passive solar, exterior shading devices, and active double wall systems, often have significant impacts on the design of the HVAC system.
- On-site energy systems that produce waste heat, such as fuel cells, engine-driven generators, or microturbines, will affect the design of HVAC systems in order for waste heat to be most effectively utilized.

Beyond these form-giving elements, there are many other specific features of a green building that affect (or are affected by) HVAC systems to achieve the best overall performance. Some of these strongly impact HVAC system conceptual design, and some require only minor adjustments to HVAC specifications. Such features might include:

- Using the commissioning process to document the owner's defined sustainable/green objectives and criteria and assist the project team to deliver.
- Effective use of ventilation (and IAQ sensors linked to the ventilation system) to improve IAQ.
- Provision of user controls for temperature and humidity control.
- Reduced system capacities to reflect lower internal loads and building envelope loads.
- Utilization of TES systems can reduce the overall size of the chiller plant equipment, such as chillers, cooling towers, and pumps. Reduction in chiller plant equipment size will save capital costs, ongoing energy, and operational costs and can reduce outdoor noise levels during the daytime.
- Selection of non-ozone-depleting refrigerants.
- Reduction and optimization of building energy usage below the levels of ASHRAE Standard 90.1-based codes or other applicable state and local energy codes. (Levels of reduction of as much as 40% to 50% below Standard 90.1 are becoming more common and are encouraged.)
- Use of reclaimed water for cooling tower makeup, and minimization of cooling tower blowdown discharge to the sanitary sewer system.
- Testing of the key systems, especially the HVAC systems.

## Key Steps

The IDP includes the following elements:

- Ensuring that as many of the interested parties as possible are represented on the design team as early as possible—this includes not only architects, engineers, and the owner (client), but also CxA, construction specialist (contractor), cost estimator, operations/maintenance person, and other specialists (outlined below).

- Interdisciplinary work among architects, engineers, costing specialists, operations people, and other relevant persons right from the beginning of the design process.

- Discussion and documentation by the owners and the design team of the relative importance of various performance and cost issues and the establishment of a consensus on these matters between client and designers and among the designers themselves.

- Provision of a design facilitator (or EDC) to suggest strategies for the team to consider, as well as a CxA to raise performance issues throughout the process and to bring specialized knowledge to the table.

- Addition of an energy specialist to test various design assumptions through the use of energy and daylight simulations throughout the process, to provide relatively objective information on a key aspect of performance.

- Addition of subject specialists (e.g., for daylighting, thermal storage) for short consultations with the design team.

- Clear articulation of performance targets and strategies to be updated throughout the process by the owner and the design team.

## Iterative Design Refinement

The design process requires the development of design alternatives. To come up with the most effective combination, these alternatives must be evaluated, refined, evolved, and finally optimized. This is the concept of *iterative design*, wherein the design is progressively refined over time, as shown in Figure 5-2.

Often fee and schedule pressures lead the designer to want to lock in a single design concept at the beginning of the project and stick with it throughout. But this precludes the opportunity to come up with a "better" system that reflects the unique combination of loads and design integration opportunities for this specific building. This better design usually evolves during schematics and early design development in the iterative process.

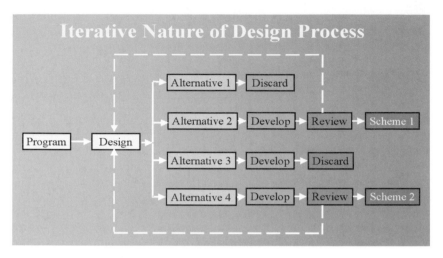

Figure 5-2   The iterative design process.

## CONCEPT DEVELOPMENT

### The Big Picture

Designers should always keep in mind the three major steps for achieving a sustainable/green design:

- Reduce the loads.
- Apply the most efficient systems.
- Look for synergies.

**Reduce the Loads.** If you have a building with a normal 500 ton cooling load, is it possible to provide a more comfortable environment with, say, only 300 tons? The team would have to really work to achieve this: solar loads on the building would have to be reduced; lighting loads to the space would also have to be lowered; maybe the building could use daylight rather than electric light during the day, so the building shape would be influenced; and the site for the building, its shape, its thermal mass, and its orientation could all work together to reduce the cooling load. Early and quick modeling can provide interesting information to assist decisions.

Similar things could be done with the heating load. Does the winter sun provide much of the building's heating needs during the day? With design changes, could the sun do more?

These considerations in a hierarchal design process are not typically brought to the attention of the HVAC engineer, who could help significantly improve the energy efficiency instead of just calculating the loads and could apply energy-efficient systems. Significant reducing utility consumption and environmental impact cannot occur by simply doing the same old job just a little bit better.

The HVAC or energy engineer can make a positive contribution to the success of a sustainable/green design. The value of the engineer to the project is significantly increased, and the results are reduced heating and cooling loads, as well as overall utility consumption.

Someone once suggested an attitudinal approach to building design that says the designer should strive toward making the building inherently "work by itself." Building systems are there simply to fine tune the operation and pick up the extreme design conditions. In contrast, buildings are often traditionally designed like advanced fighter aircraft: if the flight computers are lost, the pilot cannot fly the plane.

**Apply the Most Efficient Systems.** This is the world of the energy-efficient engineer. This is the area where ASHRAE generally operates. While it is very important, it is not enough by itself.

**Look for Synergies.** The preceding two major steps have the potential of increasing capital costs. Therefore, you might have a wonderful, energy-efficient building that will not get built due to high first cost, or the "value-engineering knives" come out and cut the project back to a traditional, affordable project—proof perhaps that "green cannot work." Part of the solution to get around this syndrome is to look for synergies of how building elements can work *together*. This also relates to the cost transfer mentioned earlier.

If a building has a large amount of southern exposure, exterior-shading devices might significantly reduce the summer solar load while still admitting lower angle winter sun. Daylight (but not direct sun) would allow shutting off the electric lights on sunny days. The HVAC system for the south perimeter zones could be significantly reduced in size and cost as the simultaneous solar and electric lighting loads are reduced. Indeed, the very nature of the HVAC system might well be simplified due to the significant load reduction. Resulting cost savings can be used to pay for some or all of the additional treatments.

A major benefit to an integrated design that is on budget is that you avoid wasting a lot of time on elemental payback exercises and value engineering (and cost cutting) because you are on budget. Many of the integrated solutions work so that if you save by cutting out an element such as the exterior shades, there is an additional cost in another area such as the size of the HVAC system.

## The Nitty-Gritty

The success of green design starts with establishing the project's goals and objectives, defining roles and responsibilities of each team member, creating communication between design team members, developing a decision-making process, and clarifying the level of effort that will be required by each member of the team. It is recommended that a workshop be conducted to introduce the team to the process of integrating sustainable development principles and determining how decisions establish objectives and criteria that are documented in the OPRs and provide guidance to the team in the delivery of the project. Objectives and criteria are tracked and used to form concise benchmarks to gauge and document the team's success in meeting the established goals. Commercially available software can assist with cost-benefit analysis, design coordination, energy and daylight analysis, and organizing process design.

The creation of documentation supporting both the decision-making process and the results of decisions is important in guiding and determining the success of green design efforts as well as establishing what was or was not successful and the reasons why. Like a business plan or construction plan, it is important to measure milestones so that adjustments can be made to correct course deviations in reaching the goals. Sustainable/green design objectives and criteria defined in the OPR document should also identify the criteria on which life-cycle cost analyses are based; also, any assumptions made must be recorded in the basis-of-design document so comparison against actual performance can be measured. Learning from the deviations that occur will allow teams as well as individuals to grow from the experience.

The owner's sustainable/green design objectives and criteria documented in the OPRs provide the organizational structure required for successful projects. Software tools available today also increase communication within a team, help stimulate innovative thinking, and help teams optimize design trade-offs by grouping related issues.

The team must develop consensus criteria such as:

- Selecting a site that minimizes environmental impacts.
- Utilizing existing infrastructure to the maximum extent possible to avoid building additional infrastructure to support the project.
- Minimizing the impact of automobiles and the infrastructure required to support them, such as parking, roads, and highways.
- Developing high-performance buildings that enhance occupant productivity and comfort, minimize energy and water consumption, and are durable and recyclable at the end of their useful service life.

Based on the consensus criteria selected, identify potential goals. Once goals are identified, develop tasks necessary to obtain these goals, including studying the impacts these goals will have upon

- project cost, schedule, and energy and water usage;
- indoor environmental quality, operational and maintenance costs, life of the building, and occupant productivity; and
- environmental impacts at the end of the building's or whole facility's useful service life.

Next, assign roles and responsibilities by identifying who is responsible for each task, when the task must be completed, and in what chronological order tasks must be completed so as to facilitate the tracking and management of the sustainable/green design process. The design and construction commissioning plans and checklists can significantly assist a team in accomplishing their goals and minimizing wasted effort.

A good OPR document provides the team the information needed to guide the team's decisions, measure the team's success, and document changes and the reasons for change in the project.

## EXPRESSING AND TESTING CONCEPTS

Expressing concepts is very important in green design because that is the way ideas and intentions are communicated to the owner and others on the design team. This is especially true since green design requires the close and active participation of many different parties.

There are three ways of expressing concepts in the design of buildings; two of these are the traditional *verbal* means and the other is the *diagrammatic* or *pictorial* means. The third has come of age more recently along with computers: *modeling*.

## VERBAL

Both the written and the spoken word play an especially important part in green design. Because there are many meetings or charettes where ideas are explored and intentions voiced, getting across what is expressed accurately assumes significance. Then, succinctly and clearly putting down on paper what has been expressed (memorializing it) is also critical to the various team members as they each go about filling their respective roles. To illustrate, one need only read Chapter 3, "Commissioning," for it to become apparent how important a written record of "what happened" during design (i.e., design intent, assumptions, etc.) is to the successful follow-through of a well-executed green design.

## DIAGRAMMATIC/PICTORIAL

The use of diagrams, sketches, photos, renderings, etc., a tried and true method of communicating a lot of information, continues to be an essential part of green design (Figure 5-3). The old adage "one picture is worth a thousand words" most certainly applies here. But there is now available a relatively new way of "creating a picture" (Figure 5-4) of a building, an energy system, or a year of operations—*modeling*.

## MODELING

This computer-age technique plays such an important part in green design because of its speed, accuracy, and comprehensiveness.

Everyone is familiar with how speedy computers are, once the input data are entered. The "slow" part in this process is the human analyst, the one who converts intentions and ideas into computer-modeling program input, which is why it is *so* important for that analyst to be very conversant with the modeling process. This is especially true for load and energy calculations that impact HVAC&R systems. The team has an idea, and they want an answers *fast* as to how well that idea would work!

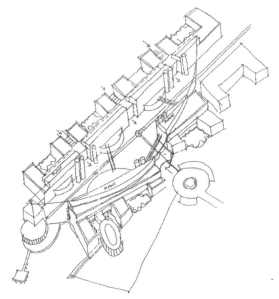

IES VE Conceptual Design Presentation

**Figure 5-3   Example of diagrammatic building sketch.**

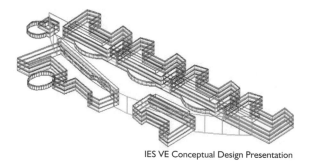

IES VE Conceptual Design Presentation

**Figure 5-4   Example of computer simulation model derived from Figure 5-3.**

Most would also acknowledge that computers are accurate: they do not make "careless" mistakes. Again, if there are inaccuracies, they usually come from the human side, which is why the analyst must be an expert at avoiding "garbage in."

Modeling programs have another advantage, especially the more sophisticated ones: they are comprehensive in what they can analyze simultaneously. The human mind can only accommodate so many ideas or concepts at once without getting confused and bogged down; a properly conceived model will not get confused and can provide answers that may be counterintuitive. As an example, a good modeling program can track heat gain from lights, plug loads, and solar energy, along with heat loss from the building envelope and infiltration, do it for every hour or every day of the year in whatever weather conditions are assumed, take into account mass effects of the structure, and still yield an accurate answer. It would be impossible for the human mind to do this in a reasonable time, unless it was very good at guessing!

See Figure 5-5 for an example of daylighting analyses output.

## BUILDING INFORMATION MODELING

The approach to building design is moving away from conventional CAD software to follow the way design software has evolved in the manufacturing sector. Your building can now be a working digital prototype! Sustainable design is driving this solution to ensure that the buildings we are erecting are designed, constructed, and operated in a manner that minimizes their environmental impact and are as close to self-sufficient as possible. This technology is known as *building information modeling* (BIM).

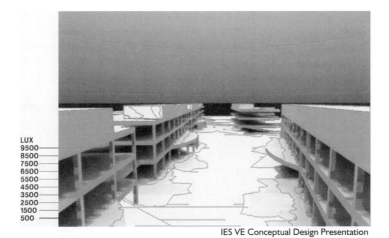

IES VE Conceptual Design Presentation

**Figure 5-5  Example of daylight analysis software output—conceptual design.**

BIM is a software tool available today that uses a relational database together with a behavioral model that captures and presents building information dynamically. In the same way that a spreadsheet is automatically updated, a change in the parametric building modeler is immediately reflected everywhere. This means, for example, revising windows from one type to another not only produces a visually different graphic representation in all views of your building but the insulation value of the glazing (R-values) is also revised. Due to the integration of BIM with existing tools of analysis, running energy calculations is greatly simplified. Visualization tools are sophisticated and allow three-dimensional views and walkthroughs of the building.

Multiple design options can therefore be developed and studied within a single model early in the design process to not only see the building and provide conventional documentation for construction but also interact with other software to perform energy analysis and lighting studies. (See Figure 5-6 for an example.)

Using BIM helps with the demanding aspects of sustainable design, such as solar applications and daylight harvesting, and also automates routine tasks such as documentation.

Schedules are generated directly from the model; if the model changes, so do the schedules. Architects are able to filter and sort material quantities automatically, bypassing the manual extraction/calculation process required. Determining the percentages of material reuse, recycling, or salvage can be tracked and studied for various sustainability design options.

You can perform year-round sun studies to understand when your building is provided natural shading and, thus, optimize the orientation of the building to maximize afternoon shading from the hot summer sun and properly size roof overhangs to minimize solar heat gain. Engineers can then reduce the capacity of cooling systems, demonstrating the building is exceeding the "baseline" building energy requirements.

| | H/L | Wt | Case 1 | Case 2 | Case 3 | Case 4 | Case 5 | Case 6 | Case 7 | Case 8 | Case 9 |
|---|---|---|---|---|---|---|---|---|---|---|---|
| 1 Solar gains (peak kW) | L | 1 | -40.91 | 28.13 | 12.33 | 0.74 | 8.12 | -27.02 | -17.43 | 0.17 | 0.17 |
| 2 Conduction gains (peak kW) | L | 1 | 2.92 | 1.31 | 0.61 | 0.07 | -9.85 | 13.89 | 2.29 | 10.11 | 16.02 |
| 3 Conduction losses (peak kW) | L | 1 | -23.28 | 20.12 | 11.96 | 3.75 | 5.59 | -17.76 | -5.90 | -2.82 | -5.46 |
| 4 Infiltration gains (peak kW) | L | 1 | 0.50 | 0.05 | 0.05 | 0.05 | 0.05 | 0.05 | 0.05 | 0.05 | 0.05 |
| 5 Infiltration losses (peak kW) | L | 1 | 0.45 | 0.00 | 0.00 | 0.00 | 0.00 | 0.00 | 0.00 | 0.00 | 0.00 |
| 6 Avg. daylight level (Lux) | H | 1 | 65.26 | -30.74 | -5.05 | -6.74 | -8.84 | 31.16 | 16.84 | 0.00 | 0.00 |
| 7 Heating requirement (Peak kV) | L | 1 | -2.10 | -0.03 | -0.02 | 0.00 | 1.64 | -7.89 | -2.00 | -3.65 | -5.97 |
| 8 Cooling requirement (peak kW) | L | 1 | -17.06 | 7.82 | 3.23 | 0.06 | 2.88 | -9.39 | -7.15 | -2.22 | -3.53 |
| Total | | | -14.22 | 26.66 | 23.11 | -2.07 | -0.41 | -16.96 | -13.30 | 1.64 | 1.28 |

IES VE Conceptual Design Presentation

**Figure 5-6 Example of multiple design options analysis.**

An energy analysis using a two-dimensional CAD file requires manually taking off the building values/areas from the floor plans and entering said data into an energy simulation application. The data for supporting green design are captured during the design process in BIM and are extracted as necessary.

The energy performance for the baseline model averages the results of four simulations of one year of operation. One simulation is based on the actual orientation of the building on the site; the others rotate the entire building by 90°, 180°, and 270°, which enables the proposed design to receive credit for a well-sited building. This is easily accomplished when using BIM.

The software carries all the data required for building a structure; it "understands" the data. It quantifies building materials so designers can move walls or insert windows and almost instantly have the building's data on energy performance, daylight harvesting, and, perhaps most importantly, costs shift accordingly in real time. Designers can test the life-cycle performance of brick walls versus concrete walls. The digital representations behave like buildings and not just drawings.

BIM is transforming the way we work and will enable further endeavors in the practice of creating sustainable, cost-effective buildings.

### Evaluating Alternative Designs

Modeling of alternative designs is made easier by the plethora of modeling tools available today. The chosen model should meet specific requirements depending on the level of accuracy needed. Large error can be introduced into modeling if users forget the "garbage in/garbage out" rule. To reduce the chance of inappropriate or misunderstood input, modeling programs can employ input and output formats that allow the user quick "reality checks."

Various stages of design, requirements, and associated tools are shown in Table 5-1.

### Parametric Analysis

See the discussion of this subject in Chapter 7, "Conceptual Engineering Design—Load Determination."

**Table 5-1    Summary of Available Analysis/Modeling Tools**

| Stage | Requirements | Tools | Reality Checks |
|---|---|---|---|
| Scoping | –Quick analysis<br>–Comparative results<br>–Reduce alternatives to consider<br>–Control strategy modeling | –System Analyzer™<br>–Modified bin analysis (where load is not entirely dependent on ambient conditions)<br>–eQUEST<br>–IES VE | –Operation cost per $ft^2$ ($m^2$)<br>–Payback or other financial measure |
| System design | –Accurate output<br>–Industry-accepted methods | –HAP<br>–TRACE 700<br>–Elite Design<br>–IES VE | –$cfm/ft^2$ ($L/min/m^2$)<br>–cfm/ton (L/s/ton) |
| Energy/cost analysis | –Accurate<br>–Industry-accepted methods<br>–Flexible<br>–Allows modeling of complex control strategies<br>–Complies with ECB method requirements of ASHRAE 90.1-2001 | –EnergyPlus<br>–DOE<br>–HAP<br>–TRACE 700<br>–SUNREL<br>–IES VE | –Btu/h-$ft^2$ ($kWh/m^2$) per year<br>–Operation cost per $ft^2$ ($m^2$)<br>–Payback or other financial measure |
|  | –Works for existing building and systems |  |  |
| Monitoring | –Simplicity<br>–Intuitive interface<br>–Systemwide<br>–Interoperable | –BacNET®<br>compatible automation systems | –Trended operating characteristics<br>–Benchmark comparisons (such as system kW/ton [kW/kWR]) |

## NREL's Nine-Step Process for Low-Energy Building Design

1.  Create a base-case building model to quantify base-case energy use and costs. The base-case building is solar neutral (equal glazing areas on all wall orientations) and meets the requirements of applicable energy efficiency codes such as ASHRAE Standards 90.1 and 90.2.

2.  Complete a parametric analysis to determine sensitivities to specific load components. Sequentially eliminate loads from the base-case building, such as conductive losses, lighting loads, solar gains, and plug loads.

3.  Develop preliminary design solutions. The design team brainstorms possible solutions that may include strategies to reduce lighting and cooling loads by incorporating daylighting or to meet heating loads with passive solar heating.

4.  Incorporate preliminary design solutions into a computer model of the proposed building design. Energy impact and cost effectiveness of each variant are determined by comparing the energy with the original base-case building and with the other variants. Those variants having the most favorable results should be incorporated into the building design.

5.  Prepare preliminary set of construction drawings. These drawings are based on the decisions made in Step 3.

6.  Identify an HVAC system that will meet the predicted loads. The HVAC system should work with the building envelope and exploit the specific climatic characteristics of the site for maximum efficiency. Often, the HVAC system is much smaller than in a typical building.

7.  Finalize plans and specifications. Ensure that the building plans are properly detailed and that the specifications are accurate. The final design simulation should incorporate all cost-effective features. Savings exceeding 50% from a base-case building are frequently possible with this approach.

8.  Rerun simulations before design changes are made during construction. Verify that changes will not adversely affect the building's energy performance.

9.  Commission all equipment and controls. Educate building operators. A building that is not properly commissioned will not meet the energy-efficiency design goals. Building operators must understand how to properly operate the building to maximize its performance.

## One Firm's Green Building Design Process Checklist

☐ Create an integrated, cross-disciplinary design team, committed to sustainability and aware of environmental issues, which include all those impacted by the building.

☐ Pre-charette and charette meetings should include the project owner, CxA, architects and landscape architects, engineers, an energy engineer with experience in computer simulation of building energy consumption, environmental consultant, facility occupants and users (including purchasing, human resources, and managers), facility manager, contractors when hired, interior designer, local utility representatives, cost consultant, and other specialty consultants.

☐ Review how occupants will use the project and any of the client's unique operational characteristics. Based on the this information, assess and decide which sustainable/green design principles are aligned with the owner's overall objectives, criteria, and activities that will provide the most positive impact in order to provide focus and priority.

☐ Define environmental standards, goals, and strategies early in predesign and clearly state these in the OPR document.

☐ Translate the owner's objectives and criteria into the construction documents.

☐ Channel development to urban areas with existing infrastructures, protecting greenfields and preserving habitat and natural resources.

☐ Increase localized density to conform to existing or desired density goals by utilizing sites that are located within an existing minimum development density of 60,000 square feet per acre, or a 1/2 mile area, with at least ten services, such as banks, restaurants, retail businesses, etc.

☐ Channel development to areas with existing transportation infrastructure that provides non-automobile-dependent choices.

☐ Select a location for the project within 1/2 mile of a rail station (commuter rail, light rail, or subway), within 1/4 mile of two or more bus lines, or in a "live, work, walk, mixed-use environment."

☐ Reduce the overall building footprint and development area.

☐ Make important decisions on the mechanical load, daylighting, solar absorption, response to local climate and environment, and key building elements at the beginning in order to define subsequent decisions.

☐ Establish benchmarks/performance targets as a reference point.

☐ Integrate recycling systems into every aspect from reusing existing building materials to purchasing new materials.

☐ Design for disassembly at the end of the building's useful life.

☐ Educate contractor and subcontractor in sustainable practices.

☐ Evaluate and benchmark OPRs, including sustainable goals at the same intervals as budgets and schedules.

❏   Have a CxA as a team member to help ensure that owners' objectives and criteria, including sustainability goals, are met at each stage; benchmark and document success for future reference; and advocate for environmental choices during the course of the project.

❏   Tie compensation for the architect and design team to achieved building performance.

## Canada's C-2000 Program

The C-2000 Program was designed in 1993 by Natural Resources Canada, a government agency, to demonstrate the feasibility of achieving very high levels of building performance. The program's technical requirements cover energy performance,[a] environmental impacts, indoor environment, functionality, and a range of other related parameters.[b] It was, therefore, expected that incremental costs for design and construction would be substantial. After a preliminary analysis of then-prevalent project costs and an informal survey of designers, provision was made for support of incremental costs in both the design and construction phases. Contributions were provided according to a sliding scale, ranging from 7% in large projects to 12% in small projects.

Even though the program targeted a select group of clients known to have an interest in high performance, it was assumed that some level of financial incentive would be required to make the program a success. However, the extent of incentives required and the best point of intervention within the project development process were very much open to question.

The first two C-2000 projects received support according to this formula in the range of $400,000 to $750,000 CAN, and funding of this order of magnitude was also planned for subsequent projects. However, after the first six projects were designed and two of them had been completed, it was found that incremental capital costs were less than expected, partly due to the fact that designers used technologies that were less sophisticated and expensive than anticipated.[c]

A careful investigation of the first two C-2000 projects constructed, Crestwood 8[d] and Green on the Grand,[e] indicated that the marginal costs for both projects, including design and construction phases, was 7%–8% more than a conventional building, a rather modest increase. Even more interesting, the designers all agreed that application of the IDP required by the C-2000 program was the main reason why high levels of performance could still be reached. It also appeared that most of the benefit of intervention was achieved during the design process.

a. At the time, the energy requirement was 50% better than the ASHRAE Standard 90.1 (the benchmark is now the Model National Energy Code for Buildings [MNECB]). Both are North American standards for good practice.
b. Larsson, N., ed. 1996. *C-2000 Program Requirements*. Ottawa: Natural Resources Canada, October 1993, updated April 1996.
c. The conservative approach of designers is based primarily on their perception that they might face legal liability problems if they use exotic and unproven technologies.
d. CETC. 1996. *Technical Report on Bentall Corporation Crestwood 8 C-2000 Building*. Ottawa: Natural Resources Canada, April.
e. CETC. 1996. *Technical Report on Green on the Grand C-2000 Building*. Ottawa: Natural Resources Canada, April.

C-2000, now called the integrated design process (IDP), includes the key steps listed on page 87.

The design process itself emphasizes the following sequence:

- First minimize heating and cooling loads through orientation, building configuration, an efficient building envelope, and careful consideration of amount, type, and location of fenestration.
- Meet these loads through the maximum use of renewables and the use of efficient HVAC systems.
- Iterate the process to produce at least two, and preferably three, design concept alternatives.

The IDP contains no elements that are radically new but, rather, integrates well-proven approaches into a systematic total process. From an engineering perspective, the IDP permits the skills and experience of mechanical, electrical, and other engineers to be integrated at the design concept level from the very beginning of the design process. For example, reduced cooling loads will result in smaller and more economical systems, which, in turn, can reduce capital and replacement costs. When carried out in a spirit of cooperation among key persons, this results in a design that is highly efficient with minimal, and sometimes zero, incremental capital costs, along with reduced long-term operating and maintenance costs.

Most project interventions are now focused on providing advice on the design process at the very early stage. Six projects were constructed on this basis, and all have either achieved the C-2000 performance requirements or have come very close. Capital costs have been either slightly above or slightly below base budgets. The most hopeful sign that the IDP approach is taking root is that several owners have subsequently used the same process for buildings that have not benefited from any subsidy.

Simple software design support tools were produced to help design teams enrolled in the C-2000 program. One outlines generic design steps and provides a simple way for designers to record their performance targets and strategies; another facilitates the task of having the client and design team reach a consensus on the relative importance of various issues. The C-2000 IDP process is now being used as a model for development of a generic international model by Solar Heating and Cooling Task 23 of the International Energy Agency (IEA), and discussions are underway with the Royal Architectural Institute of Canada (RAIC) to see if the process can be accepted as an alternative form of delivery of professional services.

# 6

# LEED GUIDANCE FOR HVAC ENGINEERS

The LEED (Leadership in Energy and Environmental Design) Green Building Rating System® published by the US Green Building Council (USGBC) has become a major force for encouraging the integration of green building principles and techniques into building projects.[1] This chapter discusses ways in which the various LEED credits affect the HVAC engineer and how the engineer can best respond to the opportunities presented by the use of the LEED system on projects. It should be noted that this discussion is intended to be a primer—not a substitute—for the more detailed information available from the USGBC in its LEED workshops and reference manuals.[2]

## LEED CREDITS AFFECTING
## MINIMUM ENERGY PERFORMANCE (MEP) DESIGN

The breakdown of credits and points available under LEED is as follows:

- Sustainable sites                8 credits/14 points
- Water efficiency                 3 credits/5 points
- Energy and atmosphere            6 credits/17 points
- Materials and resources          7 credits/13 points
- Indoor environmental quality     8 credits/15 points

| | |
|---|---|
| *Total core points* | 64 |
| *Innovation points* | 4 |
| *Design process points* | 1 |
| ***TOTAL POSSIBLE*** | **69** |

1. Although other green building rating systems exist, LEED is the predominant one and, as such, is the focus of this discussion.
2. Throughout this chapter, references to LEED rely upon the copyrighted content of the US Green Building Council's *LEED Rating System*® *for New Construction, Version 2.2* and its associated *LEED Reference Guide*.

The percentage distribution of these credits is illustrated in Figure 6-1. It can be seen that "Energy and Atmosphere" and "Indoor Environmental Quality" are the two categories with the highest number of points available, and these are the direct purview of the HVAC engineer. The discussion in the rest of this chapter will focus on the specific credits most impacted by the HVAC engineer.

## SUSTAINABLE SITES (SS) CREDITS AFFECTING MEP DESIGN

### Prerequisite

* Erosion and Sedimentation Control
  SS Credit 1—Site Selection
  SS Credit 2—Development Density and Community Connectivity
  SS Credit 3—Brownfield Development
  SS Credit 4—Alternative Transportation
  SS Credit 5—Site Development
  **SS Credit 6—Stormwater Design**
  **SS Credit 7—Heat Island Effect**
  **SS Credit 8—Light Pollution Reduction**

The primary areas of opportunity in LEED sustainable sites are for Credits 6, 7, and 8.

**SS Credit 6—Stormwater Management.** The focus in Credit 6 is minimizing the amount of stormwater runoff from the building site and maximizing the time lag between the stormwater impinging on the site and leaving the site. Often the techniques used are architectural (such as vegetative roofs), landscaping (pervious pavements, bioswales), and civil (detention basins, filtration). These systems may have added impacts on MEP in terms of design of roof drain systems and irrigation systems for roof vegetation.

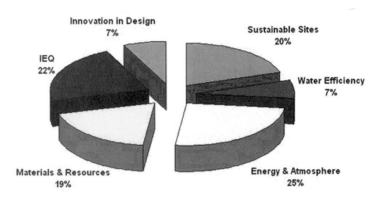

**Figure 6-1    The percentage distribution of LEED-NC Version 2.2 credits.**

However, there are also some MEP-focused techniques that have attained wide use, which focus on collecting the stormwater and storing it for future use on-site. An example of this is the on-site stormwater collection, storage, and reuse system for the Santa Monica Main Library project (Figure 6-2). The system collects stormwater through the roof drain and area drain system and stores it in a cistern system of large-diameter high density polyethylene culvert pipes under the building. The water stored in the cistern is then used for landscape irrigation. One of the most challenging aspects of the design was sizing the storage reservoirs to handle the wide variations in annual rainfall, so as to minimize the amount of stormwater that has to be allowed to run off the site in a peak rainfall year in which more rain falls than the reservoir can store. An example of the sizing calculation performed by the designer is shown in Figure 6-3.

**SS Credit 7—Heat Island Reduction.** As every mechanical engineer knows, the color and reflectance of the building roof and of the surrounding paving can have a significant impact on cooling loads inside the building and on the local microclimate through the heat island effect. To the extent that the design team wants input on the options and impacts of improving reflectance and emissivity values of these materials, the mechanical engineer can provide assistance. These values should also be integrated into the load calculations for the building, to take advantage of the improved exterior environment to reduce cooling loads and installed tonnage.

**SS Credit 8—Light Pollution Reduction.** The impact of this credit is borne by the lighting designer and/or electrical engineer, and it involves both exterior lighting (on the building and site—see Figure 6-4) and interior lighting (which may project to the exterior).

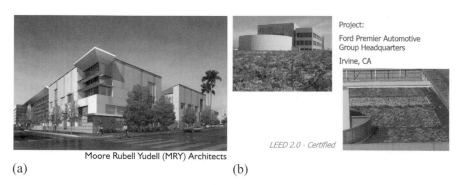

(a)                                   (b)

Moore Rubell Yudell (MRY) Architects

Project:
Ford Premier Automotive Group Headquarters
Irvine, CA

LEED 2.0 - Certified

**Figure 6-2   (a) Santa Monica Main Library project and (b) Ford Premier Automotive Group project.**

## 200,000 Gallon Storm-water Collection Cistern - Provides Water for Site Irrigation

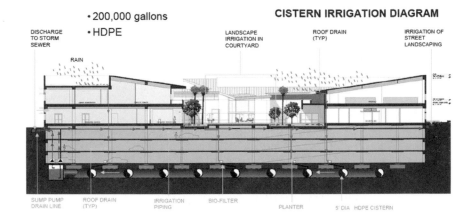

## Rainwater Collection Cistern Sizing – Drought Year

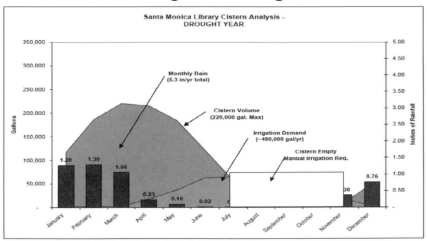

Source: Syska Hennessy Group

**Figure 6-3   Stormwater sizing calculation.**

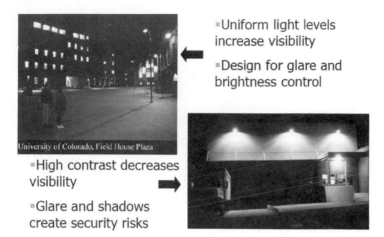

▪Uniform light levels increase visibility

▪Design for glare and brightness control

▪High contrast decreases visibility

▪Glare and shadows create security risks

**Figure 6-4   Light pollution reduction.**

## WATER EFFICIENCY (WE) CREDITS AFFECTING MEP DESIGN

WE Credit 1—Water Efficient Landscaping
**WE Credit 2—Innovative Wastewater Technologies**
**WE Credit 3—Water Use Reduction**

**WE Credit 3—Water Use Reduction.** The opportunities here primarily impact the plumbing engineer in terms of reducing the total potable water used by the building and of minimizing the impact of that water use on the sanitary and stormwater systems. Traditionally, the building plumbing system and cooling towers use about half the total annual water usage, and the irrigation system uses the rest (depending upon locale and development density). The building plumbing and cooling towers have many opportunities for potable water reductions, both through more efficient fixtures and through the use of reclaimed water (where it is available). An example of this or the Toyota South Campus Project (LEED Gold in Torrance, CA) is shown in Figure 6-5, where the total usage of potable water was reduced by 80% through the combination of efficiency and reclaimed water use.

The reduction of potable water usage through more water-efficient plumbing fixtures is relatively straightforward, as there are many fixtures and technologies available. Some of the water-efficient plumbing fixture types include:

- Low-flow lavatory and shower aerators
- Auto-controls
- Dual-flush water closets
- Ultra-low-flow urinals
- Waterless urinals (see Figure 6-6 for an example)

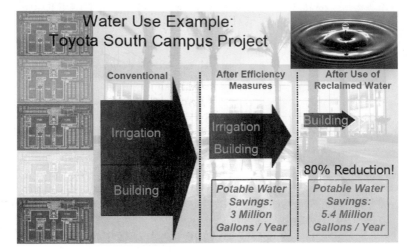

**Figure 6-5   Water use reduction.**

**Figure 6-6   Waterless No Flush™ Sonora Waterfree urinal.**

Water closets and urinals are the biggest users of water in a typical office building, so focusing on these fixtures is the most effective. LEED establishes a methodology for calculating the reduction of water usage for these fixtures that is a function of the number of building occupants and characteristics of the plumbing fixtures.

One of the technologies with the most impact is that of waterless urinals. This is an innovative concept that uses no water for flushing but instead uses a nontoxic chemical to form the trap. The urine flows through the chemical to enter the drain. These fixtures are gradually attaining code approval and user acceptance in areas across the country.

Water Efficiency: On-Site Waste
Water Treatment:
Graywater System

NRDC Santa Monica Office
LEED Platinum, 2004

**Figure 6-7   Graywater system.**

**WE Credit 2—Innovative Wastewater Technologies.** Once the water usage is reduced to a minimum through earning LEED Credit 3, the opportunity exists to earn LEED Credit 2 by treating the wastewater on-site. This can have the effect of reducing the load on the municipal sanitary sewer system and offers the potential for using the treated wastewater within the project site. There are two basic approaches to this: a *graywater treatment system*, which clarifies that portion of the sewage that does not contain human fecal matter, and a *tertiary treatment system* that treats sewage regardless of its content. Both systems produce effluent, which can be used as reclaimed water for performing certain functions, such as toilet flushing or landscape irrigation. (*Note:* Both of these uses are subject to rulings of the local building and health departments that have jurisdiction over the project, and these vary based on location.)

The graywater system collects the drainage from plumbing fixtures, such as lavatories and drinking fountains, as well as condensate from HVAC systems. The graywater is filtered and disinfected and then stored in a cistern or tank until needed. It is then piped in a special separate piping system for reclaimed water to the points of use, which can be toilets, urinals, or irrigation systems. An example of such a system is shown in Figure 6-7.

The on-site sewage treatment system approach can include traditional septic systems or more modern biological treatment systems that create a local natural wetland ecosystem that purifies wastewater after a biological digestion process is applied to the sewage. An example is shown in Figure 6-8.

## ENERGY AND ATMOSPHERE CREDITS AFFECTING MEP DESIGN

The single biggest area of potential LEED points is in the "Energy and Atmosphere" (EA) section of LEED. The credits and prerequisites for this section are as follows.

### LEED Prerequisites

- **Fundamental building systems commissioning**
- **MEP**
- **Chlorofluorocarbon (CFC) reduction in HVAC equipment**

    **EA Credit 1—Optimize Energy Performance**
    **EA Credit 2—On-Site Renewable Energy**
    **EA Credit 3—Enhanced Commissioning**
    **EA Credit 4—Enhanced Refrigerant Management**
    **EA Credit 5—Measurement and Verification**
    EA Credit 6—Green Power

**Prerequisites.** The LEED prerequisites for EA support several areas in which ASHRAE is active: building commissioning, building energy performance, and CFC reductions in refrigerants. The prerequisites provide no LEED points but are required for a project to become LEED certified.

# Water Efficiency: On-Site Waste Water Treatment…Living Machine

- On-site treatment of wastewater
- Natural wetland ecosystem process
- Water can be re-used for flushing

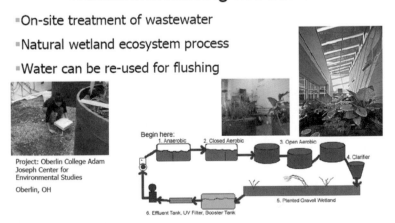

Project: Oberlin College Adam Joseph Center for Environmental Studies

Oberlin, OH

**Figure 6-8  Biological water treatment system.**

One of the most profound effects of LEED in the areas of interest to the HVAC engineer is the requirement for building systems commissioning as a prerequisite for LEED certification. This is crucial because it ensures a focus on the actual performance of the building systems as installed and goes a long way toward ensuring that the MEP systems will provide the intended performance once the building is constructed. This is also supportive of ASHRAE's focus on commissioning through Technical Committee 7.9 and the publication of Guidelines 0 and 1. Likewise, the prerequisite for MEP requires for LEED-NC-2.1 that the building comply with the applicable local energy code, which is typically based on ASHRAE Standard 90.1 in most jurisdictions, and for LEED-NC-2.2 that the building comply with ASHRAE Standard 90.1-2004. Finally, LEED requires compliance with the Montreal Protocol for CFC reduction.

**EA Credits and Sustainable Design Measures.** Energy efficiency is inextricably tied up in the sustainability of a building design, and this is illustrated by Figure 6-9, which shows how various energy measures impact the ability to earn points under the various LEED EA credits. Numerous building energy efficiency measures (envelope, HVAC, lighting, daylighting, etc.) impact the ability to earn points under EA Credit 1, but so does the use of renewable sources of energy, such as photovoltaic solar power and solar thermal heating. Because of the structure of LEED, which tends to encourage renewable energy, such renewable sources contribute to more than one credit, so they can be used to earn more points in multiple credits. The specifics of earning points under each credit are described in the sections that follow.

**EA Credit 1—Optimize Energy Performance.** This credit provides the single largest opportunity for earning LEED points in the entire LEED rating system: ten points are available for this single credit. Many projects earn substantial points under this credit, as shown in Figure 6-10.

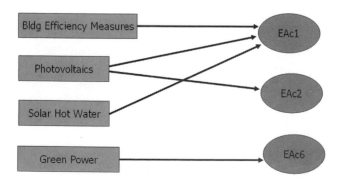

**Figure 6-9   Energy efficiency is inextricably tied up in the sustainability of a building design.**

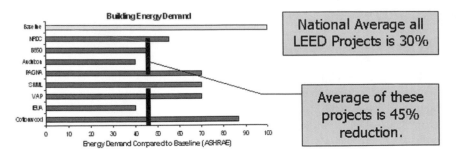

**Figure 6-10 Points earned under EA Credit 1.**

To earn points under EA Credit 1, the project's energy performance must be simulated using an hourly energy simulation program (such as DOE-2.1). There is a sliding scale for points, based upon the extent to which the simulated design projects annual energy usage lower than the minimum required by the applicable local energy code. This provides a mechanism for assessing all of the energy features of the proposed design (building envelope, HVAC system, lighting, materials colors, etc.) in terms of their integrated impact on total energy usage and on the sizing of each of the systems. It is very important that this simulation process be used as an iterative design tool to optimize the overall building design and not just as an accounting tool for computing the LEED points earned by this credit. With this in mind, the simulation model should be developed in the early design phases and used to refine the design, rather than just at the end of the design to compute the percentage better than energy code.

There are numerous specific techniques that can contribute to earning points under EA Credit 1, including daylighting, high-efficiency lighting, high-efficiency HVAC, optimized building envelopes and orientations, and renewable energy. Each of these concepts is dealt with elsewhere in this book, so the primary point to be made here is that each of these has an impact on the other systems' energy usage as well so that the design needs to be optimized as an integrated whole. This provides an important opportunity for the HVAC engineer to contribute to the early evolution and optimization of the design but requires more involvement in the early design than is often done. As a result, it requires a more proactive approach on the part of the HVAC engineer to offer ideas and to perform energy analyses that can answer formative questions about the design. This is one of the profound changes of green building design that is most challenging, as it requires redefinition of roles and scopes of work among the design team participants, yet it offers an exciting opportunity to the HVAC engineer to make a difference in the sustainability of the final design, so it is a highly desirable change in the process.

LEED requires a somewhat unique approach to energy simulation modeling in terms of the method of computing the actual energy savings on which the EA Credit 1 calculation is based. Conventionally, energy modeling computes the "percent better than energy code" as measured in Btu per ft$^2$ per year (or similar annual energy metrics). By contrast, LEED requires the calculation of the "percent annual energy cost reduction" from ASHRAE Standard 90.1, using the energy cost budget (ECB) method defined in ASHRAE Standard 90.1.

The methodology for this is as follows:

1. First input the proposed building design (including HVAC and lighting systems).
2. Create the "base-case" energy model using the prescriptive values for each system, as set by ASHRAE Standard 90.1.
   - In LEED, version 2.1 and prior, this required excluding nonregulated loads, such as plug loads, process energy, garage ventilation, elevators, exterior lighting, and any other loads not included in the standard, using ASHRAE Standard 90.1-1999. In LEED, version 2.2, these loads are included, using Appendix C of ASHRAE Standard 90.1-2004, along with the regulated loads (heating, cooling, pumps, fans, water heating, and interior lighting).
3. Create the "proposed design" energy model, which is the same building but with different inputs, reflecting the proposed energy-efficiency measures that exceed the prescriptive requirements of ASHRAE Standard 90.1.
   - Generally use the same systems in the budget building and the proposed design building. The energy savings come from load reduction, equipment efficiencies, and any on-site energy production from renewables.
4. Compute the ECB using the form in ASHRAE Standard 90.1 for both the base-case building and the proposed design.
5. The percentage different in the ECB for the two models is the percentage used to compute the number of LEED points earned under EA Credit 1.

## LEED COMMISSIONING REQUIREMENTS

LEED's perspective on commissioning is that it is essential to achieving a functional green building and, as a result, is a required prerequisite (LEED EAp1) for LEED certification. The required activities are termed *fundamental building systems commissioning*, as described below. In addition, LEED EA Credit 3 offers the opportunity to earn one point for "additional commissioning" activities, as described below.

Commissioning is typically the largest added cost for professional services related to complying with LEED, yet it really has nothing to do with the cost of LEED certification per se. It is well known that the savings that accrue from commissioning the building systems pay back quickly (generally in the three to five year range[1]), and this is for all buildings, whether LEED-rated or not. Despite this, commissioning is not generally performed on new buildings due to the traditional pressure on "first cost." So, by mandating commissioning, LEED is promoting a practice that contributes to the better long-term energy and IAQ performance of the building systems. It produces a return on its cost, and its cost is independent of the process of LEED certification.

The intent and requirements of the commissioning prerequisite and credit are quoted here in full due to the detailed nature of the requirements, and are self-explanatory:

## LEED Prerequisite EAp1:
## Fundamental Commissioning of the Building Energy Systems

- INTENT: Verify that the building's energy related systems are installed, calibrated and perform according to the OPRs, basis of design, and construction documents.
- REQUIREMENT: The following commissioning process activities shall be completed by the commissioning team, in accordance with the LEED-NC 2.2 Reference Guide.
  - Designate an individual as the CxA to lead, review, and oversee the completion of the commissioning process activities.
    - The CxA shall have documented CxA experience in at least two building projects with technical and managerial complexity similar to this project.
    - The individual serving as the CxA shall be independent of the project's design and construction teams, though they may be employees of the firms providing those services. The CxA may be a qualified employee or consultant of the Owner.
    - The CxA shall report directly to the Owner.
    - For projects smaller than 50,000 gross square feet, the CxA may include qualified persons on the design or construction teams who have the required experience).
  - The Owner shall document the OPRs. The design team shall develop the basis of design. The CxA shall review these documents for clarity and completeness. The Owner and design team shall be responsible for updates to their respective documents.
  - Develop and incorporate commissioning requirements into the construction documents.
  - Develop and implement a commissioning plan.
  - Verify the installation and performance of the systems to be commissioned.
  - Complete a commissioning report.

---

1. Mills, et al. 2004. *The Cost-Effectiveness Of Commercial-Buildings Commissioning.* Berkeley, CA: Lawrence Berkeley National Laboratory, LBNL – 56637 (Rev.).

- COMMISSIONED SYSTEMS: Commissioning process activities shall be completed for the following energy-related systems, at a minimum:
  - Heating, ventilating, air conditioning, and refrigeration (HVAC&R) systems (mechanical and passive) and associated controls
  - Lighting and daylighting controls
  - Domestic hot water systems
  - Renewable energy systems (photovoltaics, wind, solar etc.).

## LEED EA Credit 3: Enhanced Commissioning

- INTENT: Begin the commissioning process early during the design process and execute additional activities after systems performance verification is completed.
- REQUIREMENT: Implement, or have a contract in place to implement, the following additional commissioning process activities in addition to the requirements of EA prerequisite 1 and in accordance with the LEED-NC 2.2 Reference Guide:
  - Prior to the start of the construction documents phase, designate an independent CxA to lead, review, and oversee the completion of all commissioning process activities. The CxA shall, at a minimum, perform Tasks 2, 3. and 6. Other team members may perform Tasks 4 and 5.
    - The CxA shall have documented commissioning authority experience in at least two building projects with similar technical and managerial complexity as this project.
    - The individual serving as the CxA shall be:
      - Independent of the design and construction process,
      - Not an employee of the design team, though they may be contracted through them, and
      - Not an employee of, or contracted through, a contractor or construction manager holding construction contracts.
      - The CxA may be a qualified employee or consultant of the Owner.
    - The CxA shall report directly to the Owner.
    - This requirement has no deviation for project size.
  - The CxA shall conduct, at a minimum, one commissioning design review of the OPRs, basis of design, and design documents prior to mid-construction documents phase and back-check the review comments following design submission.
  - The CxA shall review contractor submittals applicable to systems being commissioned for compliance with the OPRs and basis of design. This review shall be concurrent with A/E reviews and submitted to the design team and the Owner.

– Develop a systems manual that provides future operating staff the information needed to understand and optimally operate the commissioned systems.
– Verify that the requirements for training operating personnel and building occupants are completed.
– Assure the involvement by the CxA in reviewing building operation within 10 months after substantial completion with O&M staff and occupants. Include a plan for resolution of outstanding commissioning-related issues.

### LEED EA Credit 5—Measurement and Verification (M&V)

In an effort to encourage efficient operation of LEED-certified buildings, LEED provides a credit for providing the building with the ability to measure its usage of energy and water over time. The requirements of this credit are for the permanent installation of sensors and a monitoring system to measure and track key HVAC system performance metrics, including airflow and IAQ, energy consumption of major HVAC system components (fans, pumps, chillers), and troubleshooting diagnostics. The LEED credit uses the International Performance Measurement And Verification Protocol (IPMVP) as a guide.

There are numerous benefits to the owner from the installation of this capability, including potentially increasing building value (due to reduced energy operating costs) and improving tenant/occupant satisfaction (due to improved system performance). Implementing this credit requires an increase in the control system's capability in the building to provide the added sensors. However, this is typically earned back in relatively short order if the information collected by the M&V system is used to optimize the building's operations. It also provides opportunities to the HVAC engineer for potential increased follow-up business in terms of helping to interpret the system performance data and optimizing system performance over time. These activities represent opportunities for providing value-added services to the building owner and for maintaining service relationships with that owner over time.

## MATERIALS AND RESOURCES CREDITS AFFECTING HVAC DESIGN

### Prerequisite

• Storage and Collection of Recyclables
MR Credit 1—Building Reuse
MR Credit 2—Construction Waste Management
MR Credit 3—Materials Reuse
MR Credit 4—Recycled Content
MR Credit 5—Regional Materials
MR Credit 6—Rapidly Renewable Materials
MR Credit 7—Certified Wood

Interestingly, *none* of the LEED materials and resource credits directly affect the typical activities of the HVAC engineer.

# INDOOR ENVIRONMENTAL QUALITY (IEQ) CREDITS AFFECTING MEP DESIGN

## Prerequisites

- **Minimum IAQ Performance (ASHRAE Standard 62.1-2004)**
- Tobacco Smoke Control

**EQ Credit 1—Outdoor Air Delivery Monitoring**

**EQ Credit 2—Increased Ventilation**

**EQ Credit 3—Construction IAQ Management Plan**

**EQ Credit 4—Low-Emitting Materials**

**EQ Credit 5—Indoor Chemical and Pollutant Source Control**

**EQ Credit 6—Controllability of Systems**

**EQ Credit 7—Thermal Comfort**

EQ Credit 8—Daylight and Views

After the EA credits, the IEQ requirements of LEED provide the HVAC engineer with the greatest opportunity to contribute to the green rating of a building. Many aspects of the indoor environment are governed by ASHRAE standards, which are cited by LEED in this group of credits. The emphasis of the IEQ credits is on the health and comfort of the building occupants.

## Indoor Air Quality

The first IEQ prerequisite requires that the building mechanical systems comply with ASHRAE Standard 62.1.[1] This governs a number of HVAC system attributes, including:

- Placement of intakes vs. exhausts
- Reduced face velocities at coils
  - To eliminate condensate entrainment
  - Also increase heat transfer efficiency
- Access to condensate pans
- Ductwork quality and workmanship
  - Cleanliness
  - Liner integrity
- IAQ Monitoring and Controls
  - $CO_2$, volatile organic compounds (VOCs), Particulates
  - Maintainability of IAQ over time

The first IEQ prerequisite also deals with the quality of the HVAC system installation to ensure that it is well built and functions as intended. These issues include such construction quality practices as:

---

1. See also: S. Taylor, LEED and Standard 62.1, *ASHRAE Journal Sustainability Supplement*, September, 2005.

- Duct sealing
- Condensate drains
- Functionality of dampers
- Accurate testing
- Collaboration and integration with commissioning activities

### EQ Prerequisite 1—Minimum IAQ Performance

In its effort to ensure good IAQ in LEED buildings, LEED mandates compliance with ASHRAE Standard 62.1, as described in the intent and requirements quoted below:

- INTENT: Establish minimum IAQ performance to enhance IAQ in buildings, thus contributing to the comfort and well-being of the occupants.
- REQUIREMENTS: Meet the minimum requirements of voluntary consensus standard *ASHRAE Standard 62.1-2004, Ventilation for Acceptable Indoor Air Quality.* Mechanical ventilation systems shall be designed using the ventilation rate procedure. Naturally ventilated buildings shall comply with ASHRAE Standard 62.1-2004, Section 5.1.

### EQ Prerequisite 2—Environmental Tobacco Smoke Control

LEED mandates that smoking should not be allowed in buildings receiving a LEED rating (with some exceptions for residential buildings). While this does not directly affect the HVAC engineer, it is interesting to note how radically the market has moved in the past five years since LEED was introduced. At the time of its initial publication, this prerequisite was a major challenge to accepted practices. However, today it is the norm in most markets, as codes in most local jurisdictions no longer allow smoking in buildings.

### EQ Credit 1—Outside Air Delivery Monitoring

The focus of the first credit is on ensuring that adequate fresh (outdoor) air is delivered to maintain the $CO_2$ levels of the indoor air below accepted thresholds. The intent and requirements for this credit are as follows:

- INTENT: Provide capacity for ventilation system monitoring to help sustain occupant comfort and well-being.
- REQUIREMENTS: Install permanent monitoring systems that provide feedback on ventilation system performance to ensure that ventilation systems maintain design minimum ventilation requirements. Configure all monitoring equipment to generate an alarm if underventilation is detected, via either a BAS alarm to the building operator or via an alarm that alerts building occupants.

The requirements go into detail for both mechanically and naturally ventilated buildings.

## EQ Credit 2—Increased Ventilation

LEED provides a credit for providing ventilation strategies that enhance occupant access to fresh air. This contemplates either mechanically ventilated or naturally ventilated spaces. It should be noted that this is a significant change for LEED-NC, version 2.2, because prior to this version, this credit had focused on the concept of ventilation effectiveness (referencing ASHRAE Standard 129) and had largely been limited to applications of underfloor air distribution.

The specific intent of this credit is as follows:

- INTENT: Provide additional outdoor air ventilation to improve IAQ for improved occupant comfort, well-being, and productivity.

The requirements go into detail for both mechanically and naturally ventilated buildings, with the following primary elements:

- *For Mechanically Ventilated Spaces:*
  – Increase breathing zone outdoor air ventilation rates to all occupied spaces by at least 30% above the minimum rates required by ASHRAE Standard 62.1-2004 as determined by EQ Prerequisite 1.

- *For Naturally Ventilated Spaces:*
  – Design natural ventilation systems for occupied spaces to meet the recommendations set forth in the Carbon Trust "Good Practice Guide 237" (1998). Determine that natural ventilation is an effective strategy for the project by following the flow diagram process shown in Figure 1.18 of the Chartered Institution of Building Services Engineers (CIBSE) *Applications Manual 10: 2005, Natural Ventilation in Non-Domestic Buildings.*

The design of effective natural ventilation systems is an area of potential professional growth for the HVAC engineer, and one that is in keeping with the general passive natural-energy approach to green buildings.

## EQ Credit 3—Construction IAQ Management Plan

In keeping with LEED's focus on actual construction and delivery processes (as exemplified by the commissioning requirements), there are credits that focus on controlling construction impacts on IAQ. This is done through a combination of a "construction IAQ management plan" for the periods during construction and the flushing or testing of the indoor air to minimize the level of pollutants prior to occupancy. This has obvious advantages to the occupants in terms of improved IAQ and is thus also a benefit to the HVAC engineer in terms of reduced liability and improved occupant satisfaction relative to IAQ.

For both of the credits (EQc3.1 and EQc3.2) related to construction IAQ, the intent of LEED is the same:

- INTENT: Reduce IAQ problems resulting from the construction/renovation process in order to help sustain the comfort and well-being of construction workers and building occupants.

Credit EQc3.1 focuses on the construction phase, and Credit EQc3.2 deals with the period prior to the end of construction activities but prior to occupancy of the building. Both require the development and implementation of a construction IAQ management plan, which coordinates the efforts to optimize IAQ. (See Figures 6-11 and 6-12 for illustrations of bad construction IAQ management and good construction IAQ management, respectively.)

For the construction phase credit, "EQc3.1 Construction IAQ Management Plan—During Construction," elements of the construction IAQ management plan

**Figure 6-11 Bad construction IAQ management.**

**Figure 6-12 Good construction IAQ management.**

will include a variety of good housekeeping procedures and protection of materials during construction:

- Protect all HVAC ducts and materials from deposition of dust and debris.
- Responsible use of chemicals (and limiting off-gassing of VOCs in the space).
- Use of filters (minimum MERV-8 efficiency) on all return air grilles if air handlers are operated during construction.

For the pre-occupancy credit, "EQc3.2 Construction IAQ Management Plan—Before Occupancy," the important elements are that the indoor air in the building be purged of contaminants, which can be demonstrated by one of two methods:

1. Thorough flushing of the air in the building interior with outdoor air (at a specified cumulative rate that will typically take eight to ten days to accomplish).
2. Through air quality testing to ensure that contaminant levels are below acceptable threshold levels (as specified by LEED).

The LEED EQc4 credits require the use of low-emitting materials for construction to minimize or eliminate the emission of VOCs in the building interior. While this is not directly the purview or responsibility of the HVAC engineer, it has a definite positive impact on the building's IAQ. It is also one of the striking features of most green buildings: when you walk into the new green building, there is no "new building smell," which is typical of other buildings in which multiple materials and finishes are emitting VOCs.

The final IAQ-related LEED credit is "EQc5 Indoor Chemical and Pollution Source Control," which has the following intent:

- INTENT: Minimize exposure of building occupants to potentially hazardous particulates and chemical pollutants.

This is typically accomplished with the following techniques:

- Entryway grates at exterior doors to reduce entrainment and tracking in of dirt and dust (see Figure 6-13).
- Deck-to-deck partition with separate exhaust system in areas of chemical use.
- Provision of exhaust systems and enclosures that keep copy machine rooms (and similar "dirty" environments) at negative pressures to reduce migration of contaminants into office areas.
- Use of MERV-13 or better filters on all outdoor air and return air intakes of HVAC systems. (*Note:* this requires attention at the time of HVAC equipment selection to ensure that the filters will fit, as they are typically thicker than lower-efficiency filters.)

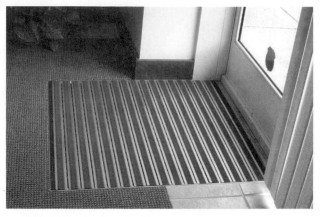

**Figure 6-13 Sample entryway grate.**

### EQ Credit 6.2—Controllability of Systems—Thermal Comfort

Research by ASHRAE and others has documented the desirability of providing building occupants with the ability to control the HVAC system variables (temperature, airflow, humidity) affecting their thermal comfort. LEED has honored this finding by providing a credit for systems that can provide this capability, with the specific credit intent as follows:

- INTENT: Provide a high level of thermal comfort system control by individual occupants or by specific groups in multi-occupant spaces (i.e., classrooms or conference areas) to promote the productivity, comfort, and well-being of building occupants.

This intent is typically met by providing zoning of HVAC systems that, at a minimum, permits the separate control of interior and perimeter zones, as depicted in Figure 6-14. It can be even better met through the use of underfloor air distribution (which allows a user-controlled diffuser for each workstation) or other work-station-level zoning techniques.

### EQ Credit 7—Thermal Comfort

The next LEED IEQ credits require the design and validation of the HVAC system to meet the requirements of ASHRAE Standard 55, with specific intent and requirements as follows:

- INTENT: Provide a comfortable thermal environment that supports the productivity and well-being of building occupants.

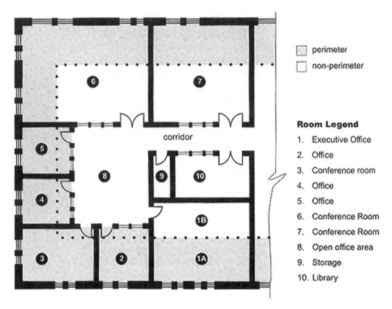

**Figure 6-14 HVAC system interior and perimeter zones.**

### EQ Credit 7.1—Thermal Comfort—Compliance

Credit EQc7.1 covers the *design* of the HVAC system and its ability to meet Standard 55.

- REQUIREMENTS: Demonstrate that the project design complies with *ASHRAE Standard 55-2004, Thermal Comfort Conditions for Human Occupancy.* Demonstrate design compliance in accordance with Section 6.1.1, Documentation (of ASHRAE Standard 55).

### EQ Credit 7.2—Thermal Comfort—Monitoring

Credit EQc7.2 covers the *actual performance* of the HVAC system to validate that it meets comfort criteria established by ASHRAE Standard 55.

- REQUIREMENTS: Provide validation of the desired comfort criteria as determined by EQ Credit 7.1 using either of the two methods described by ASHRAE Standard 55-2004 (analysis of environmental variables or occupant survey).

Combined, these credits make it more likely that the HVAC system will satisfy the comfort needs of the occupants, which is a positive benefit to the HVAC engineer. The validation also provides a source of potentially valuable feedback to the designer regarding the actual performance of the HVAC systems designed.

### EQ Credit 8—Daylight and Views

Credit EQc8 encourages the use of daylighting techniques to bring natural light into the interior of the building (see Figure 6-15). While this is not the direct purview of the HVAC engineer, the HVAC design is certainly affected by the introduction of daylighting into the building interior. It will potentially affect heating and cooling loads from solar insolation, as well as the internal heat gains from lighting (which may be dimmed or off when daylight is available). Thus, this credit becomes a point of coordination for the HVAC engineer with the architect and lighting designer as the daylighting strategy is implemented in the building design.

### Innovation in Design Credits

ID Credit 1—LEED Certified Professional
ID Credit 2—Innovation in Design (four possible points)

**Figure 6-15 Daylighting example.**

- INTENT: To provide design teams and projects the opportunity to be awarded points for exceptional performance above the requirements set by the LEED-NC Green Building Rating System® and/or innovative performance in green building categories not specifically addressed by the LEED-NC Green Building Rating System®.

It is interesting to note that almost no LEED innovation credits have been awarded for innovative HVAC systems in the projects certified by the USGBC to date. This seems to represent a worthy challenge to the HVAC engineering community!

## LEED REGISTRATION AND CERTIFICATION PROCESS

The process by which LEED projects are registered, reviewed, and approved is described below. The most important aspect of the LEED system is that it is a *third-party certification process* rather than a self-certification process, which gives added significance to the LEED rating conferred on a project.

The LEED project process consists of three steps:

- Step 1: Project registration
  - Register the project and project team online
  - Receive welcome packet and online project listing
- Step 2: Technical support
  - Receive *LEED Reference Guide* package
  - Obtain access to LEED credit interpretation rulings (CIRs)
- Step 3: Building certification
  - Submit documentation at end of construction
    - *Note:* it is now possible to submit the design-related credits at the end of the design phase and the construction-related credits at the end of the construction phase. This reduces uncertainty about which credits will be earned and minimizes the length of time the design team must be involved in the certification process.
  - After USGBC review and approval, the project receives a LEED plaque and certificate commemorating the LEED rating awarded.

Traditionally, the LEED rating was thought of in terms of filling out the "LEED scorecard" (see Figure 6-16), summarizing the LEED credits being sought for the project. Although this process is now largely outmoded (given the online submittal process), the same basic concept applies: selecting the credits for the project. One of the most significant aspects of the LEED rating is that it is largely tailored to the individual project. Thus, no two projects need utilize exactly the same LEED credits nor earn them in the same exact way. This allows the project to evolve to meet its needs rather than to "chase credits," and this is an important point: the LEED strategy

**LEED™ Scorecard**

Purpose of Form: Use this form to score your project against the LEED™ Green Building Rating System. Fill it out at the time of registration. It will help you, and us, to keep track of the prerequisites and applicable credits on your project. It will also be used to track compliance when that documentation is submitted to the US Green Building Council.

**14 Sustainable Sites**

Prerequisite: Erosion and Sedimentation Control
Credit 1: Site Selection
Credit 2: Urban Redevelopment
Credit 3: Brownfield Redevelopment
Credit 4: Alternative Transportation

Credit 5: Reduced Site Disturbance
Credit 6: Stormwater Management
Credit 7: Landscape and Exterior Design to Reduce Heat Islands
Credit 8: Light Pollution Reduction

**5 Water Efficiency**

Credit 1: Water Efficient Landscaping
Credit 2: Innovative Wastewater Technologies
Credit 3: Water Use Reduction

**17 Energy and Atmosphere**

Prerequisite 1: Fundamental Building Systems Commissioning
Prerequisite 2: Minimum Energy Performance
Prerequisite 3: CFC Reduction in HVAC&R Equipment

Credit 1: Optimize Energy Performance
Credit 2: Renewable Energy
Credit 3: Additional Commissioning
Credit 4: Elimination of HCFC's and Halons
Credit 5: Measurement and Verification
Credit 6: Green Power

**13 Materials and Resources**

Prerequisite: Storage & Collection of Recyclables
Credit 1: Building Reuse
Credit 2: Construction Waste Management
Credit 3: Resource Reuse

Credit 4: Recycled Content
Credit 5: Local/Regional Materials
Credit 6: Rapidly Renewable Materials
Credit 7: Certified Wood

**15 Indoor Environmental Quality**

Prerequisite 1: Minimum IAQ Performance
Prerequisite 2: Environmental Tobacco Smoke (ETS) Control
Credit 1: Carbon Dioxide (CO2) Monitoring
Credit 2: Increase Ventilation Effectiveness
Credit 3: Construction IAQ Management Plan

Credit 4: Low-Emitting Materials
Credit 5: Indoor Chemical and Pollutant Source Control
Credit 6: Controllability of Systems
Credit 7: Thermal Comfort
Credit 8: Daylight and Views

**64 Total Core LEED Rating System Points**

**5 Innovation and Design Process Points**

LEED Innovation Credits
LEED Accredited Professional

Total Points Scored

**LEED Green Building Certification Levels**

| LEED Certified | = 26 - 32 Points |
| LEED Certified Silver Level | = 33 - 38 Points |
| LEED Certified Gold Level | = 39 - 51 Points |
| LEED Certified Platinum Level | = 52+ Points |

Source: LEED Reference Guide

**Figure 6-16 The LEED scorecard.**

for a project should evolve from the whole-building design process for the project, selecting design concepts that help meet the project goals rather than have the design be driven primarily by a desire to earn a certain number of LEED points.

## SUMMARY

The LEED Green Building Rating System® provides an opportunity for the design and development team for a project to strive for a level of accountability as to the effectiveness of their efforts to produce a building that is truly "green." Some aspects of this accountability of the rating include:

- Third-party validation of achievement.
- Standardized sets of measures of green features.
- Ever-evolving set of criteria to keep pace with changes in technology.

Some of the benefits of participating in this rating effort include:

- Qualifying for the growing array of state and local government incentives.
- Contributing to growing knowledge base.
- Receiving a LEED certification plaque and certificate commemorating the LEED rating.
- Receiving marketing exposure through the USGBC Web site, case studies, and media announcements.
- Receiving recognition in the marketplace as a LEED building.
- Receiving recognition from building occupants.

It is clear that there is a very important role here for the HVAC engineer in designing a LEED-rated building.

- 18 LEED credits (totaling 34 LEED points) are in the purview of the mechanical, electrical, and plumbing engineers.
- 11 LEED credits (totaling 21 LEED points) focus specifically on issues typically handled by the HVAC engineer.

By applying the principles described in the other chapters of this *GreenGuide* and using the metrics of LEED described in this chapter to measure the impact, the HVAC engineer can make a significant contribution to delivering a green building.

## REFERENCES

ASHRAE. 2000. *ANSI/ASHRAE Standard 52.2-1999, Method of Testing General Ventilation Air-Cleaning Devices for Removal Efficiency by Particle Size.* Atlanta: American Society of Heating, Refrigerating and Air-Conditioning Engineers, Inc.

ASHRAE. 2004. *ANSI/ASHRAE Standard 55-2004, Thermal Environmental Conditions for Human Occupancy.* Atlanta: American Society of Heating, Refrigerating and Air-Conditioning Engineers, Inc.

ASHRAE. 2004. *ANSI/ASHRAE Standard 62.1-2004, Ventilation for Acceptable Indoor Air Quality.* Atlanta: American Society of Heating, Refrigerating and Air-Conditioning Engineers, Inc.

ASHRAE. 2004. *ANSI/ASHRAE/IESNA Standard 90.1-2004, Energy Standard for Buildings Except Low-Rise Residential Buildings.* Atlanta: American Society of Heating, Refrigerating and Air-Conditioning Engineers, Inc.

Taylor, S.T. 2005. LEED and Standard 62.1. *ASHRAE Journal Sustainability Supplement*, September.

US Green Building Council, LEED-NC v 2.2 Rating System.

US Green Building Council, *LEED-NC v 2.2 Reference Guide.*

# 7

# CONCEPTUAL ENGINEERING DESIGN— LOAD DETERMINATION

The traditional load determination methods, such as the cooling load temperature difference method or rules of thumb, are rough first approximations based on old correlations or simplified heat transfer calculations. Designing low-energy buildings requires the engineer to have a thorough understanding of the dynamic nature of the interactions of the building with the environment and the occupants. To optimize the design, detailed computer simulations allow the engineer to accurately model the major loads and interactions.

Loads can be divided into those stemming from the envelope and those from internal sources. Envelope loads include the impacts of the architectural features; heat and moisture transfer through the walls, roof, floor, and windows; and infiltration. Internal loads include lights, equipment, people, and process equipment. Examine all of the loads in two ways: separately, to determine their relative impacts, and together, to determine their interactions.

The engineer must also understand the energy sources and flows in the building and their location, magnitude, and timing. Once the engineer understands these sources and flows, he/she can be creative in coming up with solutions. The charts in Figure 7-1 show how average energy use breaks down in typical office buildings at three different climatic locations in the United States. As designs are developed, breakdowns such as these should be kept in mind so that the energy-using areas that matter most are given priority in the design process.

When trying to minimize energy use in buildings, the first step is to identify which aspects of building operation offer the greatest energy-saving opportunities. For example, as shown in the Figure 7-1 pie charts, for Chicago a reduction of space heating energy would be the first priority. In Miami and Philadelphia, however, the first priority would be reduction of energy used in lighting. In Philadelphia the second priority would be space heating, but in Miami it would be space cooling.

To find energy end-use statistics for many types of buildings (office, education, health care, lodging, retail, etc.) based on building location, age, size, and principal energy sources, consult the *EN4M Energy in Commercial Buildings* tool found at www.eere.energy.gov/buildings/tools_directory/software.cfm/ID=299/.

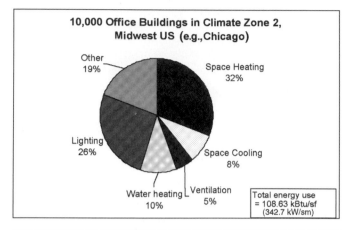

10,000 Office Buildings in Climate Zone 2, Midwest US (e.g., Chicago)

Other 19%
Space Heating 32%
Lighting 26%
Space Cooling 8%
Water heating 10%
Ventilation 5%

Total energy use = 108.63 kBtu/sf (342.7 kW/sm)

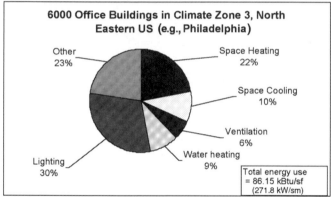

6000 Office Buildings in Climate Zone 3, North Eastern US (e.g., Philadelphia)

Other 23%
Space Heating 22%
Space Cooling 10%
Ventilation 6%
Water heating 9%
Lighting 30%

Total energy use = 86.15 kBtu/sf (271.8 kW/sm)

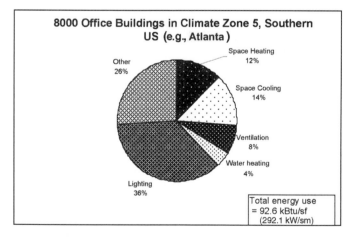

8000 Office Buildings in Climate Zone 5, Southern US (e.g., Atlanta)

Other 26%
Space Heating 12%
Space Cooling 14%
Ventilation 8%
Water heating 4%
Lighting 36%

Total energy use = 92.6 kBtu/sf (292.1 kW/sm)

Figure 7-1   Average office building energy use in US.

## ENERGY IMPACTS OF ARCHITECTURAL FEATURES

For economic reasons, it is very important to focus your efforts in the following sequence: *reduce loads first.* Reducing loads will have a compounded effect on reducing overall energy consumption due to lower HVAC requirements.

Work with architects to improve the thermal envelope, especially in areas of external shading of glazing; spectrally selective glazing, appropriately placed and sized; good roof and wall insulation with a radiant barrier; and orientation of building (east-west long axis).

To gain an understanding of how the proposed design of the building affects energy use, the engineer should perform a series of parametric simulations on the building in the specific location. Sequentially take each load out of the energy balance individually and note the effect on overall energy use.

For instance, to effectively remove wall conductive heat transfer from the energy balance, set the wall thermal resistance to a very high value (such as R-100 [R-17.6]). Run the simulation model and note the building energy requirements. Then, reset the wall insulation back to the actual R-value and proceed to the next parameter. Continue to do this individually with the floors, roofs, and windows. Then remove the window solar gain, daylighting, and other site-specific shading that may be involved. The final envelope load is infiltration, which may or may not be important depending on whether the building is pressurized by outdoor air ventilation. Then look at the effect of outdoor air by setting it to zero. Plot the results for each case by annual energy use or by peak load to compare the impacts, as illustrated in Figure 7-1. A "code" building would be the building that meets the minimum requirements of the current ASHRAE/IESNA Standard 90.1, and the "design" building is a proposed building design.

Look for creative solutions to minimize the impact of each load, starting with those with the largest impact or those that are the easiest to implement. From the parametric analysis example in Figure 7-2, outdoor air has the largest impact on the annual energy use. One solution to this problem may be to monitor $CO_2$ levels to reduce the ventilation requirements. If solar gain through the windows is a large problem, look at orientation, size, and location of the windows, glass type, and possibly external shading.

## THERMAL/MASS TRANSFER OF ENVELOPE

Basic, steady-state energy transfer through building envelopes is well known, well understood, and easy to calculate (see Figure 7-3). Increased R-values are certainly beneficial in heating climates and can also help in cooling-load-dominated climates. Use as a minimum the recommended amounts spelled out in ASHRAE's energy standard (Standard 90.1, latest approved edition). While values above those recommended can be beneficial in simple structures, high values can be counterproductive over a heating/cooling season in more complex structures. It is always

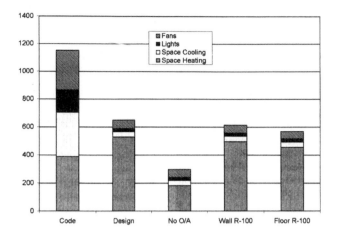

**Figure 7-2 Sample parametric analysis to determine relative impacts of building envelope.**

- U-factor
- R-values
- Thermal Mass
- Heat Capacity

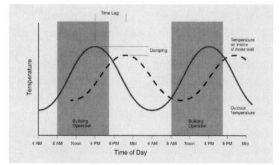

Source: US DOE Energy Efficiency and Renewable Energy

**Figure 7-3 Heat transfer through building envelope.**

wise to evaluate R-value benefits through the application of load and energy simulation programs.

The effects of thermal mass are sometimes not as easy to gauge intuitively. (See Figure 7-4.) Therefore, the above-mentioned simulation programs, properly applied, are very useful in this regard. If thermal mass is significant in the building being planned (or if increased mass would be easy to vary as optional design choices), then such programs are essential for evaluating the "flywheel" effects of thermal mass on both loads and longer-term energy use.

GreenTip #6 describes a technique for combining nighttime ventilation with a building's thermal mass to achieve load and/or energy savings.

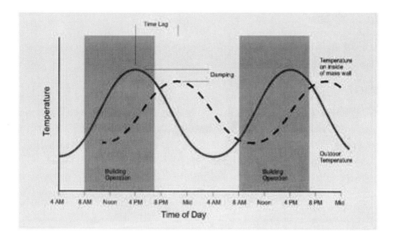

**Figure 7-4 The effects of thermal mass.**

## ENGINEERING LOAD-DETERMINING FACTORS

Parametric simulations should be completed for the internal loads as well, in a process similar to that described in the architectural features section above. Set the lighting load to zero, set the equipment load to zero, and then take all the people out of the building. In this manner, the engineer can understand what is driving the energy use in the building. For office buildings, lights are usually the major culprits. Therefore, optimize the daylighting and electric lighting design.

Likewise, evaluate office equipment loads. Make recommendations about the effects of the choices of computers, monitors, printers, and other types of equipment. In many offices, this equipment is left on all night. An office building should have very few loads when the building is unoccupied. Leaving an office full of equipment on all night and on weekends can easily add up to large energy consumption.

For example, assume one 34,000 ft$^2$ (3,159 m$^2$) office building has nighttime plug loads of 10 kW, and assume the building is unoccupied for 14 hours a day during the week and 24 hours a day on the weekend. This adds up to around 6000 hours per year and 60,000 kWh—or $4200 (at $0.07/kWh)—of electricity that could have easily been reduced. It is important for the engineer to bring up these issues during the design stage (and later, during the operation) of the building because no one else may be paying attention to such details. Designing and building a great building is only half the job; operating it in the correct manner is the other—and often more important—half. The engineer can make better operation possible by designing piping, wiring, and controls capable of easily turning things off when not being used.

Educate owners about efficient office equipment and appliances to reduce plug loads, covering such things as flat screen computer monitors, laptops vs. desktop central processing units, copy machines, refrigerators, and process equipment. Consider measuring usage in one of the clients' existing buildings to get an accurate picture of load distribution and population profiles. Attempt to develop a total building electric load profile for every minute of one week.

## LIGHTING

Good sustainable designs should not increase first costs over typical designs. More expensive equipment, glazing, and lighting should yield lower capacity requirements to offset those extra costs. Smaller equipment translates into smaller wiring, transformers, fuses, switchgear, etc.

The paragraphs below briefly outline some key considerations in lighting design, especially those involved in daylighting. Chapter 13 contains additional information for HVAC&R designers on lighting systems.

### Electric Lighting

Work with electric lighting designers to lower connected and actual lighting loads while improving visual acuity. Collectively set the budget at less than the levels set in the current ASHRAE Standard 90.1 (or local energy codes, whichever is more stringent). Typically this will mean values less than 1 watt per gross square foot connected load. Consider the following items: indirect, low ambient lighting levels; using task lights at workstations to allow lower area lighting levels; and using accent lighting in lieu of high area lighting levels to highlight relief, color, and areas of interest.

### Daylight Harvesting

**Studies on Daylight Harvesting.** One of the principal rationales for the use of daylighting is the beneficial impact it has on the occupants of the building in terms of improved productivity and well-being. An example of the benefits is shown in one study of daylight harvesting impacts on retail sales, which showed a 40% improvement in sales. Another study of daylighting impacts on school performance[1] showed a 20% improvement in math scores and a 26% improvement in reading scores.

**Daylight Harvesting Design Process: Some Key Points.**

*Design Process/Design Options*
- Prospect in the schematic design phase.
- Integrate the design of all systems:
    - Envelope
    - HVAC
    - Lighting
    - Interiors
- Take credit for HVAC downsizing to "pay" for daylight harvesting.
- Physical modeling can be useful.

---

1.  Heschong Mahone Group for Pacific Gas and Electric Company, www.h-m-g.com.

- Address *all* the problems of daylit buildings:
  - Direct insolation and glare control
  - Daylighting central cores
  - Heat gain—glazing selection, low-E glass
  - Heat loss—glazing selection, double-pane, interior shading,
  - Owner awareness of design assumptions

### Energy Conservation Regulation Compliance
- Prescriptive path vs. performance path (energy budget)
- A well-designed daylit building will comply with Title 24 encrgy budgets

### Daylighting Considerations
- Envelope
- HVAC
- Lighting/controls
- Interiors/operations
- Implementation

### Envelope
- Siting and orientation
- Fenestration/effective aperture area
  - Glass type
  - Area
- Shading and window management
  - External shading wherever possible
- Fenestration location

### HVAC Loads
- Ensure realistic load assumptions among the design team
- Design for actual load, not "imaginary worst case"
- Take advantage of reduced internal loads to downsize
- Select appropriate HVAC system
- Capture skin loads before they enter the space
  - Return air grilles over windows
  - Returns or exhaust fans over skylights

### HVAC Zonation
- Upsize cooling capacity in zones near daylighting fenestration, if warranted
- May need to upsize heating in such zones as well
- Perimeter HVAC zones should be sized to be similar to daylighting zones (15–18 ft deep)

### Implementation
- Entire design has to be implemented
- Cannot take out daylighting and not upsize HVAC to compensate

- Cannot delete shading devices
- Cannot delete controls
- Cannot paint interiors black
- Methods of operation need to be explained to occupants of the building.
- DON'Ts in the daylighting process:
    - (Avoid) partial design and specification
    - (Avoid) partial implementation
    - (Avoid) counterproductive building operation
- Cost-benefit considerations:
    - Lighting energy savings
    - Energy savings due to reduced HVAC loads
    - Capital savings due to reduced equipment size
    - Capital savings from SCE "Savings by Design" (more information at www.savingsbydesign.com)
    - Improved comfort
    - Improved productivity and sales
    - "Delight" factor sells the space

## SYSTEM/EQUIPMENT EFFICIENCIES

It is important to use cooling and heating equipment of the correct size. The old rule of thumb of 250–350 gross ft$^2$/ton cooling load does not apply to sustainable buildings. Recent high-performance building projects operate between 600 and 1000 GSF/ton. Set cooling equipment and system performance targets in terms of kW/ton (kW/kWR), such as:

| | |
|---|---|
| Chiller | 0.51 kW/ton (0.145 kW/kWR) |
| Cooling tower | 0.011 kW/ton (0.003 kW/kWR) |
| Chilled-water pump | 0.026 kW/ton (0.007 kW/kWR) |
| Condenser water pump | 0.021 kW/ton (0.006 kW/kWR) |
| Air-handling unit | 0.05 kW/ton (0.014 kW/kWR) |

Industry standards such as ASHRAE 90.1 (latest approved version) give minimum requirements for equipment efficiencies and system design and installation. Understand that these represent the least-efficient end of the spectrum of energy-conserving buildings that should be built! To be considered green, a building must exceed these standards.

There are a number of sources of information on energy efficiency as related to green building design in addition to ASHRAE, some of which are listed below. This Guide does not endorse any of them; the list is presented for informational purpose only. Readers should be aware that the sources use various methods to arrive at their final recommendations and that some of the guidance offered may have a hidden (or not so hidden) agenda. Some may use economics as a basis; however, the underlying economic assumptions should be understood prior to using the information. Others attempt to push energy efficiency to its technical limits. Therefore, before using any of these sources for guidance, investigate the premises used, the methods of analysis, and the background of the author.

## Governmental Agencies

*   EPA's ENERGY STAR® program, www.energystar.gov
*   DOE's Federal Energy Management Program, www.eere.energy.gov/femp
*   DOE's High Performance Buildings Institute, www.highperformance buildings.gov
*   California Energy Commission, www.energy.ca.gov

## Environmental Groups

*   American Council for an Energy Efficient Economy, www.aceee.org
*   Alliance to Save Energy, www.ase.org

## Industry Groups

*   Geothermal Heat Pump Consortium, www.geoexchange.org
*   Air-Conditioning and Refrigeration Institute, www.ari.org

Installing efficient pieces of equipment alone does not make a building green. It only creates the opportunity for the building to operate with reduced energy consumption. Integrating those pieces into a system is discussed in the following literature (use latest edition published):

*   *ASHRAE Handbook—Applications*
*   ASHRAE's *Fundamentals of Water System Design*
*   Trane's *Multiple-Chiller-System Design and Control Manual*
*   *CoolTools Chilled Water Plant Design Guide*

In addition to system integration, the comfort and process heating and cooling systems should be integrated with the building.

As the building design changes, so should the systems. For example, lighting or glazing retrofits can greatly reduce system cooling requirements. In addition, reduction in the space load may change the characteristics of the load. When the space sensible load is drastically reduced, the space sensible heat ratio becomes much steeper. In such cases, the applicability of the system must be investigated. Both air and water systems should be revisited when major building retrofits occur.

## Key Considerations in the HVAC Design Process

### DESIGN INTENT

- Set goals for performance
  - Energy performance
  - Environmental performance
  - Comfort
  - Operating cost
  - Determine how to achieve the goals
- System by system
  - Integrated design

### VERIFY THAT DESIGN INTENT IS MET

- In design
  - Verification of Cx goals in design
  - Coordination between design disciplines
  - Include commissioning in design documents
- In construction
  - Procurement of equipment and materials
  - Installation
- At start-up and testing
- In operations

### DESIGN INTEGRATION

- Integration with other disciplines
  - Architecture, lighting, interiors, structural
  - Daylighting
  - Underfloor air distribution
  - "Form-follows-function" design
- Increased emphasis on HVAC performance
  - Thermal comfort
  - IAQ
  - Energy efficiency

### HVAC SYSTEMS

- High-efficiency equipment
- Systems responsive to partial loads
  - 80% of year, system operates at <50% of peak capacity.
- Emphasis on "free" cooling and heating
  - Economizers (air, water)
  - Evaporative cooling (cooling towers, precooling)
  - Heat recovery
- Emphasis on IAQ
- Underfloor air distribution is new wave

## LOAD REDUCTION

- Reduce envelope loads
  - Solar loads
- Reduce lighting loads
  - ASHRAE Standard 90.1 or local energy codes as a design maximum
- Reduce power loads
  - Site and building type specific, perhaps 1.0 to 1.5 W/ft$^2$ (16 W/m$^2$) as a maximum
- Reduced air-conditioning tonnage
  - Can provide higher air-conditioning efficiency for same cost

## COOLING AND HEATING LOAD REDUCTION

- Envelope loads
  - Shading
  - Glass selection
  - Glass percentage
- Internal loads
  - Lighting power density (LPD)
  - Equipment loads (ENERGY STAR$^®$)
  - Controls/occupancy sensors

## ASHRAE GreenTip #6:
## Night Precooling

### GENERAL DESCRIPTION

Night precooling involves the circulation of cool air within a building during the nighttime hours with the intent of cooling the structure (see Figure 7-5). The cooled structure is then able to serve as a heat sink during the daytime hours, reducing the mechanical cooling required. The naturally occurring thermal storage capacity of the building is thereby utilized to smooth the load curve and for potential energy savings. More details on the concept of thermal mass on building loads are included in Chapter 4, "Architectural Design Impacts."

There are two variations on night precooling. One, termed *night ventilation precooling*, involves the circulation of outdoor air into the space during the naturally cooler nighttime hours. This can be considered a passive technique except for any fan power requirement needed to circulate the outdoor air through the space. The night ventilation precooling system benefits the building IAQ through the cleansing effect of introducing more ventilation air. With the other variation, *mechanical*

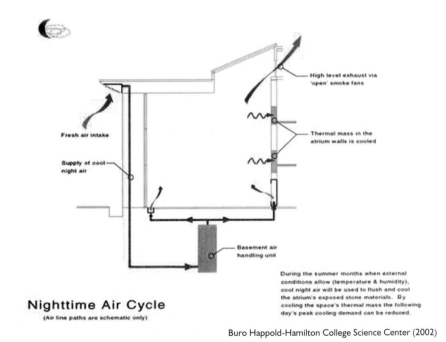

High level exhaust via 'open' smoke fans

Fresh air intake

Thermal mass in the atrium walls is cooled

Supply of cool night air

Basement air handling unit

**Nighttime Air Cycle**
(Air line paths are schematic only)

During the summer months when external conditions allow (temperature & humidity), cool night air will be used to flush and cool the atrium's exposed stone materials. By cooling the space's thermal mass the following day's peak cooling demand can be reduced.

Buro Happold-Hamilton College Science Center (2002)

**Figure 7-5 Schematic example of nighttime air cycle.**

*precooling*, the building mechanical cooling system is operated during the nighttime hours to precool the building space to a setpoint usually lower than that of normal daytime hours.

Consider these key parameters when evaluating either concept:

- local diurnal temperature variation
- ambient humidity levels
- thermal coupling of the circulated air to the building mass

The electric utility rate structure for peak and off-peak loads also is important to determine the cost-effectiveness, in particular for a mechanical precooling scheme.

A number of published studies show significant reductions in overall operating costs by the proper precooling and discharge of building thermal storage. The lower overall costs result from load shifting from the day to the nighttime with its associated off-peak utility rates. For example, Braun (1990) showed significant energy cost savings of 10% to 50% and peak power requirements of 10% to 35% over a traditional nighttime setup control strategy. The percent savings were found to be most significant when lower ambient temperatures allowed night ventilation cooling to be performed.

For a system incorporating precooling to be considered a truly green design concept, the total energy used through the entire 24-hour day should be lower than without precooling. A system that uses outdoor air to do the precooling only requires the relatively lower power needed to drive the circulation fans, compared to a system that incorporates mechanical precooling. Electrical energy provided by the utility during peak demand periods also may be "dirtier" than that provided during normal periods, depending on the utility and circumstances.

The system designer needs to be aware of the introduction of additional humidity into the space with the use of night ventilation. Thus, the concept of night ventilation precooling is better suited for drier climates. A mechanical nighttime precooling system will prevent the introduction of additional humidity into the space by the natural dehumidification it provides, but at the expense of greater energy usage compared to night ventilation alone.

Both variations (night ventilation and night mechanical precooling) are not 100% efficient in the thermal energy storage (TES) in the building mass, particularly if the building is highly coupled (thermally) with the outside environment. Certain building concepts used in Europe are designed to increase the exposure of the air supply or return with the interior building mass (see, for example, Andersson et al. [1979]). This concept will increase the overall efficiency of the thermal storage mass.

For either type of system, the designer must carefully analyze the structure and interaction with the HVAC system air supply using transient simulations in order to assess the feasibility with their particular project. A number of techniques and commercially available computer codes exist for this analysis (Balaras 1995).

## WHEN/WHERE IT'S APPLICABLE

Night precooling would be applicable in the following circumstances:

- When the ambient nighttime temperatures are low enough to provide sufficient opportunity to cool the building structure through ventilation air. Ideally, a low ambient humidity level would also occur. A hot, dry environment, such as the southwestern United States, is an ideal potential area for this concept.
- When the building occupants would be more tolerant of the potential for slightly cooler temperatures during the morning hours.
- When the owner and design team are willing to include such a system concept and to commit to (1) a proper analysis of the dynamics of the building thermal performance and (2) the refinement of the control strategy upon implementation to fine-tune the system performance.
- More massive buildings, or those built with heavier construction materials such as concrete or stone as compared to wood, have a greater potential for benefits. Just as important is the interaction of the building mass with the building internal and HVAC system circulating air. This interaction may allow for more efficient transfer of thermal energy between the structure and the airspace.

## PROS AND CONS

### Pro

1. Night ventilation precooling has good potential for net energy savings because the power required to circulate the cooler nighttime air through the building is relatively low compared to the power required to mechanically cool the space during the daytime hours.
2. Mechanical precooling could lead to net energy savings, although there will likely be a net increase in total energy use due to the less-than-100% TES efficiency in the building mass.
3. Both variations require only minor, if any, change to the overall building and system design. Any changes required are primarily in the control scheme.
4. Night ventilation can provide a better IAQ environment due to increased circulation of air during the night. A greater potential exists with the ventilation precooling concept. Both will be better than if the system were completely shut off during unoccupied hours.

### Con

1. Temperature control should be monitored carefully. The potential exists for the building environment to be too cool for the occupant's comfort during the early hours of the occupied period. This will result in increased service calls or complaints and may end with the night precooling being bypassed or turned off.

2.  The increased runtime on the equipment could lead to lower equipment life expectancy or increased frequency in maintenance. Careful attention should be given to the resulting temperature profile through the day during the commissioning process. Adjustments may be needed to the control schedule to keep the building within the thermal comfort zone.
3.  Proper orientation must be given to the building operator to understand how the control concept affects the overall system operation throughout the day.
4.  Future turnovers in building ownership or operating personnel could negatively affect how successfully the system performs.
5.  Occupants would probably need at least some orientation so that they would understand and be tolerant of the differences in conditions that may prevail with such a system. Future occupants may not have the benefit of such orientation.

## KEY ELEMENTS OF COST

The following provides a possible breakdown of the various cost elements that might differentiate a nighttime precooling scheme from a conventional one and gives an indication of whether the net cost for the precooling option is likely to be lower (L), higher (H), or the same (S). This assessment is only a perception of what might be likely, but it may not be correct in all situations. **There is no substitute for a detailed cost analysis as part of the design process.** The listings below may also provide some assistance in identifying the cost elements involved.

### First Cost

| | |
|---|---|
| • Mechanical ventilation system elements | S |
| • Architectural design features | S |
| • System controls | H |
| • Analysis and design fees | H |

### Recurring Cost

| | |
|---|---|
| • Energy for mechanical portion of system | |
|    Ventilation precooling | L |
|    Mechanical precooling | S/H |
| • Total cost to operate cooling systems | L |
| • Maintenance of mechanical ventilation and cooling system | S/H |
| • Training of building operators | H |
| • Orientation of building occupants | H |
| • Commissioning cost | H |
| • Occupant productivity | S |

## SOURCES OF FURTHER INFORMATION

The following is a sampling of representative papers that can provide further background information.

Andersson, L.O., K.G. Bernander, E. Isfält, and A.H. Rosenfeld. 1979. Storage of heat and coolth in hollow-core concrete slabs. Swedish experience and application to large, American style buildings. Second International Conference on Energy Use and Management, Lawrence Berkeley National Laboratory, LBL-8913.

Balaras, C.A. 1995. The role of thermal mass on the cooling load of buildings. An overview of computational methods. *Energy and Buildings* 24(1):1–10.

Braun, J.E. 1990. Reducing energy costs and peak electrical demand through optimal control of building thermal storage. *ASHRAE Transactions* 96(2):876–88.

Keeney, K.R., and J.E. Braun. 1997. Application of building precooling to reduce peak cooling requirements. *ASHRAE Transactions* 103(1):463–69.

Kintner-Meyer, M., and A.F. Emery. 1995. Optimal control of an HVAC system using cold storage and building thermal capacitance. *Energy and Buildings* 23:19–31.

Ruud, M.D., J.W. Mitchell, and S.A. Klein. 1990. Use of building thermal mass to offset cooling loads. *ASHRAE Transactions* 96(2):820–29.

# 8

# SPACE THERMAL/COMFORT DELIVERY SYSTEMS

Occupant comfort and health are important in green building design. Sacrificing the quality of the indoor environment in the name of green design is not a viable strategy, since maintaining maximum human productivity and performance is paramount.

## THERMAL

To provide for a thermally comfortable environment that supports the productive performance of the building occupants,

- comply with ASHRAE Standard 55 (latest approved edition) plus approved addenda for thermal comfort, including humidity control, within established ranges for comfort, and
- install a permanent temperature and humidity monitoring system configured to provide operators control over thermal comfort performance and effectiveness of humidification and/or dehumidification systems in the building. *Individual control has green advantages by allowing variation to suit individual preference as well as shutting down when individual spaces are unoccupied.*

## INDOOR AIR QUALITY (IAQ)

Reduce the amount of indoor and outdoor air contaminants that have adverse impacts on the environment and human health with the following steps:

- Evaluate and preferentially specify materials that are low-emitting, nontoxic, and chemically inert.
- Do not install combustion appliances unless they are sealed—combustion or power-vented; avoid gas ranges.
- Prevent exposure of building occupants and systems to environmental tobacco smoke.
- Utilize effective moisture control to curb humidity and prevent mold problems.
- Avoid using hard-to-seal building cavities, such as dropped ceiling plenums, for air movement unless these areas can be properly constructed and sealed.

Given the fact that ceiling plenums will be used on many projects, attention should be paid to ensure that exterior elements (e.g., fluted roof deck to exterior wall) are properly sealed.

- Establish minimum IAQ performance to prevent the development of IAQ problems in the building, maintaining the health and well-being of the occupants.
- Meet the minimum requirements of voluntary consensus standard *ASHRAE Standard 62, Ventilation for Acceptable Indoor Air Quality* (latest approved edition) and approved addenda.
- Provide IAQ monitoring to sustain long-term occupant health and comfort.
- Power-vent equipment that emits VOCs and ozone.
- Consider natural or LP gas with vented hoods, as gas ranges are less GHG-emitting than grid-supplies electric ranges.
- Have permanent minimum IAQ performance monitoring for $CO_2$, CO, $O_3$, $NO_x$, and VOCs.

## ENERGY EXCHANGE

Distributing energy throughout a building is usually accomplished through the flow of steam or a hydronic fluid, air, electrons (electricity), and sometimes a refrigerant. Air and hydronic flows in particular also serve the function of disposition of used air (exhaust) or liquid waste. That the air and hydronic media being so moved are often at different thermal levels (i.e., warm, cold), opportunities are offered to incorporate green design techniques. There are several practical techniques whereby energy from one flow stream can be transferred usefully to another.

Seven separate GreenTips on energy recovery systems are contained within this chapter.

### Cooling and dehumidifying

**Chilled water (CHW).** Circulating CHW through is generally the least energy-efficient process for refrigerated air conditioning but in many cases is the most cost effective, particularly for large buildings. The relatively low thermal efficiency results from it being a two-step process. The most common process uses a refrigerant using an electrical recycling vapor-compression system to a low temperature. The CHW then chills a secondary refrigerant (water) to the temperature needed for cooling or both cooling and dehumidifying the air supplied to the occupied zones.

CHW can also be produced by a heat-driven absorption process that uses a mixture of water with lithium bromide or ammonia.

Some installations can benefit from energy reduction by using thermal storage using either CHW or ice, if further reduced in temperature and melted for later use.

The CHW is pumped through air-conditioning (A/C) coils that may be in a central plant room, in several building zones, or in individual rooms.

Further energy losses result from heat gained through insulation on the CHW piping.

**Liquid desiccant (LD).** Circulating LD is a single-step process that uses heat such as that from natural gas or from LP, solar, or waste heat sources when available. LD generally does not need piping to be insulated.

**Synthetic refrigerant (SR).** Circulating SR from an electrical recycling vapor-compression system can be directly used in A/C coils that are in the same housing or plant room. Alternatively, the SR can be distributed to A/C coils in several building zones or in individual rooms, such as with variable refrigerant flow or volume (VRF or VRV) systems.

**Natural refrigerant (NR).** Ammonia, hydrocarbons, and carbon dioxide should also be considered.

## Leakage

There are differing hazards and costs associated with refrigerant leaks, and all should be considered serious faults.

- *Chilled water.* Leaks mainly cause building damage. Pressure testing procedures must be specified.
- *Liquid desiccant.* Some can have minor toxicity and also can cause building damage. Piping, being uninsulated, allows ready inspection.
- *Synthetic refrigerant.* Different types of SR variously damage the ozone layer and increase global warming potential. These leaks can be hard to detect, especially in systems with extended piping, and may develop long after installation. As Pearson (2003) noted, high losses have been reported, leading to concern that SR maybe should be confined to plant room or factory certified leak-free systems.
- *Natural refrigerant.* Hydrocarbons are highly flammable, and ammonia is toxic and has limited flammability. These refrigerants are required by relevant codes to be confined to plant rooms having appropriate ventilation and spark protection.

## ENERGY DELIVERY METHODS

### Media Movers (Fans/Pumps)

**Basics: Power, Flow and Pressure.** If air conditioning (heating/cooling) could be produced exactly where it is needed throughout a building, overall system efficiency would increase because there would be no additional energy used to move (distribute) conditioned water or air. For acoustic, aesthetic, logistic, and a variety of other reasons, this ideal seldom is realized. Therefore, fans and pumps are used to move energy in the form of water and air. Throughout this process, the goal is to minimize *system* energy consumption.

Minimization of a media mover's power at full-load and part-load conditions is the goal. Understanding how fan and pump power change with flow and pressure is imperative. (*Note:* Refer also to Chapter 2, "Background and Fundamentals.")

$$Power \sim Flow \times \Delta Pressure$$

As flow drops, so does the pressure differential through pipes, chillers, and coils. Pressure drop through these devices varies approximately with (flow)$^{1.85}$. This, in turn, means that the power required to cause flow through these devices varies approximately as (flow)$^{2.85}$. (In an ideal world, power changes with the cube of the flow. While the 2.0 and 3.0 exponent is not fully achieved in practical application, it is nevertheless clear that reducing flow can drastically reduce energy use.)

However, there are some pressure drops in typical systems that do *not* change as flow decreases. These are:

- The pressure differential setpoint that many system controls use
- Cooling tower static lift
- Pressure drop across balancing devices

Therefore, for flow through these system components, power will vary *directly* with flow.

To reduce pressure drop, the pipe or duct size should be maximized and valve and coil resistance minimized. (Duct and pipe sizes are discussed in the "distribution" section below.) Coil sizes should be maximized within the space allowed to reduce pressure drop on both the water side and the air side. The ideal selection will require striking an economic balance between first cost and projected energy savings (operating cost).

**Chilled-Water Pumps.** Historically a design chilled-water temperature differential ($\Delta T$) across AHU cooling coils of 10°F (5.5°C) was used, which results in a flow rate of 2.4 gpm/ton (2.6 L/min per kW). In recent years, the 60% increase in required minimum chiller efficiency from a 3.8 COP (ASHRAE Standard 90-75) to 6.1 COP (ASHRAE Standard 90.1-2004) has led to a reexamination of the assumptions used in designing hydronic media flow paths and in selecting movers (pumps) with an eye to reducing energy consumption.

The CoolTools team came to the following conclusion:

> …the trend for most applications is that higher chilled water delta-Ts result in lower energy costs, and they will always result in the same or lower first costs. (*CoolTools Chilled Water Plant Design Guide*).

Simply stated, increase the temperature difference in the chilled-water system to reduce the chilled-water pump flow rate and increase chiller efficiency with warmer return water. This reduces installed cost and operating costs. The *CoolTools Chilled Water Plant Design Guide* recommends starting with a chilled-water temperature difference of 12°F to 20°F (7°C to 11°C). It is important to understand that in order for the chiller plant to utilize a higher chilled-water $\Delta T$, you must start at the load coils (AHUs).

**Condenser Water Pumps.** In the same manner, design for condenser water flow has traditionally been based on a 10°F (5.6°C) $\Delta T$, which equates to 3 gpm/ton (3.2 L/min per kW). Today's chillers will give approximately a 9.4°F (5.2°C) $\Delta T$ with that flow rate. The CoolTools guide states, "Higher delta-Ts will reduce first costs (because pipes, pumps, and cooling towers are smaller), but the net energy-cost impact may be higher or lower depending on the specific design of the chillers and tower."

The CoolTools team, in their summary, state:

> In conclusion, there are times you can "have your cake and eat it too." In most cases larger $\Delta T$'s and the associated lower flow rates will not only save installation cost but will usually save energy over the course of the year. This is especially true if a portion of the first cost savings is reinvested in more efficient chillers. With the same cost chillers, at worst, the annual operating cost with the lower flows will be about equal to "standard" flows but still at a lower first cost (*CoolTools Chilled Water Plant Design Guide*).

The *CoolTools Chilled Water Plant Design Guide* recommends a design method that starts with a condenser water temperature difference of 12°F to 18°F (7°C to 10°C).

Thus, reducing chilled- and condenser-water flow rates (conversely, increasing the $\Delta T$s) can not only reduce operating cost but, more important, can free funds from being applied to the less efficient infrastructure and allow them to be applied toward increasing overall efficiency elsewhere.

**Variable Flow Systems.** The above discussion suggests that variable-flow air and water systems are an excellent way to reduce system energy consumption. Variable flow (either air or water) is required by ASHRAE Standard 90.1 for many applications, but it may be beneficial in even more applications than the standard requires. Today's most used technology for reducing flow is the VFD.

## Distribution Paths (Ducts/Pipes/Wires)

Ducts, pipes, and wires are used to move media. Proper sizing is a balancing act between energy use and cost and between material use and first cost. In terms of space consumed in running these carriers, in terms of the energy carried for the cross section of the carrier involved, wires (electricity) have the capacity for carrying the most, followed by pipes (hydronics), and followed, in turn, by ducts (air). Another consideration is that the different energy-carrying media have different characteristics and capabilities in terms of meeting the requirements of the spaces served. The above factors will have some influence on determining the type of HVAC system to be used.

**Sizing Considerations.** The previous section, "Media Movers," stated that reducing flow rates may reduce both installed cost (by reducing duct, pipe, fan, and pump sizes) and operating cost (by reducing pump and fan energy use). Decreased duct and pipe sizes also lead to less insulation. However, the design professional may want to leave pipe and duct sizes larger to minimize energy cost if the incremental installed cost savings would be relatively small. The best designs begin with generalized ranges (as were stated above for chilled and condenser water) that are fine tuned for the specific application. This fine tuning may be done with commonly available analysis and design software.

To reduce pressure drop, pipes and ductwork should be laid out prior to locating pumps, chillers, and air handlers.

**Hydronic Fluid Selection.** Fluid properties can greatly affect system performance. Antifreeze generally increases pressure drop and decreases heat transfer effectiveness. This leads to reduced system efficiency and perhaps increased cost due to the need for larger components. So the design professional should first examine the system to determine if antifreeze is an absolute necessity or whether water, with proper antifreeze safeguards, could not be used.

If antifreeze is truly necessary:

• Determine whether it is *freeze* or *burst* protection that is being sought. If an affected component (such as a chiller) does not need to be operated during freezing conditions, perhaps only burst protection is necessary. The amount of antifreeze needed can be greatly reduced, although slush may form in the pipes.

• Use only the minimum antifreeze necessary to provide protection; higher concentrations will simply reduce performance.

• Balance all environmental aspects of the antifreeze. Understand that while some antifreeze solutions are viewed as "less toxic," they can significantly increase system installed and operating costs. Often the greatest "environmental cost" of a particular antifreeze is the increased energy consumption.

• For more information on burst and freeze protection, consult the manufacturer. (e.g., www.dow.com).

## Air Quality Considerations

• Determine the condition of the ductwork system and develop methodologies to meet current standards for occupant comfort and well-being.

• Inspect, evaluate, and document hygiene factors, microbial contamination, leakage, and thermal qualities according to ACGHI's *Bioaerosols: Assessment and Control*. If the standard is not met, proceed with required remediation.

• Clean ductwork in accordance with the National Air Duct Cleaning Association's (NADCA) *General Specification for the Cleaning of Commercial Heating, Ventilating and Air Conditioning Systems*.

• Achieve ductwork leakage class as described in Table 7 of the *2005 ASHRAE Handbook—Fundamentals*, Chapter 35. This can also be referenced with Sheet Metal and Air Conditioning Contractors' National Association (SMACNA) *HVAC Duct Construction Standards, Third Edition 2005,* Table 1-3, Chapter 1.

## OCCUPIED-SPACE ENERGY DELIVERY MEANS
## (AIR TERMINALS/CONDUCTIVE/CONVECTIVE/RADIANT DEVICES)

Employ architectural and HVAC design strategies to increase ventilation effectiveness and prevent short-circuiting of airflow delivery.

For mechanically or naturally ventilated spaces, design ventilation systems that result in an air change effectiveness ($E$) of at least 0.9 under both heating and cooling modes using accepted reference standards of testing, design, or analysis. This requires special attention when heating from overhead.

Install a permanent $CO_2$ or an HVAC flow rate monitoring system that provides visual and/or auditory feedback on space ventilation performance by indicating when indoor $CO_2$ levels are greater than outdoor levels by more than 530 parts per million at any time. Or monitor ventilation rates to not drop below those specified in ASHRAE Standard 62.1-2004, Table 6-1, and calculated using the ventilation rate procedure. (*Note:* This reference is for a typical office building. Other building types or activity levels may require different values. Refer to ASHRAE Standard 62.1-2004, appendices or the references.) It is always important to consider the minimum outdoor airflow required to maintain the proper building pressurization.

## PROVISION OF OUTDOOR
## AIR VENTILATION MECHANICAL VENTILATION

Design the HVAC system so that the rate of outside air intake can be measured.

- In buildings with VAV ventilation systems, special controls may be required to maintain minimum outside air intake at all times. VAV control units must have a minimum open position to ensure required distribution of outside air to all portions of the building.
- Locate air intakes away from sources of pollution.

### Natural Ventilation

- Provide natural ventilation in combination with air conditioning, using operable windows, cross-ventilation, and the stack effect. Provide interlocking controls to avoid simultaneous operation. (See GreenTip #13 on hybrid ventilation later in this chapter.)
- Consider designing the building with no mechanical cooling and provide windows or other natural means such as evaporative cooling, earth, or groundwater contact without refrigeration.
- Install operable windows; include window frame switches to shut off mechanical air conditioning to rooms with open windows.

## ASHRAE GreenTip #7:
## Air-to-Air Energy Recovery—Heat Exchange Enthalpy Wheels

### GENERAL DESCRIPTION

There are several types of air-to-air energy systems for recovering energy from building exhaust air, and consideration should be given to using the type most appropriate for the application. The table at the bottom of the page indicates typical ranges in performance.

A heat exchange enthalpy wheel, also known as a *rotary energy wheel*, has a revolving cylinder filled with an air-permeable medium with a large internal surface area. Adjacent airstreams pass through opposite sides of the exchanger in a counterflow pattern. Heat transfer media may be selected to recover heat only or sensible plus latent heat. Because rotary exchangers have a counterflow configuration and normally use small-diameter flow passages, they are quite compact and can achieve high transfer effectiveness.

Cross-contamination, or mixing, of air between the airstreams occurs in all rotary exchangers by one of two methods: carryover or leakage. Carryover occurs as air is entrained within the medium and is carried into the other airstream. Leakage occurs because the differential pressure across the two airstreams drives air from the high-pressure to the low-pressure airstream. Because cross-contamination can be detrimental, a purge section can be installed to reduce carryover.

Two control methods are commonly used to regulate wheel energy recovery. In the first, an air bypass damper controlled by a wheel supply air temperature sensor regulates the proportion of air that is permitted to bypass the exchanger. The second,

| Type | Cross Leakage, % | Pressure Drop, % | Effective Sensible, % | Effective Latent, % |
|---|---|---|---|---|
| Plate—dry air streams | 0–5 | 100–400 | 50–75 | — |
| Plate—wetted exhaust air[a] | 0–5 | 100–500 | 50–75 | 50–85 |
| Plate—membrane[b] | 0–5 | 100–500 | 50–75 | 50–85 |
| Heat wheel | 0.5–10 | 100–300 | 50–75 | — |
| Enthalpy wheel[c] | 0.5–10 | 100–300 | 50–75 | 50–85 |
| Heat pipe | 0–1 | 100–500 | 50–65 | — |
| Runaround loop | 0 | 100–500 | 50–65 | — |
| Thermosyphon | 0 | 100–500 | 50–65 | — |
| Twin towers | 0 | 100–300 | 50–65 | — |

a. Indirect evaporative—total energy transfer.
b. Porous membrane—total energy transfer.
c. Desiccant coated—total energy transfer.

and more common, method regulates the energy recovery rate by varying the wheel's rotational speed.

## WHEN/WHERE IT'S APPLICABLE

In general, rotary air-to-air energy recovery systems can be used in process-to-process, process-to-comfort, and comfort-to-comfort applications, where energy in the exhaust stream would otherwise be wasted. Regions with higher energy costs favor higher levels of energy recovery; however, the economics of scale often favor larger installations. Energy recovery is most economical when there are large temperature differences between the airstreams, the source of supply is close to the exhaust, and they are both relatively constant throughout the year. Applications with a large central energy source and a nearby waste energy use are more favorable than applications with several scattered waste energy sources and uses. Rotary energy wheels are best applied when cross-contamination is not a concern.

## PROS AND CONS

### Pro

1.  The total HVAC system installed cost may be lower because central heating and cooling equipment may be reduced in sized.
2.  With a total energy wheel, humidification costs may be reduced in cold weather and dehumidification costs may be lowered in warm weather.
3.  Plate energy exchangers are the simplest type due to having no moving parts (compared to rotary types). Some are sealed to prevent leakage and cross-contamination and have plates made from hygienic materials in a cleanable cross-flow arrangement.
4.  Rotary type exchangers have cross-contamination because they rotate between the exhaust and supply air ducts. Cross-contamination may be reduced by inserting a purge section in the wheel between entering and leaving the ducts. This uses some supply air.
5.  Rotary wheels require little maintenance and are simple to operate.
6.  Heat pipe systems have variable tilt for performance control. This requires some maintenance and allows minor cross-contamination.
7.  The other systems listed are suitable for installations that are unable to bring the exhaust outlet duct and ventilation air inlet duct close together. This can be the case for existing building A/C systems.

### Con

1.  Energy recovery requires that the supply and exhaust airstreams be within close proximity.
2.  Cross-contamination can occur between the airstreams.
3.  Energy wheel adds pressure drop to the system.

## KEY ELEMENTS OF COST

The following provides a possible breakdown of the various cost elements that might differentiate the above system from a conventional one and an indication of whether the net cost for the alternative option is likely to be lower (L), higher (H), or the same (S). This assessment is only a perception of what might be likely, but it obviously may not be correct in all situations. **There is no substitute for a detailed cost analysis as part of the design process.** The listings below may also provide some assistance in identifying the cost elements involved.

### First Cost

* Central equipment costs          L
* Co-locating exhaust and supply sources   S/H
* Ductwork             S/H
* Design fees            S

### Recurring Cost

* Overall energy cost           L
* Maintenance of system          S/H
* Training of building operators       S/H
* Filters             H

## SOURCES OF FURTHER INFORMATION

ASHRAE. 2004. *2004 ASHRAE Handbook—HVAC Systems and Equipment,* chapter 44, Air-to-Air Energy Recovery. Atlanta: American Society of Heating, Refrigerating and Air-Conditioning Engineers, Inc.

Trane Company. 2000. Energy conscious design ideas—Air-to-air energy recovery. *Engineers Newsletter* 29(5). Publication ENEWS-29/5. Lacrosse, WI: Trane Company.

## ASHRAE GreenTip #8:
## Air-to-Air Energy Recovery—Heat Pipe Systems

### GENERAL DESCRIPTION

A heat pipe heat exchanger is a completely passive energy recovery device with an outward appearance of an ordinary extended surface, finned tube coil. The tubes are divided into evaporator and condenser sections by an internal partition plate. Within the permanently sealed and evacuated tube filled with a suitable working fluid, there is an integral capillary wick structure. The working fluid is normally a refrigerant, but other fluorocarbons, water, and other compounds are used in applications with special temperature requirements.

Heat transfer occurs when hot air flowing over the evaporator end of the heat pipe vaporizes the working fluid. A vapor pressure gradient drives the vapor to the condenser end of the heat pipe tube, where the vapor condenses, releasing the latent energy of vaporization. The condensed fluid is wicked back to the evaporator where it is revaporized, thus completing the cycle. Using this mechanism, heat transfer along a heat pipe is 1000 times faster than through copper.

Heat pipes typically have zero cross-contamination, but constructing a vented double-wall partition can provide additional protection. Changing the slope or tilt of a heat pipe controls the amount of heat it transfers. Operating the heat pipe on a slope with the hot end below (or above) the horizontal improves (or retards) the condensate flow back to the evaporator. By utilizing a simple temperature sensor-controlled actuator, the output of the exchanger can be modulated by adjusting its tilt angle to maintain a specific leaving temperature.

### WHEN/WHERE IT'S APPLICABLE

In general, air-to-air energy recovery systems can be used in process-to process, process-to-comfort, and comfort-to-comfort applications, where energy in the exhaust stream would otherwise be wasted. Regions with higher energy costs favor higher levels of energy recovery; however, the economics of scale often favor larger installations. Energy recovery is most economical when there are large temperature differences between the airstreams, the source of supply is close to the exhaust, and they are both relatively constant throughout the year. Applications with a large central energy source and a nearby waste energy use are more favorable than applications with several scattered waste energy sources and uses.

### PROS AND CONS

#### Pro

1. The total HVAC system installed cost may be lower because central heating and cooling equipment may be reduced in sized.

2. They require little maintenance and are simple to operate.
3. Cross-contamination is not a significant concern.

## Con

1. The system requires that the supply and exhaust airstreams be within close proximity.
2. Heat pipe adds pressure drop to the system.
3. Decomposition of the thermal fluid can deteriorate performance.

## KEY ELEMENTS OF COST

The following provides a possible breakdown of the various cost elements that might differentiate a building with a heat pipe system from one without and an indication of whether the net cost is likely to be lower (L), higher (H), or the same (S). This assessment is only a perception of what might be likely, but it obviously may not be correct in all situations. **There is no substitute for a detailed cost analysis as part of the design process.** The listings below may also provide some assistance in identifying the cost elements involved.

### First Cost

| | |
|---|---|
| • Central equipment costs | L |
| • Co-locating exhaust and supply sources | S/H |
| • Ductwork | S/H |
| • Design fees | S |

### Recurring Cost

| | |
|---|---|
| • Overall energy cost | L |
| • Maintenance of system | S/H |
| • Training of building operators | S/H |
| • Filters | H |

## SOURCES OF FURTHER INFORMATION

ASHRAE. 2003. *2003 ASHRAE Handbook—HVAC Systems and Equipment.* Atlanta: American Society of Heating, Refrigerating and Air-Conditioning Engineers, Inc.

Trane Company. 2000. Energy conscious design ideas—Air-to-air energy recovery. *Engineers Newsletter* 29(5). Publication ENEWS-29/5. Lacrosse, WI: Trane Company.

## ASHRAE GreenTip #9:
## Air-to-Air Energy Recovery—Run-Around Systems

### GENERAL DESCRIPTION

A typical coil energy recovery loop places extended surface, finned tube coils in the supply and exhaust airstreams of a building or process. The coils are connected in a closed loop via counterflow piping through which an intermediate heat transfer fluid (typically water or an antifreeze solution) is pumped. An expansion tank must be included to allow fluid expansion and contraction.

The coil energy recovery loop cannot transfer moisture from one airstream to another. However, indirect evaporative cooling can reduce the exhaust air temperature, which significantly reduces cooling loads. And in comfort-to-comfort applications, the energy transfer is seasonally reversible. Specifically, the supply air is preheated when the outdoor air is cooler than the exhaust and precooled when the outdoor air is warmer.

Complete separation of the airstreams eliminates cross-contamination as a concern, but freeze protection must be considered. A dual-purpose three-way control valve can prevent freeze-ups by controlling the temperature of the solution entering the exhaust coil to 30°F or above. This condition is maintained by bypassing some of the warmer solution around the coil. This valve can also ensure that a prescribed air temperature from the supply coil is not exceeded.

### WHEN/WHERE IT'S APPLICABLE

In general, air-to-air energy recovery systems can be used in process-to-process, process-to-comfort, and comfort-to-comfort applications, where energy in the exhaust stream would otherwise be wasted. Regions with higher energy costs favor higher levels of energy recovery; however, the economics of scale often favor larger installations. Energy recovery is most economical when there are large temperature differences between the airstreams, the source of supply is close to the exhaust, and they are both relatively constant throughout the year. Runaround loops are highly flexible and well suited to renovation and industrial applications. The loop accommodates remote supply and exhaust ducts and allows the simultaneous transfer of energy between multiple sources and uses.

### PROS AND CONS

#### Pro

1. The total HVAC system installed cost may be lower because central heating and cooling equipment may be reduced in sized.
2. The loop accommodates remote supply and exhaust duct locations.
3. Cross-contamination is not a concern.

## Con

1. It requires a pump, which offsets some energy recovery savings.
2. It adds pressure drop to the system.
3. Relative to passive air-to-air heat exchangers (i.e., heat wheels or heat pipes), it requires more maintenance and controls.

## KEY ELEMENTS OF COST

The following provides a possible breakdown of the various cost elements that might differentiate a building with a run-around coil system from one without and an indication of whether the net cost is likely to be lower (L), higher (H), or the same (S). This assessment is only a perception of what might be likely, but it obviously may not be correct in all situations. **There is no substitute for a detailed cost analysis as part of the design process.** The listings below may also provide some assistance in identifying the cost elements involved.

### First Cost

- Central equipment costs       L
- Hydronics (piping, pumps, and controls)    H
- Ductwork       S
- Design fees       S

### Recurring Cost

- Overall energy cost       L
- Maintenance of system       S/H
- Training of building operators       S/H
- Filters       H

## SOURCES OF FURTHER INFORMATION

ASHRAE. 2003. *2003 ASHRAE Handbook—HVAC Systems and Equipment.* Atlanta: American Society of Heating, Refrigerating and Air-Conditioning Engineers, Inc.

Trane Company. 2000. Energy conscious design ideas—Air-to-air energy recovery. *Engineers Newsletter* 29(5). Publication ENEWS-29/5. Lacrosse, WI: Trane Company.

## GENERAL DESCRIPTION

With a ceiling supply and return air system, the ventilation effectiveness may be compromised if sufficient mixing does not take place. While there are no data suggesting that cold air supplied at the ceiling will short circuit, it is possible that a fraction of the supply air may bypass directly to the return inlet without mixing at the occupied level when heating from the ceiling. For example, when heating with a typical overhead system with supply temperatures exceeding 15°F (8.3°C) above room temperature, ventilation effectiveness will approach 80% or less. In compliance with Table 6.2, ASHRAE Standard 62.1-2004, zone air distribution effectiveness is only 0.8, so ventilation rates must be multiplied by 1/0.8 or 1.25. While proper system design and diffuser selection can alleviate this problem, another potential solution is displacement ventilation.

In displacement ventilation, conditioned air with a temperature slightly lower than the desired room temperature is supplied horizontally at low velocities at or near the floor. Returns are located at or near the ceiling. The supply air is spread over the floor and then rises by convection as it picks up the load in the room. Displacement ventilation does not depend on mixing. Instead, you are literally displacing the stale polluted air and forcing it up and out the return or exhaust grille. Ventilation effectiveness may actually exceed 100%, and Table 6.2 of ASHRAE Standard 62.1-2004 indicates a zone air distribution effectiveness of 1.2 shall be used.

Displacement ventilation is common practice in Europe, but its acceptance in North America has been slow primarily because of the conventional placement of ductwork at the ceiling level and more extreme climatic conditions.

## WHEN/WHERE IT'S APPLICABLE

Displacement ventilation is typically used in industrial plants and data centers, but it can be applied in almost any application where a conventional overhead forced air distribution system could be utilized and the load permits.

Because the range of supply air temperatures and discharge velocities is limited to avoid discomfort to occupants, displacement ventilation has a limited ability to handle high heating or cooling loads if the space served is occupied. Some designs use chilled ceilings or heated floors to overcome this limitation. When chilled ceilings are used, it is critical that building relative humidity be controlled to avoid condensation. Another means of increasing cooling capacity is to recirculate some of the room air.

Some associate displacement ventilation solely with underfloor air distribution and the perceived higher costs associated with it. In fact, most underfloor pressurized plenum, air distribution systems do not produce true displacement ventilation but, rather, well-mixed airflow in the lower part of the space. It can, however, be a viable alternative when considering systems for modern office environments where data cabling and flexibility concerns may merit a raised floor.

## PROS AND CONS

### Pro

1. Displacement ventilation offers improved thermal comfort and IAQ due to increased ventilation effectiveness.
2. There is reduced energy use due to extended economizer availability associated with higher supply temperatures.

### Con

1. It may add complexity to the supply air ducting.
2. It is more difficult to address high heating or cooling loads.
3. There are perceived higher costs.

## KEY ELEMENTS OF COST

The following provides a possible breakdown of the various cost elements that might differentiate a system utilizing displacement ventilation from one that does not and an indication of whether the net cost is likely to be lower (L), higher (H), or the same (S). This assessment is only a perception of what might be likely, but it obviously may not be correct in all situations. **There is no substitute for a detailed cost analysis as part of the design process.** The listings below may also provide some assistance in identifying the cost elements involved.

### First Cost

- Controls                              S
- Equipment                           S
- Distribution ductwork            S/H
- Design fees                          S

### Recurring Cost

- Energy cost                         L
- Maintenance of system          S
- Training of building operators    S/H
- Orientation of building occupants  S/H
- Commissioning cost              S

## SOURCES OF FURTHER INFORMATION

Advanced Buildings Technology and Practice, www.advancedbuildings.org.

ASHRAE. 2004. *2004 ASHRAE Handbook—Fundamentals*. Atlanta: American Society of Heating, Refrigerating and Air-Conditioning Engineers, Inc.

ASHRAE. 2004. *ASHRAE Standard 62.1-2004, Ventilation for Acceptable Indoor Air Quality*. Atlanta: American Society of Heating, Refrigerating and Air-Conditioning Engineers, Inc.

Bauman, F., and T. Webster. 2001. Outlook for underfloor air distribution. *ASHRAE Journal* 43(6). Atlanta: American Society of Heating, Refrigerating and Air-Conditioning Engineers, Inc.

Interpretation IC-62-1999-30 of *ANSI/ASHRAE Standard 62-1999, Ventilation for Acceptable Indoor Air Quality*, August 2000.

Public Technology Inc., US Department of Energy and the US Green Building Council. 1996. *Sustainable Building Technical Manual—Green Building Design, Construction and Operations*.

ASHRAE GreenTip #11:
Dedicated Outdoor Air Systems

## GENERAL DESCRIPTION

ASHRAE Standard 62 describes in detail the ventilation required to provide a healthy indoor environment as it pertains to IAQ. Traditionally designers have attempted to address both thermal comfort and IAQ with a single mixed air system. But ventilation becomes less efficient when the mixed air system serves multiple spaces with differing ventilation needs. If the percentage of outdoor air is simply based on the critical space's need, then all other spaces are overventilated. In turn, providing a separate dedicated outdoor air system (DOAS) may be the only reliable way to meet the true intent of ASHRAE Standard 62.

A DOAS uses a separate air handler to condition the outdoor air before delivering it directly to the occupied spaces. The air delivered to the space from the DOAS should not adversely affect thermal comfort (i.e., too cold, too warm, too humid); therefore, many designers call for systems that deliver neutral air. However, there is a strong argument for supplying cool dry air and decoupling the latent conditioning as well as the IAQ components from the thermal comfort (sensible only) system.

The only absolute in a DOAS is that the ventilation air must be delivered directly to the space from a separate system. Control strategy, energy recovery, and leaving air conditions are all variables that can be fixed by the designer.

## WHEN/WHERE IT'S APPLICABLE

While a DOAS can be applied in any design, it is most beneficial in a facility with multiple spaces with differing ventilation needs. A DOAS can be combined with any thermal comfort conditioning system, including, but not limited to, all-air systems, fan-coil units, and hydronic radiant cooling. Note, however, that a design incorporating a separate 100% outdoor air unit delivering air to the mixed air intakes of other units is not a DOAS as defined here. While this type of system may have benefits, such as using less energy or providing more accurate humidity control, it still suffers from the multiple space dilemma described above.

## PROS AND CONS

### Pro

1. A DOAS ensures compliance with ASHRAE 62.1-2004 for proper multiple space ventilation and adequate IAQ.
2. It reduces a building's energy use when compared to mixed air systems that require overventilation of some spaces.
3. It allows the designer to decouple the latent load from the sensible load, hence providing more accurate space humidity control.
4. It allows easy airflow measurement and balance and keeps ventilation loads off main HVAC units.

## Con

1. Depending on overall design (thermal comfort and IAQ), it may add additional first cost associated with providing parallel systems.
2. Depending on overall design, it may require additional materials with their associated embodied energy costs.
3. Depending on overall design, there may be more systems to maintain.
4. With two airstreams, proper mixing may not occur when distributed to the occupied space.
5. The total airflow of two airstreams may exceed airflow of a single system.

## KEY ELEMENTS OF COST

The following provides a possible breakdown of the various cost elements that might differentiate a building with a DOAS from one with another system and an indication of whether the net cost is likely to be lower (L), higher (H), or the same (S). This assessment is only a perception of what might be likely, but it obviously may not be correct in all situations. **There is no substitute for a detailed cost analysis as part of the design process.** The listings below may also provide some assistance in identifying the cost elements involved.

### First Cost

- Controls     H
- Equipment     S/H
- Distribution ductwork     S/II
- Design fees     S/H

### Recurring Cost

- Energy cost     S/L
- Maintenance of system     S/H
- Training of building operators     S/H
- Orientation of building occupants     S
- Commissioning cost     S/H

## SOURCES OF FURTHER INFORMATION

ASHRAE. 2004. *ASHRAE Standard 62.1-2004, Ventilation for Acceptable Indoor Air Quality.* Atlanta: American Society of Heating, Refrigerating and Air-Conditioning Engineers, Inc.

Coad, W.J. 1999. Conditioning ventilation air for improved performance and air quality. *Heating/Piping/Air Conditioning*, September.

Morris, W. 2003. The ABCs of DOAS. *ASHRAE Journal* 45(5).

Mumma, S.A. 2001. Designing dedicated outdoor air systems. *ASHRAE Journal* 43(5).

### GENERAL DESCRIPTION

A significant component of indoor environmental quality (IEQ) is the indoor air quality (IAQ). ASHRAE Standard 62.1-2004 describes in detail the ventilation required to provide a healthy environment. However, providing ventilation based strictly on the peak occupancy using the ventilation rate procedure (Section 6.1) will result in overventilation during periods. Any positive impact on IAQ brought on by overventilation will be outweighed by the costs associated with the energy required to condition the ventilation air.

$CO_2$ can be used to measure or control the per-person ventilation rate and, in turn, allow the designer to introduce a ventilation demand control strategy. Simply put, the amount of $CO_2$ present in the air is an indicator of the number of people in the space and, in turn, the amount of ventilation air that is required. $CO_2$-based ventilation control does not affect the peak design ventilation capacity required to serve the space as defined in the ventilation rate procedure, but it does allow the ventilation system to modulate in sync with the building's occupancy.

The key components of a $CO_2$ demand-based ventilation system are $CO_2$ sensors and a means by which to control the outdoor fresh air intake, i.e., a damper with a modulating actuator. There are many types of sensors, and the technology is evolving while, at the same time, costs are dropping. Sensors can be wall-mounted or mounted in the return duct, but it is recommended that the sensor be installed within the occupied space whenever possible.

### WHEN/WHERE IT'S APPLICABLE

$CO_2$ demand control is best suited for buildings with a variable occupancy. The savings will be greatest in spaces that have a wide variance, such as gymnasiums, large meeting rooms, and auditoriums. For buildings with a constant occupancy rate, such as an office building or school, a simple nighttime setback scenario may be more appropriate for ventilation demand control, but $CO_2$ monitoring may still be utilized for verification that high IAQ is achieved.

### PROS AND CONS

#### Pro

1. $CO_2$ demand control reduces a building's energy use as it relates to overventilation.
2. It assists in maintaining adequate ventilation levels regardless of occupancy.

#### Con

1. There is an added first cost associated with the sensors and additional controls.

2. There are additional materials and their associated embodied energy costs.
3. Evolving sensor technology may not be developed to full maturity.

## KEY ELEMENTS OF COST

The following provides a possible breakdown of the various cost elements that might differentiate a building utilizing a $CO_2$ ventilation demand control strategy from one that does not and an indication of whether the net cost is likely to be lower (L), higher (H), or the same (S). This assessment is only a perception of what might be likely, but it obviously may not be correct in all situations. **There is no substitute for a detailed cost analysis as part of the design process.** The listings below may also provide some assistance in identifying the cost elements involved.

### First Cost

- Controls                              H
- Design fees                           S/H

### Recurring Cost

- Energy cost                           L
- Maintenance of system                 S/H
- Training of building operators        S/H
- Orientation of building occupants     S/H
- Commissioning cost                    S/H

## SOURCES OF FURTHER INFORMATION

Advanced Buildings Technologies and Practices, www.advancedbuildings.org.

ASHRAE. 2004. *ANSI/ASHRAE Standard 62.1-2004, Ventilation for Acceptable Indoor Air Quality.* Atlanta: American Society of Heating, Refrigerating and Air-Conditioning Engineers, Inc.

ASTM. 1998. *ASTM D 6245-1998: Standard Guide for Using Indoor Carbon Dioxide Concentrations to Evaluate Indoor Air Quality and Ventilation.* American Society for Testing and Materials.

Lawrence, T.M. 2004. Demand-controlled ventilation and sustainability. *ASHRAE Journal* 46(12):117–21.

Schell, M., and D. Int-Hout. 2001. Demand control ventilation using $CO_2$. *ASHRAE Journal* 43(2). Atlanta: American Society of Heating, Refrigerating and Air-Conditioning Engineers, Inc.

Trane Company. *A Guide to Understanding ASHRAE Standard 62-2001.* Lacrosse, WI: Trane Company. http://trane.com/commercial/issues/iaq/ashrae2001.asp.

US Green Building Council. 2005. *LEED Reference Guide*, Version 2.2. www.usgbc.org/.

## ASHRAE GreenTip #13:
## Hybrid Ventilation

### GENERAL DESCRIPTION

A hybrid ventilation system allows the controlled introduction of outdoor air ventilation into a building by both mechanical and passive means; thus, it is sometimes called mixed-mode ventilation. It has built-in strategies to allow the mechanical and passive portions to work in conjunction with one another so as to not cause additional ventilation loads compared to what would occur using mechanical ventilation alone. It thus differs from a passive ventilation system, consisting of operable windows alone, which has no automatic way of controlling the amount of outdoor air load.

Two variants of hybrid ventilation are the *changeover* (or *complementary*) type and the *concurrent* (or *zoned*) type. With the former, spaces are ventilated either mechanically or passively, but not both simultaneously. With the latter variant, both methods provide ventilation simultaneously, though usually to zones discrete from one another.

Control of hybrid ventilation is obviously an important feature. With the changeover variant, controls could switch between mechanical and passive ventilation seasonally, diurnally, or based on a measured parameter. In the case of the concurrent variant, appropriate controls are needed to prevent "fighting" between the two ventilation methods.

### WHEN/WHERE IT'S APPLICABLE

A hybrid ventilation system may be applicable in the following circumstances:

- When the owner and design team are willing to explore employing a nonconventional building ventilation technique that has the promise of reducing ongoing operating costs as well as providing a healthier, stimulating environment.
- When it is determined that the building occupants would accept the concept of using the outdoor environment to determine (at least, in part) the indoor environment, which may mean greater variation in conditions than with a strictly controlled environment.
- When the design team has the expertise and willingness—and has the charge from the owner—to spend the extra effort to create the integrated design needed to make such a technique work successfully.
- Where extreme outside conditions—or a specialized type of building use—do not preclude the likelihood of the successful application of such a technique.

Buildings with atriums are particularly good candidates.

## PROS AND CONS

### Pro

1. Hybrid ventilation is an innovative and potentially energy-efficient way to provide outdoor air ventilation to buildings and, in some conditions, to cool them, thus reducing energy otherwise required from conventional sources (power plant).
2. Corollary to the above, it could lead to a lower building life-cycle cost.
3. It could create a healthier environment for building occupants.
4. It offers a greater sense of occupant satisfaction due to the increased ability to exercise some control over the ventilation provided.
5. There is more flexibility in the means of providing ventilation; the passive variant can act as backup to the mechanical system and vice versa.
6. It could extend the life of the equipment involved in providing mechanical ventilation since it would be expected to run less.

### Con

1. Failure to integrate the mechanical aspects of a hybrid ventilation system with the architectural design could result in a poorly functioning system. Some architectural design aspects could be constrained in providing a hybrid ventilation system, such as building orientation, depth of occupied zones, or grouping of spaces.
2. Additional first costs could be incurred since two systems are being provided where only a single one would be provided otherwise, and controls for the passive system could be a major portion of the added cost.
3. If automatic operable window openers are utilized, these could result in security breaches if appropriate safeguards and overrides are not provided.
4. If integral building openings are utilized in lieu of, or in addition to, operable windows, pathways for the entrance of outside pollutants and noise or of unwanted insects, birds, and small animals would exist. If filters are used to prevent this, they could become clogged or could be an additional maintenance item to keep clear.
5. Building operators may have to have special training to understand and learn how best to operate the system. Future turnovers in building ownership or operating personnel could negatively affect how successfully the system performs.
6. Occupants would probably need at least some orientation so that they would understand and be tolerant of the differences in conditions that may prevail with such a system. Future occupants may not have the benefit of such orientation.
7. Special attention would need to be given to certain safety issues, such as fire and smoke propagation in case of a fire.

8. Although computer programs (such as computational fluid dynamics) exist to simulate, predict, and understand airflow within the building from passive ventilation systems, it would be difficult to predict conditions under all possible circumstances.

9. Few codes and standards in the US recognize and address the requirements for hybrid ventilation systems. This would likely result in local code enforcement authorities having increased discretion over what is acceptable or not.

## KEY ELEMENTS OF COST

The following provides a possible breakdown of the various cost elements that might differentiate a hybrid ventilation system from a conventional one and an indication of whether the net cost for the hybrid option is likely to be lower (L), higher (H), or the same (S). This assessment is only a perception of what might be likely, but it obviously may not be correct in all situations. **There is no substitute for a detailed cost analysis as part of the design process.** The listings below may also provide some assistance in identifying the cost elements involved.

### First Cost

- Mechanical ventilation system elements             S
- Architectural design features                      H/L
- Operable window operators                          H
- Integral opening operators/dampers                 H
- Filters for additional openings                    H
- Controls for passive system/coordination with mechanical system H
- Design fees                                        H

### Recurring Cost

- Energy for mechanical portion of system            L
- Maintenance of above                               L
- Energy used by controls, mechanical operators      H
- Maintenance of passive system                      H
- Training of building operators                     H
- Orientation of building occupants                  H
- Commissioning cost                                 H
- Occupant productivity                              H

### SOURCE OF FURTHER INFORMATION

Kosik, W.J. 2001. Design strategies for hybrid ventilation. *ASHRAE Journal* 43(10). Atlanta: American Society of Heating, Refrigerating and Air-Conditioning Engineers, Inc.

# 9

# INTERACTION WITH THE LOCAL ENVIRONMENT

## WHY A CHAPTER ON THIS?

The *GreenGuide* is not intended as a complete compilation of all interactions that buildings, and HVAC systems in particular, have on and with the environment. Many of the issues are common knowledge as being important to "green" building design among engineers and lay people alike; such areas as energy consumption, location of buildings, and the construction process are prime examples. This Guide is primarily intended to convey ideas on how to improve buildings and their systems. There are, however, some areas that are either not intuitively obvious as being potential impacts of HVAC systems or are items that some may not consider to be true "sustainability" issues. Regardless of your definition of *sustainability* or the various labels and compartmentalization, it is assumed that the reader of this Guide *is* interested in designing buildings and their systems that provide for the needs of the occupants while minimizing the adverse impacts. Therefore, this chapter provides examples of several areas that the HVAC engineer may not initially think are important when minimizing environmental impacts but that truly are significant.

In this chapter, we describe how building HVAC systems may impact the local environment and methods for mitigating or reducing the impact. We particularly focus on issues possibly not normally considered by HVAC engineers but still HVAC related. The issues discussed include the interaction with the local outdoor as well as indoor environment.

## INDOOR ENVIRONMENTAL QUALITY (IEQ)

The terms *indoor environmental quality* (IEQ) and *indoor air quality* (IAQ) are sometimes confused as being one and the same. In reality, IEQ is a broader, more encompassing concept that includes IAQ as one of the key factors. Five areas are considered key to providing good IEQ.

- Air quality and ventilation
- Thermal comfort

- Acoustics and noise
- Lighting levels
- Visual perception

### Air Quality and Ventilation

This is one area where the outdoor environment can have a negative impact on the building HVAC and the indoor environment, or vice versa, depending on the specific situation. Location of outdoor air intakes near a known contamination source (such as a loading dock with potential idling engines) can seriously degrade the IAQ by introducing, rather than removing, contaminants. Similarly, building exhausts can contaminate the local area near the exhaust discharge, making this air unsuitable for human exposure or re-intake into the building. Chapter 44, "Building Air Intake and Exhaust Design," of the *2003 ASHRAE Handbook—Applications* contains more information on this topic.

Assuming no contamination of the local air surrounding the building, then good IAQ is possible by providing adequate ventilation and distribution within the space, as discussed in Chapter 8 and ASHRAE Standard 62.2.

### Thermal Comfort

Similar to lighting levels, thermal comfort affects the occupants and overall building IEQ. Thermal comfort of the occupied space is covered in Chapter 8 of this Guide. The interaction with the local environment has minimal impact on thermal comfort.

### Acoustics

The acoustical environment can also be an important factor in determining good IEQ. This is treated in detail in a separate section later in this chapter.

### Lighting Levels

Adequate lighting levels are required for the building occupants. The lighting levels required vary according to the design purpose of the room or building zone. The local environment, in the form of trees, landscaping, or other buildings, can influence the lighting that may enter the space and, hence, affect the lighting levels inside. Lighting and its impact on HVAC load determination is discussed in more detail in Chapter 13 of this Guide.

### Visual

This is another area that influences how a person perceives the IEQ. Rarely would the HVAC system interact with visual perception of the indoor space, with one possible aspect being exposed ductwork. Any HVAC system interaction with visual perception will likely be dealt with by the project architect.

## COOLING TOWER SYSTEMS

Cooling towers are a very efficient method of cooling. Cooling towers remove heat by evaporation and can cool close to the ambient wet-bulb temperature. The wet-bulb temperature is always lower than the dry-bulb; thus, water cooling allows more efficient condenser operation (as much as a 50% energy savings) than air cooling. Cooling towers also conserve water. In a typical cooling tower operation, only 1% of the recirculated water is evaporated. This evaporation will cool the remaining 99% of the water for reuse. In addition to evaporation, some recirculated water must be bled from the system to prevent soluble and semi-soluble minerals from reaching too high a concentration. This bleed or blowdown is usually sent to a publicly owned treatment works (POTW).

### Water Treatment

The water in the evaporative cooling loop must be treated to minimize biological growth, scaling, and corrosion. Typically a combination of biocides, corrosion inhibitors, and scale inhibitors are added to the system.

Corrosion inhibitors are usually phosphate or nitrogen based (fertilizers) or molybdenum or zinc based (heavy metals). These inhibitors are more effective when added in combinations. These materials have low vapor pressure and are not "used" by the system. The inhibitors simply need to remain in the solution at the proper concentration to maintain a protective film on the metal components. Their only loss is through bleed and drift.

Most scaling inhibition is done by polymer-based chemicals, organic phosphorous compounds (phosphonates), or by acid addition. The acid reacts with the alkalinity in the water to release $CO_2$ and is used up. The polymer and phosphonate scale inhibitors remain in the solution to delay scaling; their major loss is through bleed and drift. Some polymers are designed to be biodegradable, easily broken down by bacteria in the environment, while others are not.

There are very wide assortments of biocides. A typical system will maintain an oxidizing biocide such as bromine or chlorine at a constant level and slug feed a non-oxidizing biocide once a week. Chlorine and bromine have a high vapor pressure in water. Much of the chlorine and bromine added to the tower is stripped from the water into the air; a small quantity actually reacts with organics in the tower. Drift and bleed will contain all of the non-oxidizing biocides and a small quantity of oxidizing biocides and the reaction products of the biocides.

### Drift

To promote efficient evaporation, cooling towers force intimate contact of outdoor air with warm water. Besides removing heat by evaporation, dust, pollen, and gas components of the air will become entrained in the water, while some high-vapor components of the water (bromine, chlorine), as well as entrained water drops, will migrate to the air. Airborne dust and pollen that are captured by the water can promote biological growth in the tower.

Small water droplets entrained and carried out with the air passing through the tower are called *drift*. Drift is always present when operating a cooling tower. Since drift is generated by small droplets of the cooling tower water, it contains all of the dissolved minerals, microbes, and water treatment chemicals in the tower water. Drift is a source of $PM_{10}$ emission (particulate matter less than 10 µ in diameter) and is as well a suspected vector in *Legionella* transmission.

Drift is usually reported as a percentage of the recirculating water, though it is more accurately described in terms of the parts per million (ppm) of the air passing through the tower. Tower designs use drift eliminators to capture some of this entrained water. A typical value for drift from cooling towers is 0.005% of the recirculating water; many tower designs have drift values as low as 0.001%.

To put these values in perspective, an example of a 400-ton cooling system with a particular treatment program is useful. A 400-ton system would circulate approximately 1200 gpm (4542 L/min) through the tower and chiller. At nominal rates, 12 gpm would be lost to evaporation, 0.06 gpm would be lost to drift (0.005%), and, at four cycles of concentration, 4 gpm (15.1 L/min) would be intentionally bled from the system. Chlorine addition would be 0.09 gal/h (0.34 L/h) of 12.5% liquid bleach (0.06 lb/h or 0.027 kg/h $Cl_2$). This chlorine addition should maintain about 0.2 ppm free chlorine and 0.2 ppm combined chlorine. The combined chlorine is from the reaction of chlorine with organic molecules and may include some hazardous by-products, such as chloroform. For corrosion and scale protection, the water would be maintained with 2 ppm zinc, 3 ppm triazole, and 20 ppm polyphosphate. Once a week four pounds of a 1.5% solution of isothiazoline, a non-oxidizing biocide, will be fed to the system for biofilm control. The drift and bleed will contain the same quantity of minerals and chemicals as are maintained in the recirculated water.

Table 9-1 shows monthly results for operating this tower at an assumed 75 hours per week (300 hours per month).

## Table 9-1   Material Release from Example Cooling Tower Operation

|  | Release Rate | Total per Month |
|---|---|---|
| Evaporation | 12 gpm (45.4 L/min) | 216,000 gal (818 m) |
| Bleed at four cycles | 4 gpm (15.1 L/min) | 72,000 gal (273 m) |
| Drift | 0.060 gpm (0.005%)<br>0.012 gpm (0.001%) | 1,080 gal (4088 L)<br>216 gal (818 L) |
| Chlorine addition | 0.06 lb/h (0.027 kg/h) | 18.0 pounds $Cl_2$ (8.2 kg) |
| Chlorine in bleed | Free @ 0.2 ppm<br>Combined @ 0.2 ppm | 0.12 lb free (0.055 kg)<br>0.12 lb combined |
| Unaccounted chlorine |  | 17.7 lb $Cl_2$ (8.0 kg) |

From the Table 9-1, it is seen that most of the chlorine used in this tower is unaccounted for. Some of this loss is due to oxidation of organic and inorganic material in the cooling system, resulting in nonhazardous chloride ions; much of the chlorine is released into the atmosphere as chlorine gas. While it is hard to be quantitative, over the course of one year this tower could release over 100 lb (45.5 kg) of chlorine gas into the immediate building environment. If less effective drift eliminators were used, 12,000 gal (45.4 m) of contaminated tower water would also be released every year and 800,000 gal (3028 m) of water containing heavy metals, phosphates, and biocides will be sent to a POTW system. Most POTW systems are designed to handle only organic waste; much of these cooling tower chemicals will pass through the system untreated or will be released later as gaseous emissions at the POTW.

Over the lifetime of the building, these releases could be among the most significant impacts on the local environment that the building will cause. This example highlights the magnitude to which the issue may come if not addressed.

### Green Choices—Water Treatment

The water treatment plan illustrated above is not the only choice. There are many ways to treat the system that will have a less negative impact on the environment. Besides being rapidly stripped into the air, chlorine and bromine may react with organic molecules to produce very hazardous daughter products; however, other oxidizing biocides, such as hydrogen peroxide, do not have this issue. By continuous monitoring of the cooling system, chemical additions can be added only when needed. This technique can yield equivalent performance with less added chemicals. Also, the EPA maintains a Web site on green chemistry, www.epa.gov//greenchemistry/, which contains criteria on how to evaluate the life-cycle environmental impact of a particular chemical.

Non-chemical water treatment has the potential to be a powerful method for water treatment; however, its successful use depends on the water chemistry, operating procedures, and degree of pollution of the specific system. There are several different nonchemical technologies available, including those based on pulsed electric fields, mechanical agitation, and ultrasound. Each of these technologies has developed a widespread following. ASHRAE has investigated the scale prevention effectiveness of some of these nonchemical technologies and has published the results from ASHRAE's Research Project RP-1155 (Cho et al. 2003). These technologies offer the promise of eliminating the storage and handling of toxic chemicals at the site, eliminating the risk caused by any spills or leaks of cooling tower water, eliminating the issues of the bleed water at the POTW, and eliminating much of the concern with drift. Bleed water, instead of being sent to the POTW, could be used on-site for irrigation or other nonpotable needs.

### Green Choices—Tower Selection

All cooling tower designs are efficient at removing heat from water through evaporation; however, not all designs perform as well environmentally. Some tower designs are more prone to splashout, spills, drift, and algae growth than others. Splashout involves tower water splashing from the tower. This happens most often in no-fan conditions (circulating tower water with the fans off) when there are strong winds. Cross-flow towers are more prone to this issue since, in no-fan condition, some water will fall outside of the fill.

Spills can happen from the cold-water basin from overflowing at shutdown when all of the water in the piping drains into the basin. Proper water levels will prevent this. Some tower designs use hot-water basins to distribute water at the top of the tower, while other designs use a spray header pipe. The hot-water basin design can overflow if the nozzles clog; a spray header pipe never overflows.

Algae are a nuisance problem in basins and can contribute to microbial growth. Algae control requires harsher chemical treatment than typical biological control. Since algae are plants, they need sunlight to grow. Some tower designs are light-tight, which completely eliminates algae as an issue in cooling towers, while other designs are more open and algae growth can be an issue.

The amount of drift varies extensively in tower design. Some tower designs have very little drift, and the less the drift from a tower, the lower the amount of water containing minerals, water treatment chemicals, and microbes that will be released into the surrounding environment.

Fan-power draw also varies between designs. There are two general types of fans: axial and centrifugal. With axial fans the fan blades are mounted perpendicular to the axis like propellers on a plane or boat. All induced draft cooling towers and some forced draft designs use axial fans. With centrifugal fans, the fan blades are mounted parallel to the axis as along the outside of a drum. Centrifugal fans are only used in forced draft towers and typically use twice the energy to achieve the same amount of airflow as an axial fan.

## Maintenance

An often overlooked method to minimize environmental impact is maintenance. Cooling towers operate outdoors under changing conditions. Wind damage to inlet air louvers, excessive airborne contamination, clogging of water distribution nozzles, and mechanical problems can best be prevented and quickly corrected with periodic inspections and maintenance.

## Sources of Further Information on Cooling Tower Impacts

EPA Green Chemistry Web site, www.epa.gov/greenchemistry/.

Young, I.C., S.H. Lee, and W. Kim. 2003. Physical water treatment for the mitigation of mineral fouling in cooling tower water applications. *ASHRAE Transactions* 109(1):346–57.

## DISTRICT ENERGY SYSTEMS

District energy (DE) systems involve the provision of thermal energy (heating and/or cooling) from one or more central energy plants to multiple buildings or facilities via a network of interconnecting thermal piping. Generally, district heating (DH) systems deliver heat as steam or hot water, while district cooling (DC) systems deliver cooling as chilled water or chilled secondary coolant (such as an aqueous glycol or an aqueous sodium nitrite solution) or even as a refrigerant.

DE systems often deliver multiple significant positive impacts to the local building environments that they serve. These typical impacts include the areas outlined in following sections.

### Energy Consumption

Heating and/or cooling buildings using DE systems can affect overall energy consumption in various ways, from modest increases or decreases to very dramatic decreases in fuel and energy consumption. The energy consumed within the boundaries of DE-served buildings will, of course, be dramatically reduced compared to a baseline building with its own dedicated boilers and chillers. This energy reduction within the buildings will be offset by the energy consumed in the central DE plants and in distributing the thermal energy from the central plants to the customer buildings. If the central DE plant utilizes similar technology (e.g., gas boilers and electric chillers) as otherwise used in the individual buildings, there may be little or no net reduction in energy use. However, the larger (and generally more efficient) DE plant equipment more than offsets extra energy consumed in distribution of the thermal energy to the buildings for at least a slight net reduction in overall energy consumption. Reductions in overall fuel and energy consumption are achieved through the ability of DE plants to more readily and more economically utilize alternative technologies than is the case for individual buildings. These technologies include, but are not limited to, dual-fuel boilers; alternative fuel boilers (including renewable fuels such as low Btu landfill gas, municipal solid waste, wood waste, etc.); high-efficiency boilers; high-efficiency chillers; alternative energy-efficient refrigerants (e.g., ammonia); non-electric chiller plants (e.g., absorption chillers, engine-driven chillers, or turbine-driven chillers); hybrid chiller plants (with various combinations of electric and non-electric chillers); energy-efficient series or series-parallel chiller configurations for high Delta-T systems; thermal energy storage (TES); cogeneration of combined heat and power (CHP); trigeneration or combined cooling, heating, and power (CCHP); and the use of natural renewable thermal energy (such as geothermal heat for DH systems and cold deep water sources, such as lakes or oceans, for DC systems).

### Emissions

As is the case with energy consumption, DE serves to eliminate many emissions from the local building environment, such as boiler exhausts and chiller plant heat rejection. Some emissions are of course relocated to the site of the central DE plant. However, just as DE plants tend to have at least incrementally higher levels of energy efficiency, they tend to have at least incrementally lower levels of emissions, versus those associated with otherwise individually heated and cooled buildings. And for DE systems utilizing one or more of the alternative technologies (as cited above on energy consumption), the overall emissions can be significantly reduced in terms of air pollutants ($SO_X$, $NO_X$, and precipitates) and greenhouse gases ($CO_2$, $NO_X$, and some refrigerants).

### Noise and Vibration

Through the avoidance of needing boilers and/or chillers in the buildings, the occupants of DE-served buildings experience a local building environment that is free from the potential noise and vibrations associated with such equipment.

### Chemical Supplies and Blowdown

Due to the avoidance of needing boilers and/or chiller plants in each building, DE systems can eliminate or greatly mitigate the presence of potentially harmful fuel and chemicals to be handled within the occupied buildings. With the use of DE, the storage, handling, and disposal of fuel, boiler water treatment chemicals, refrigerants, condenser water treatment chemicals, and chilled-water treatment chemicals can all be removed to the location of the central DE plants. Thus, the potential for related chemical spills, disruptions, and associated hazards are avoided within the occupied building environments.

### Efficiency and Reliability

Central energy plants associated with DE systems, compared to the alternative of dispersed multiple smaller boilers and chiller plants within individual buildings, generally have higher levels of operational efficiency and reliability. This is because larger DE plants can more easily justify sophisticated design, automated optimized control systems, more attentive maintenance programs, and more highly trained and focused operations and maintenance personnel.

### Space Utilization and Aesthetics

Through the avoidance of needing heating and/or chilling plants in the buildings, DE systems provide improved space utilization for the occupants of the individual buildings. Also, there is no longer a need for local boiler exhaust stacks and/or for local chiller plant heat rejection such as via roof-mounted cooling towers. In addition to the improved aesthetics of having buildings without such stacks and towers, multiple and sometimes unsightly exhaust plumes from stacks and towers are also removed from the local building environment.

## Where DE Systems are Used

DE systems are routinely utilized on university campuses; DE systems are also often used for other institutional applications (including schools, hospital and medical facilities, airports, military installations, and other federal, state, and local government facilities), for privately owned multi-building commercial/industrial facilities, and for thermal "utilities" serving urban business districts. DE systems serve as few as two buildings or as many as many hundreds of buildings. The ideal times to consider utilizing DE for serving the heating and/or cooling needs of a building are either during master planning and new construction or during expansion or renovation of buildings or their HVAC systems.

## Source of Further Information on DE Systems Impacts

ASHRAE. 2004. *2004 ASHRAE Handbook—Systems and Equipment*, Chapter 11, District heating and cooling. Atlanta: American Society of Heating, Refrigerating and Air-Conditioning Engineers, Inc.

## WATER CONSUMPTION DUE TO COOLING SYSTEM OPERATION

Trade-offs must be carefully considered between energy consumption and water consumption. In many cases, site energy can be saved at the expense of using site water. Examples of this include using evaporative cooling compared with direct expansion cooling. Likewise, chillers with water-based cooling towers are more efficient than air-based systems—but this energy saving comes with using water. Further complicating the analysis is that water does not arrive at the site without energy impacts of treating and pumping. Much of the electricity production is produced with thermal electric plants that use evaporative cooling for the condenser part of the Rankine cycle. As additional environmental pressures are applied, more power plants are using water evaporation through towers, rather than discharging the heat to rivers, lakes, and oceans. Even hydroelectric power plants evaporate water because of the large lake surface areas compared to free-running rivers.

We have just begun to understand these interactions and know that the numbers vary considerably by climate and location—many of the variations are because of variations in power plant designs. On a national average in the US, 0.50 gal of water evaporate for every kilowatt-hour of electricity produced. Additional breakdown by region and for energy mixes is provided in the references.

Because of the variations in climate and HVAC design, no "rules of thumb" yet exist to compare water and electrical consumption. Hourly computer simulations can be used to compare systems with the water consumption calculated based on water and energy balances.

Other uses or sources of water should also be considered. For example, condensate from cooling coils can be used for irrigation. In some cases, blowdown from cooling towers can also be used depending on the levels of dissolved solids and chemical treating.

### Source of Further Information on Water Consumption Impacts of HVAC

Torcellini, P.A., M. Longm, and R. Judkoff, 2004. Consumptive water use for U.S. power production. *ASHRAE Transactions* 110(1):96–100.

## ACOUSTICS

Sound and vibration are the often unheralded contributors to occupant comfort and health that should be an integral part of green building design and should not be forgotten. While noise is not always an obvious problem, current research into the effect of noise on human productivity and performance has shown noise to be an important parameter affecting a worker's performance (Waye 2002). In addition to worker productivity, relatively low levels of indoor noise can also adversely affect workers' well-being. Outdoors, increasing environmental noise levels lower the general quality of life and degrade the environment. Unforeseen effects due to sound reflections can have serious adverse affects after a project is completed.

### Noise Criteria

There are three basic criteria for indoor noise in wide use today: noise criteria, room criteria, and A-weighting (dB[A]). A full description of these criteria and how they are used can be found in the *2007 ASHRAE Handbook—Applications*, Chapter 47. Figures 9-1a and 9-1b show the sound pressure levels associated with various sound frequency levels for both the noise and room criteria.

There is one basic criterion used for rating outdoor noise: A-weighting or dB(A). The criterion can be stated in terms of a maximum sound level ($L_{max}$) or as an energy average sound level, the equivalent sound level ($L_{eq}$).

### Indoor Sound Sources

*Fan and mechanical equipment noise and vibration.*

1. Paths—There are two paths by which sound travels to people: airborne and structureborne. Airborne sound travels through the air, through windows, and is how we hear other people's voices. Structureborne sound is how we hear ourselves (through the bones in our skull) and is what happens when you hear someone walking on the floor above or an elevator passing by or the hum of a pump on the floor above. In Figure 9-2, a typical floor fan room is shown, along with the many paths by which sound reaches the occupants of the space it serves.

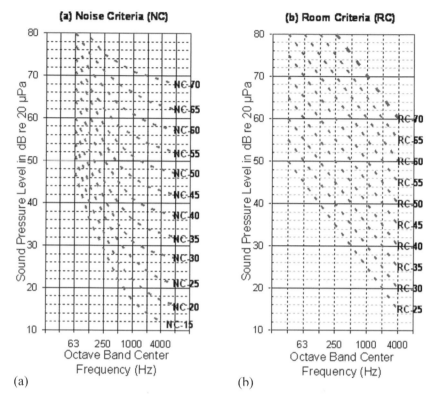

Figure 9-1 Noise and room criteria levels.

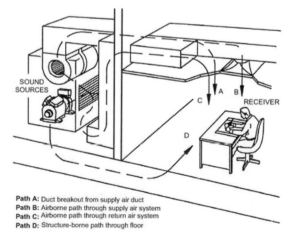

Path A: Duct breakout from supply air duct
Path B: Airborne path through supply air system
Path C: Airborne path through return air system
Path D: Structure-borne path through floor

Figure 9-2 Typical floor fan room sound transmission paths.

Paths A, B, and C are airborne transmission paths, while Path D is a structureborne path. All are important in securing the comfort of the occupants. See *ASHRAE Handbook—Applications*, Chapter 47, for more information.

*Sources—There are many sources of mechanical noise in a building.*

1. Equipment—fans, pumps, chillers, compressors, vacuum pumps.
2. Duct system airflow—regenerated noise due to high flow velocities, close spacing of successive duct elements, too rapid transitions. Anytime static pressure is lost, there is noise generated. The greater the loss, the higher the noise.
3. Pipe system flow noise and vibration—excessive flow velocities; air in the fluid, closely spaced successive valves, turns, and junctions can all contribute to noise in occupied spaces.

*Noise and vibration control.*

1. There is extensive information in the ASHRAE publications referenced below for controlling noise and vibration generated by building mechanical equipment.

## Outdoor Sound Sources

*Cooling towers.*

1. This equipment comes in a variety of sizes from small units to great behemoths that service industrial plants. There are three sources of noise from this equipment: fans, motors, and water. The two primary types are the cross-flow, induced draft tower, and forced draft centrifugal fan towers.
   a. Fan noise can be controlled in the induced draft fans through the use of larger units with slower fan speeds. Recent innovations include the wide blade hatchet-shaped fans that move more air at slower speeds due to their larger, shaped blade surfaces and the newer multiple-blade fans that achieve much the same effect by using more blades at slower speeds to move the same air as a conventional fan.
   In forced draft towers, fan noise is controlled through the use of sound attenuators since these towers can handle the additional pressure loss without losing cooling capacity.
   b. Newer, more efficient, and quieter motors with variable-speed controls have served to eliminate the start-up whine associated with older equipment.
   c. While fan noise can be controlled through speed and innovative fan designs, water noise can still be a problem in very quiet environments. To control water noise requires interrupting the path of the noise from the tower to the listener using barriers or silencers. The

alternative is to reduce the water noise generated by reducing the height from which the water drops fall or by having them fall onto a quieter surface than the water in the sump basin.

d.  Utilization of a TES system can result in the cooling towers and associated condenser pumps to shut down during daytime operation (typically from 9:00 a.m. to 5:00 p.m.), thereby eliminating the outdoor noise levels associated with this equipment.

*Direct expansion condensing units.*

1.  This equipment is air-cooled, primarily with propeller fans that are the main noise source. The compressor noise adds annoying tonal components depending on the type of compressor.

    a.  Centrifugal compressors are the traditional type and generate noise in the lower-frequency bands (31.5, 63, and 125 Hz)

    b.  Scroll compressors have higher-frequency noise.

    c.  Screw compressors have come into wide use in recent years and have serious pure-tone issues on which manufacturers have made significant strides, but great care still needs to be taken with these types of units.

    Enclosing the compressor in a sound-sealed compartment or wrapping the compressor in a sound barrier of composite material can contain the compressor noise.

*Drycoolers.*

1.  This equipment is largely associated with computer room air conditioners (CRACs), coils, and fans. In addition to the fan noise, fan vibration often excites the casing of the units, which radiates more noise. Larger, slower moving fans can reduce much of this noise. Applying damping compounds and stiffening the casing sheet metal can reduce the casing radiated noise.

*Exhaust fans.*

1.  These fans and their ventilation supply counterparts can be inside the building, with only the discharge putting noise into the environment, or located outside, where the radiated noise also contributes.

    Low-noise fan selection and the use of duct silencers can significantly reduce this noise.

*Sound reflections.*

1.  An expected sound reflection can double or quadruple (four times) the sound energy. Each surface within about 10 ft (3 m) (in addition to the

ground or the roof) will increase the resulting noise from a source by 3 dB (see Figure 9-3). One also needs to be aware of potential sound reflecting structures that are more distant. These can reflect sound around a barrier directly to a noise-sensitive receiver.

## Potential GreenTips

1. Lower fan tip speeds result in lower noise levels; this applies generally to all fans but is most pertinent to propeller ventilation fans, small centrifugal exhaust fans, and cooling tower fans. For example, consider two fans at 10,000 cfm (4.72 m/s) and 2 in. (498 Pa) static pressure: one is a 24 in. (61 cm) backward-inclined centrifugal fan at 1610 rpm, and the other is a 22 in. (55.9 cm) fan at 1953 rpm. The difference in sound power is approximately 5 dB and lower for the larger, slower moving fan.

2. Use variable-speed drives instead of inlet guide vanes or volume dampers. This reduces energy use as well as generated noise levels. A fan without inlet guide vane can be up to 10 dB quieter.

3. The use of a VFD on cooling tower fans can reduce noise levels by 6 dB during the more sensitive nighttime periods as well as eliminate the whine associated with a motor ramping up to speed.

4. Locating sound sources near reflective surfaces can result in adverse increases in the effective sound level of equipment. Try to keep sound sources a minimum of 10 ft (3 m) from a wall if at all possible. Be aware of the relative location of a sound source and a wall and to where sound may be reflected.

5. Utilization of a thermal storage system can cause the cooling towers and associated condenser pumps to shut down during daytime operation (typically from 9:00 a.m. to 5:00 p.m.), thereby eliminating the outdoor noise levels associated with this equipment.

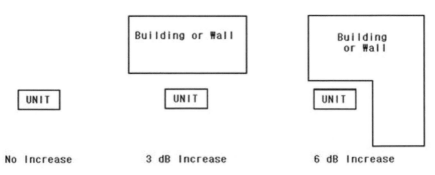

**Figure 9-3  Effect of reflecting surfaces.**

## Sources of Further Information on Acoustics

ARI. 1997. *ARI STANDARD 275-97, Standard for Application of Sound Rating Levels of Outdoor Unitary Equipment.* Arlington, VA: Air-Conditioning and Refrigeration Institute.

ASHRAE. 2003. *2003 ASHRAE Handbook—HVAC Systems and Equipment,* Chapter 47, Sound and Vibration Control. Atlanta: American Society of Heating, Refrigerating and Air-Conditioning Engineers, Inc.

ASHRAE. 2005. *2005 ASHRAE Handbook—Fundamentals,* Chapter 7, Sound and Vibration. Atlanta: American Society of Heating, Refrigerating and Air-Conditioning Engineers, Inc.

Schaffer, M. 2004. *Practical Guide to Noise and Vibration Control for HVAC Systems, Second Edition.* Atlanta: American Society of Heating, Refrigerating and Air-Conditioning Engineers, Inc.

Waye, K.P., J. Bengtsson, R. Rylander, F. Hucklebridge, P. Evans, A. Clow. 2002. Low frequency noise enhances cortisol among noise sensitive subjects during work performance. *Life Science* 70(7):745–58.

## COOLING SYSTEM HEAT SINKS

Building HVAC systems exchange a significant amount of thermal energy between the building and the surrounding environment; that is their function. Chapter 11 of this Guide describes technical details of heating and cooling systems and their interaction with the environment as a heat source or heat sink.

Green design features are available that provide high energy efficiency but have the potential for adversely impacting the environment. For example, systems have been installed that use deep water in a lake or nearby ocean as a heat sink, which results in significant energy savings. This technique has been used in Scandinavian systems for approximately 20 years and more recently in colder regions in North America. The possible net energy savings are impressive, but these must be balanced against potential adverse impacts on local aquatic environment.

## GREEN ROOF AND COOL ROOF TECHNOLOGIES

While similar in some respects, *green roof* and *cool roof* technologies can serve very different purposes.

### Green Roofs

The practice of placing a living vegetative surface onto a building rooftop (hence, a *green* roof) is not new. The famous Hanging Gardens of ancient Babylon were built around 500 BC and were built of arched stone beams waterproofed with reeds and a thick tar coating. Modern materials allow for a wide range of vegetative system concepts for a green roof. Green roof systems can be generally classified as *intensive* or *extensive*. An intensive green roof system is, in essence, a miniature ecosystem built of several layers with drainage systems and, quite often, irrigation as well. Intensive green roofs would generally have a minimum of 12 in. (30 cm) of soil depth but can include even deeper soil substrates. These types of roofs can indeed be miniature parks with large trees, shrubs, and manicured landscapes. The resulting additional structural load can be large, in the range of 80 to 150 lb/ft$^2$ (390 to 732 kg/m$^2$).

In contrast, an extensive green roof involves a much smaller soil depth, from 1 to 6 in. (2.5 to 15 cm). They are primarily installed for the environmental benefits. Extensive green roofs are commonly installed using modular plots, such as illustrated in Figure 9-4. The additional load on the building structure is not nearly as much as with an intensive green roof, with typical values being on the order of 10 to 15 lb/ft$^2$ (49 to 73 kg/m$^2$).

One of the earlier benefits touted for green roofs was the potential for reduction in building HVAC energy consumption. Green roofs can reduce the heat flow into the building in summer and out from the building roof in winter due to the insulation properties of the added soil. The potential also exists for green roofs to act as an

Green roof
modular blocks

Green roof over
parking garage

**Figure 9-4   Representative green roof installations.**

active cooling system to remove heat from the roof surface in the summer through evapotranspiration, but this generally would require the roof surface to be irrigated (and, hence, increase water consumption).

From an HVAC engineer's perspective for a green roof, the question is how exactly to analyze the green roof thermal transport properties. The green roof soil and other layers do add additional thermal insulation to the roof, and plant materials may shade the soil surface. In addition, the evapotranspiration effect of water leaving the green roof soil surface increases heat loss from the green roof, potentially keeping it cooler. Building energy simulation models are not well suited for analyzing the heat loss due to the effect of moisture evaporation from the soil. For example, in the eQUEST (DOE-2 based) simulation model, the only additional parameter available for manipulation beyond the soil thermal conductivity and capacitance is the soil absorptivity. In one study, it was proposed to adjust the absorptivity downward to account for rainwater evaporation effects (Hilten 2005). The recommended soil surface absorptivity values to use for various cities in the US when analyzing the impact of a green roof on building energy loads are listed in Table 9-2. These values are compared to a "normal" soil surface absorptivitiy of approximately 0.7.

The primary benefit of green roofs that is driving the market acceptance, particularly in Europe, is that of reduced stormwater runoff. Several major US cities have recently adapted roof vegetation requirements for new construction in order to reduce stormwater runoff. For example, the city of Chicago has requirements for green roof or cool roof technologies that vary on the building type, if public assistance is provided for the development. For example, the city of Chicago has requirements for green roof or cool roof technologies that vary with building type, and in some cases public assistance is provided for the development.

Table 9-3 summarizes potential benefits and drawbacks for green roof systems.

**Table 9-2   Recommended Soil Surface Absorptivity Values to Use in Building Energy Analysis of Green Roofs**

| City | Absorptivity |
|---|---|
| Atlanta | 0.46 |
| Denver | 0.58 |
| Honolulu | 0.56 |
| Los Angeles | 0.62 |
| New York | 0.39 |
| Phoenix | 0.67 |
| Seattle | 0.36 |

**Table 9-3    Potential Benefits and Drawbacks of Green Roofs**

| Advantages | Disadvantages |
|---|---|
| Stormwater runoff reduction | Additional structural load |
| Reduced heat gains in summer and heat loss in winter to building structure | Cost |
| Longer life for the base roofing system (may not apply to an intensive green roof) | Additional maintenance, ranging from limited for an extensive green roof with low-maintenance plants to high for a manicured landscaped intensive roof |
| Reduced noise transmission from outside | Optimal roof type, plant materials, and soil depths will vary depending on climate |
| Aesthetic benefits to people in or around the building with the additional green space | Documentation of benefits such as reduction in heat island effect has not been proven |
| Other general environmental benefits, such as reduced nitrogen runoff (source: bird droppings), air pollutant absorption, potential carbon sink, bird habitat | |

Green roofs are also considered to potentially reduce the urban heat island effect, but the exact influence and interaction of green roofs with the local thermal environment is not scientifically proven.

## Cool Roofs

The primary intent of cool roof technology is to reduce the amount of energy absorbed by a roof surface. New advanced coating materials allow for the selective absorption and reflection of various spectral wavelengths. This allows for the design of roofing systems with visual coloring that can enhance a building's character while still reflecting a good deal of the total incident solar energy, of which a significant portion extends beyond the visual wavebands to include infrared and ultraviolet light. The net result is that a lower fraction of the incident solar energy (only about 20% or less) is absorbed by the structure (Figure 9-5). This reduces the cooling load on the building's HVAC as well as significantly increases the expected life of the building's roof. The environmental benefits of cool roofs are that they can decrease the urban heat island effect by reflecting some of the incident solar energy back into space as opposed to absorbing the heat and releasing it to the surroundings.

In summary, both technologies can have a positive environmental benefit but in somewhat different ways. A comparison of the various properties of each is given in Table 9-4. Note that the *net* environmental impact of a technology should be taken into account for an individual project. For example, if adding a green roof requires additional building materials to strengthen the load-bearing capacity of the roof, then this would be a negative impact that must be considered.

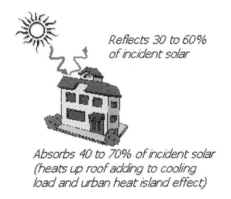

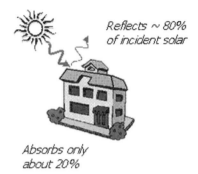

Reflects 30 to 60% of incident solar

Reflects ~ 80% of incident solar

Absorbs 40 to 70% of incident solar (heats up roof adding to cooling load and urban heat island effect)

Absorbs only about 20%

Conventional Roof

Cool Roof

**Figure 9-5   Thermal benefits of cool roof technology.**

**Table 9-4   Comparison of Green Roof and Cool Roof Technologies**

| Property | Green Roof | Cool Roof |
|---|---|---|
| Decrease roof surface temperature | Yes | Yes |
| Impact on cooling load | ↓ | ↓ |
| Impact on heating load | ↓ | ↑ |
| Building structural concern | Yes | No |
| Improved stormwater management | Yes | No |
| Reduce urban heat island impact | Yes (?) | Yes |
| Cost impact | ++ | Minor |

### Sources of Further Information on Green and Cool Roofs

City of Chicago Green Roof Policy summary, http://egov.cityofchicago.org/webportal/COCWebPortal/COC_EDITORIAL/Green_Roof_Policy_Matrix_revised.pdf

Cool Roof Rating Council (third party rating of cool roof material performance), www.coolroofs.org

Green Roof Industry Information Clearinghouse and Database, www.greenroofs.com

*Guideline for Planning, Execution and Upkeep of Green-Roof Sites* (German standard similar in nature to ASHRAE standard specifically focused on green roof design issues), www.f-l-l.de/english.html

(An overview of the above guideline is available for download at www.greenroofservice.com/downpdf/IntroductiontotheGermanFLL2.pdf.)

## DESIGNING HEALTHY BUILDINGS

The science of green building design implicitly includes the design of healthy building environments, and this concept can and should be extended to address the aerobiology of the indoor environment. The field of aerobiological engineering includes technologies that incorporate active, passive, or natural means of disinfecting air and surfaces indoors (Kowalski 2006). The field of green building design bears a lot in common with aerobiological engineering since they both aim at providing healthy habitats in the most practical or feasible manner (ASHE 2002). The USGBC has developed the LEED Green Building Rating System® for Commercial Interiors (USGBC 2002). This voluntary rating system provides guidance to building owners, occupants, interior designers, architects and others who design and install building interiors. It addresses topics related to sustainable design, such as space usage, water and energy efficiency, and IEQ.

For aerobiological engineering, several technologies fall into the category of green building design, including passive solar disinfection, vegetation air cleaning, biofiltration of air, material selectivity, and architectural design for hygienic living. These are all primarily developmental technologies and may have limitations in terms of application and effectiveness, but combined with other green building systems they may contribute to an integrated solution to IAQ problems and healthy living.

### Ventilation Air Cleaning

Currently little credit can be taken in the LEED rating system for air treatment systems (other than for air filters used during and after construction), but it is likely the matter will receive increasing attention in the future. The incorporation of air-cleaning systems in any building requires some investment in design, labor, equipment, and maintenance, and, like all building systems, the matter needs to be evaluated in economic terms. At present, however, there is insufficient information to fully establish the cost/benefits of indoor air cleaning or the actual requisite level of indoor aerobiological quality necessary for healthy living. It could be assumed, however, that the presence of any air-cleaning system is likely to provide a higher quality of health than the absence of air cleaning. Since even some minimal amount of air cleaning, a MERV 8 filter for example, will provide noticeable improvement in air quality, especially for atopic or allergic individuals, it is recommended that some modest level of air filtration be provided in all buildings, and that individuals who require higher levels of air cleanliness should consider designing and installing combined air filtration and UVGI systems insofar as budgets permit.

Incorporating sustainable air-cleaning technologies into the design of a green building is a field that requires research. There are currently no formal requirements for air-cleaning systems in commercial or residential buildings, but there are some proposed guidelines for such applications (IUVA 2005). Natural ventilation could be considered a green building technology due to the low energy consumption, but natural ventilation systems must be designed to facilitate airflow through a building. When climate permits, natural ventilation may provide an adequate number of air exchanges but cannot easily provide for filtration of outdoor spores. The use of 100% outdoor air with enthalpy recovery wheels is one green building option, but it is necessary to filter outdoor air to remove environmental spores. The minimization of biological contamination through envelope design is another green approach provided it can be accomplished economically (Rosenbaum 2002).

## Material Selectivity for Health

The appropriate selection of building materials to avoid those that may contribute to microbial growth is a developing science that begs attention today. The use of sustainable materials and renewable energy resources does not necessarily conflict with the objective of providing aerobiologically clean indoor environments, but there are areas of overlapping concerns and some considerations that need to be addressed.

Carpets often contain various chemical by-products that may act as indoor pollutants. Carpets and rugs also provide a substrate that may both collect and grow mold and mildew under moist conditions. Since mold spores tend to settle over time, they inevitably collect in carpets and the problem is exacerbated by traffic. A wet spill is sufficient to initiate mold spore germination.

## Passive Solar Disinfection

The subject of daylighting does receive some LEED credit, however, and maximizing daylighting can provide natural disinfection to building air and surfaces. Passive exposure to solar irradiation as a means of destroying airborne pathogens is based on the fact that sunlight contains some ultraviolet radiation and is lethal to airborne human pathogens (El-Adhami et al. 1994; Fernandez 1996; Beebe 1959). Sunlight in which all UVB and UVA was removed showed no significant disinfection within the first 120 minutes. Partially filtered sunlight that passes at least some of the UVA or UVB wavelengths can have a potentially significant disinfection effect over periods of days or weeks. Maximum fenestration will enhance this effect.

Generous quantities of sunlight and skylight are available even in the most northern latitudes, and careful attention to the design of indoor illumination via maximized fenestration may enhance the self-disinfection of buildings. When window space cannot be maximized, it may be possible to use UV-transmittant glass. In a hypothetical design for a building using passive solar exposure to control airborne microbes, the windows form a plenum for return air, with the outside panes being UV transmitting glass (Ehrt et al. 1994).

### Vegetation Air Cleaning

Large amounts of living vegetation can act as a natural biofilter, reducing levels of airborne microorganisms (Darlington et al. 1998; Rautiala et al. 1999). Winter gardens may act as "buffer zones" in moderating the indoor climate (Watson and Buchanan 1993). The surface area of large amounts of vegetation may absorb or adsorb microbes or dust. The oxygen generation of the plants may have an oxidative effect on microbes. The increased humidity may have an effect on reducing some microbial species, although it may favor others. The presence of symbiotic microbes such as *Streptomyces* may aid disinfection of the air. Natural plant defenses against bacteria may operate against mammalian pathogens. Finally, gardens and vegetation may have an effect on the psyche of occupants that may stimulate a sense of well-being. A number of houseplants have been identified that may contribute to passive cleaning of indoor air pollutants, but their effects on airborne microbes are not known (Wolverton 1996).

Although houseplants are often considered a source of potential fungal spores, a study by Rautiala et al. (1999) indicated that no significant increase in concentrations of fungi in air or surface samples occurred when houseplants were added to indoor environments. In a vegetation air-cleaning system, air flows through areas filled with vegetation or through entire greenhouses before entering the ventilation system. Often, such vegetation areas include waterfalls, which may also have an effect on local ionization levels. One downside to keeping large amounts of vegetation indoors is that the potting soil may include potentially allergenic fungi.

Vegetation has also been shown to have the capability of removing indoor pollutants and chemicals (Darlington et al. 2001). Atriums can provide solutions to IAQ problems in commercial and institutional buildings when large spaces are provided in multistory buildings but also present complex interfacing with ventilation and environmental control systems (Kainlauri and Vilmain 1993).

### Biofiltration

Biofiltration has been in successful use in the water industry for some time and offers a potential alternative method for controlling indoor aerobiology through the use of filtration material containing bacteria that are antagonistic to pathogens. In this sense it is similar to aspects of vegetation air cleaning. In one suggested approach, the building structure is used as the biofilter (Darlington et al. 2000). Prior to its acceptance for dealing with VOCs and $CO_2$, efforts were made to determine if this amount of biomass in the indoor space could impact IAQ. A relatively large, ecologically complex biofilter composed of a bioscrubber of about 10 $m^2$ (110 $ft^2$), 30 $m^2$ (325 $ft^2$) of vegetation, and a 3,500 L (925 gal) aquarium was incorporated in an airtight 160 $m^2$ (1720 $ft^2$) room in a newly constructed office building in downtown Toronto. This space maintained about 0.2 ach compared to the 15 to 20 ach (with 30% outdoor air) of other spaces in the same building. Total VOCs (TVOCs) and formaldehyde levels in the biofilter room were the same as, or significantly lower than, other spaces in the building. Aerial spore levels were slightly higher than other indoor spaces but were well within reported values for "healthy" indoor spaces. Levels appeared to be dependent on horticultural management practices within the space. Most of the fungal genera identified were either common indoor spores or other genera associated with living or dead plant material or soil.

## Sources of Further Information on Designing Healthy Buildings

ASHE. 2002. Green Healthcare Design Guidance Statement. American Society of Healthcare Engineering.

Beebe, J.M. 1959. Stability of disseminated aerosols of *Pastuerella tularensis* subjected to simulated solar radiations at various humidities. *Journal of Bacteriology* 78:18–24.

Darlington, A., M.A. Dixon, and C. Pilger. 1998. The use of biofilters to improve indoor air quality: The removal of toluene, TCE, and formaldehyde. *Life Support Biosph Sci* 5(1):63–69.

Darlington, A., M. Chan, D. Malloch, C. Pilger, and M.A. Dixon. 2000. The biofiltration of indoor air: Implications for air quality. *Indoor Air 2000* 10(1):39–46.

Darlington, A.B., J.F. Dat, and M.A. Dixon. 2001. The biofiltration of indoor air: Air flux and temperature influences the removal of toluene, ethylbenzene, and xylene. *Environ Sci Technol* 35(1):240–46.

Ehrt, D., M. Carl, T. Kittel, M. Muller, and W. Seeber. 1994. High performance glass for the deep ultraviolet range. *Journal of Non-Crystalline Solids* 177:405–19.

El-Adhami, W., S. Daly, and P.R. Stewart. 1994. Biochemical studies on the lethal effects of solar and artificial ultraviolet radiation on *Staphylococcus aureus*. *Arch Microbiol* 161:82–87.

Fernandez, R.O. 1996. Lethal effect induced in *Pseudomonas aeruginosa* exposed to ultraviolet-A radiation. *Photochem & Photobiol* 64(2):334–39.

IUVA. 2005. General guideline for UVGI air and surface disinfection systems. IUVA-G01A-2005 International Ultraviolet Association. Ayr, Ontario, Canada, www.iuva.org.

Kainlauri, E.O., and M.P. Vilmain. 1993. Atrium design criteria resulting from comparative studies of atriums with different orientation and complex interfacing of environmental systems. *ASHRAE Transactions* 99(1):1061–69.

Kowalski, W.J. 2006. *Aerobiological Engineering Handbook: Airborne Disease and Control Technologies*. New York: McGraw-Hill.

Rautiala, S., S. Haatainen, H. Kallunki, L. Kujanpaa, S. Laitinen, A. Miihkinen, M. Reiman, and M. Seuri. 1999. Do plants in office have any effect on indoor air microorganisms? *Indoor Air 99: Proceedings of the 8th International Conference on Indoor Air Quality and Climate, Edinburgh, Scotland*, pp. 704–09.

Rosenbaum, M. 2002. A green building on campus. *ASHRAE Journal* 44(1):41—44.

USGBC. 2002. LEED Green Building Rating System®. United States Green Building Council, www.usgbc.org.

Watson, D., and G. Buchanan. 1993. *Designing Healthy Buildings.* Washington DC: American Institute of Architects.

Wolverton, B.C. 1996. *How to Grow Fresh Air: 50 Plants That Purify Your Home of Office.* Baltimore, MD: Penguin Books.

## HVAC, BUILDING ENVELOPE, AND IAQ INTERACTION

Recent years have seen a marked increase in recognition of the impact that building materials (envelope, furniture, paints, flooring, etc.) have on IAQ. This was the primary reason for changes to the ASHRAE Standard 62.2 outdoor air ventilation requirements to include an allowance for the building area and not just total number of occupants.

The IAQ can be negatively affected by off-gassing of chemicals in building materials or chemicals used during the construction or fabrication of the components. The LEED program contains a number of credit point items that relate to reducing the introduction of potentially harmful materials into a building environment. It also describes methods to help ensure that key HVAC components, such as ductwork, do not become contaminated during construction and act as a source of indoor pollution after occupancy.

## TECHNOLOGY DESCRIPTION

Pulse-powered physical water treatment uses pulsed, electric fields (a technology developed by the food industry for pasteurization) to control scaling, biological growth, and corrosion. This chemical-free approach to water treatment eliminates environmental and health-and-safety issues associated with water treatment chemicals. Pulse-powered systems do not require pumps or chemical tanks. Pulse-powered systems tend to be forgiving of operational upsets and promote cooling tower operation at higher cycles of concentration (therefore, less blowdown and less water usage) than standard chemical treatment. Independent studies have shown not only that the method is effective for cooling towers but that the performance of pulse-powered systems is superior to standard chemical treatment in biological control and water usage. The performance results of pulse-powered technology for chemical-free water treatment, as documented by various independent evaluations, support the objectives of green buildings and have earned LEED points for certification in a number of projects.

## WHEN/WHERE IT'S APPLICABLE

Pulse-powered technology is applicable on the recirculating lines of cooling towers, chillers, heat exchangers, boilers, evaporative condensers, fluid coolers, and fountains.

The technology produces a pulsed, time-varying, induced electric field inside a PVC pipe that is fit into the recirculating water system. The electric signal changes the way minerals in the water precipitate, totally avoiding hard-lime scale by instead producing a non-adherent mineral powder in the bulk water. The powder is readily filterable and easily removed. Bacteria are encapsulated into this mineral power and cannot reproduce, thereby resulting in low bacteria populations. The water chemistry maintained by pulse-powered technology is noncorrosive, operating at the saturation point of calcium carbonate (a cathodic corrosion-inhibiting environment). The low bacteria count and reduction or elimination of biofilm reduces concern about microbial influenced corrosion. The absence of aggressive oxidizing biocides eliminates the risk of other forms of corrosion.

## PROS AND CONS

### Pro

1.  The potential for lower bacterial contamination while providing scale and corrosion control.
2.  Lower energy and water use than in traditional chemical treatment.
3.  Blowdown water is environmentally benign and recyclable.

4.   Life-cycle cost savings compared to chemical treatment.
5.   Reduction or elimination of biofilm.
6.   Removes health and safety concerns about handling chemicals.
7.   Eliminates the environmental impact of blowdown, air emissions, and drift from toxic chemicals.

### Con

1.   It does not work effectively on very soft or distilled water, since the technology is based on changing the way minerals in the water precipitate.
2.   Water with high chloride or silica content may limit the cycles of concentration obtainable to ensure optimum water savings since the technology operates at the saturation point of calcium carbonate.
3.   Energy usage is still required to operate.

### KEY ELEMENTS OF COST

The following economic factors list the various cost elements associated with traditional chemical treatment that are avoided with chemical-free water treatment. This is a general assessment of what might be likely, but it may not be accurate in all situations. There is no substitute for a detailed cost analysis as part of the design process.

*   *Direct Cost of Chemicals.* This item is the easiest to see and is sometimes considered the only cost. For cooling towers in the US, this direct cost usually runs between $8.00 and $20.00 per ton of cooling per year.
*   *Water Softener.* Water softeners have direct additional costs for salt, media, equipment depreciation, maintenance, and direct labor.
*   *OSHA and General Environmental Requirements.* Many chemicals used to treat water systems are OSHA-listed hazardous materials. Employees in this field are required to have documented, annual training on what to do in the event of a chemical release or otherwise exposed contamination.
*   *General Handling Issues.* Chemical tanks, barrels, salt bags, etc., take space. A typical chemical station requires 100 ft (9.3 m) of space.
*   *Equipment Maintenance.* Lower overall maintenance for the systems as a whole may be possible.
*   *Water Savings.* Cooling towers are often a facility's largest consumer of water. Most chemically controlled cooling towers operate at two to four cycles of concentration. Cycles of concentration can often be changed to six to eight cycles with chemical-free technology, with an annual reduction in water usage costs and the associated environmental impacts.
*   *Energy Savings.* Energy is required to operate the pulse-powered system, but overall energy usage can be lower. The reduction or elimination of biofilm (a

slime layer in a cooling tower) results in energy savings versus chemical treatment due to improved heat transfer. Biofilm has a heat transfer resistance four times that of scale and is also the breeding ground for *Legionella* amplification. Preventing this amplification thus saves costs.

## SOURCES OF FURTHER INFORMATION

Bisbee, D. 2003. *Pulse-Power Water Treatment Systems for Cooling Towers.* Energy Efficiency & Customer Research & Development, Sacramento Municipal Utility District, November 10, 2003.

*Codes and Standards Enhancement Report, Code Change Proposal for Cooling Towers.* Pacific Gas and Electric, April 8, 2002.

HPAC. 2004. Innovative grocery store seeks LEED certification. *HPAC Engineering* 27:31.

Torcellini, P.A., N. Long, and R. Judkoff. 2004. Consumptive water use for U.S. power production. *ASHRAE Transactions* 110(1):96–100.

Trane. 2005. *Trane Installation, Operation, and Maintenance Manual: Series R®  Air-Cooled Rotary Liquid Chillers*, RTAA-SVX01A-EN. Lacrosse, WI: Trane Company. www.trane.com/Commercial/Equipment/PDF/3/RTAA-SVX01A-EN.pdf.

# 10

# ENERGY DISTRIBUTION SYSTEMS

For there to be heating, cooling, lighting, and electric power throughout a building, the energy required by these functions is usually distributed from one or more central points. The most common media used to distribute energy are steam, hydronics, air, and electricity. Refrigerants are also used as a means of energy transfer between components of refrigeration equipment. Usually, except for industrial and certain specialized applications, the length of refrigerant piping runs is not great.

Discussion here will concentrate primarily on steam, hydronics, and air systems.

## STEAM

### Advantages

- Steam flows to the terminal usage without aid of external pumping.
- Steam systems are not greatly affected by the height of the distribution system, which has a significant impact on a water system.
- Steam distribution can readily accommodate changes in the system terminal equipment.
- Major steam distribution repair does not require piping draindown.
- The thermodynamics of steam utilization are effective and efficient.

### Disadvantages

- Steam traps, condensate pumps, etc., require frequent maintenance and replacement.
- Steam piping systems often have dynamic pressure differentials that are not easily controlled by typical steam controls appurtenances.
- Returning condensate by gravity is not always possible. Lifting the steam back to the source can be a challenge due to pressure differential dynamics, space constraints, wear and tear on equipment, etc.

- Venting of steam systems (PRVs, flash tanks, boiler feed tanks, condensate receivers, safety relief valves, etc.) must be done properly and must be brought to the outside of the building.
- High-pressure steam systems require a full-time boiler plant operator, adding to operating costs.

With the advantages listed, steam is often a logical choice for commercial and industrial processes and for large-scale distribution systems, such as on campuses and in large or tall buildings. However, steam traps require periodic maintenance and can become a source of significant energy waste. Condensate return venting issues can add to operation and design challenges. Venting of flash steam can also cause significant energy to be wasted. (See subsequent section on steam traps, condensate return, and pressure differential.)

### Classification

Steam distribution is either one-pipe or two-pipe. One-pipe distribution is defined to be where both the steam supply and the condensate travel through a single pipe connecting the steam source and the terminal heating units. Two-pipe distribution is defined to be where the steam supply and the steam condensate travel through separate pipes. Two-pipe steam systems are further classified as gravity return or vacuum return.

Steam systems are also classified according to system operating pressure:

- Low pressure is defined to be 15 psig (103 kPa) or less.
- Medium-pressure steam is defined to be between 15 and 50 psig (103 and 345 kPa).
- High pressure is defined to be over 50 psig (345 kPa).

Selection of steam pressure is based on the constraints of the process served. The level of system energy rises with the system pressure. Higher steam pressure may allow smaller supply distribution pipe sizes, but it also increases the temperature difference across the pipe insulation and may result in more heat loss. Higher steam pressure also dictates the use of pipes, valves, and equipment that can withstand the higher pressure. This translates to higher installation cost.

### Piping

Supply and return piping must be installed to recognize the thermodynamics of steam and to allow unencumbered steam and condensate flow. Piping that does not slope correctly—that is, it is installed with unintended water traps or has leaks—will not function properly and will increase system energy use. Careful installation will result in efficient and effective operation. If steam or condensate leaks from the piping system, additional water must be added to make up for the losses. Makeup water is chemically treated and is an operating cost.

## Control

Control of steam flow at the terminal equipment is very important. Steam control valves are selected to match the controlled process. If steam flow varies over a wide range, it may be necessary to have multiple control valves. On large terminal heat exchange units, a single control valve usually cannot provide effective control; in such cases, multiple control valves of various sizes operated in a sequential manner are used. Consult sizing data available from manufacturers.

## Condensate Return and Flash Steam

The energy conservation and operational problems that come up when utilizing steam often occur because the system design and/or installation does not properly address the issue of returning condensate to the boiler or cogeneration plant. The simplest and most efficient way to return condensate is via gravity. When this is not possible, the designer needs to clearly understand issues of lift, condensate rate, and pressure differential to ensure that operational problems such as water hammer and reduction in capacity do not occur.

In addition, the proper sizing and routing of vents for both the condensate receiver and the flash tank need to be clearly understood by the designer and the operator.

When considering recovery of flash steam, it is important to accurately calculate the expected operating pressure of the flash steam to ensure that the system or equipment served by the flash steam will be able to operate under all system conditions. Alternatively, if the flash tank is to be vented to atmosphere, the designer must fully understand the amount of energy that will be relived and wasted by venting the flash tank to the atmosphere.

## Steam Traps

Selection of steam traps is related to the function of the terminal device or pipe distribution served. Steam traps have the function of draining condensate from the supply side of the system to the condensate return side of the system without allowing steam to flow into the return piping. The flow of steam into return piping unnecessarily wastes energy, and significant energy waste can occur if periodic maintenance is not performed. Properly sized and installed steam traps allow terminal heat exchange equipment to function effectively. If condensate is not fully drained from the terminal heat exchange equipment, the heat transfer area is reduced, resulting in a loss of capacity.

## Efficient Steam System Design and Operation Tips

- Preheat boiler plant makeup water with waste heat
- Use ecofriendly chemical treatment

- Recover flash steam, making sure to understand pressure differential of flash steam compared to system pressure serving equipment where flash steam recovery is used
- Minimize use of pumped condensate return
- Do not use steam pressure for "lifting" condensate return
- Consider clean steam generators where steam serves humidification or sterilization equipment

### Sources of Further Information

Manufacturers of equipment—control valves, steam traps, and other devices—are valuable resources. There are multiple Web sites that contain system design information. Perform a search on the Internet for *steam piping design*. "Steam Systems" is Chapter 10 of the *ASHRAE Handbook—HVAC Systems and Equipment*. Other sources:

- Armstrong Intelligent Systems Solutions, www.armstrong-intl.com.
- Spirx Sarco Design of Fluid Systems—Steam Learning Module, www.spiraxsarco.com/learn/.

## HYDRONICS

Pumping heated water and chilled water is common system design practice in many buildings. Water is often diluted with an antifreeze fluid to avoid water freezing in extremely cold conditions (thus referring to these systems as *hydronic*.) There are many approaches in utilizing these systems.

### Classification

Hot-water heating systems are classified as low-temperature water, medium-temperature water, or high-temperature water.

- Low-temperature water systems operate at temperatures of 250°F (121°C) and below.
- Medium-temperature water systems operate at temperatures between 250°F and 350°F (121°C and 177°C).
- High-temperature water systems operate at temperature above 350°F (177°C).

Chilled-water systems distribute cold water to terminal cooling coils to provide dehumidification and cooling of conditioned air or cooling of a process. They can also serve cooling panels in occupied spaces. Chilled-water panels, which serve as a heat sink for heat radiated from occupants and other warm surfaces to the radiant panel, can be used to reduce the sensible load normally handled solely by mechanical air cooling. The percentage by which the load can be reduced depends upon the panel surface area and dew-point limitations, which are necessary to avoid any possibility of condensation.

Condenser water systems connect mechanical refrigeration equipment to outdoor heat dissipation devices such as cooling towers or water- or air-cooled condensers. These, in turn, reduce the temperature of the condenser water by rejecting heat to the outdoor environment.

## Piping, Flow Rates, and Pumping

Each of these systems uses two pipes—a supply and a return—to make up the piping circuit, and each uses one or more pumps to move the water through the circuit. Information on the design and characteristics of these various systems can be found in Chapters 12, 13, and 14 of the *ASHRAE Handbook—HVAC Systems and Equipment*.

Cost-effective design depends on consideration of the system constraints. Piping must be sized to provide the required load capacities and arranged to provide necessary flow at full- and part-load conditions. Design will be determined by several system characteristics and selections:

- Supply and return water temperatures
- Flow rates at individual heat transfer units
- System flow rate at design condition
- Piping distribution arrangement
- System water volume
- Equipment selections for pumps, boilers, chillers, and coils
- Temperature control strategy

Pumping energy can be a significant portion of the energy used in a building. In fact, pumping energy is roughly equal to the inverse fifth power of pipe diameter, so a small increase in pipe size has a dramatic effect on lower pumping energy (horsepower or kW). Traditionally, it was common to select heating water flow for coils based on a temperature difference of 20°F (11.1 K) between the supply and return. Flow rate in gallons per minute (gpm) was calculated by dividing the heating load in Btu per hour by 10,000 (1 Btu per lb °F × 8.33 lb per gallon × 60 minutes per hour × 20°F temperature difference) (liters per second [L/s] was calculated by dividing the kW heating load by 4.187 [specific heat capacity of water, kJ/kgK] × 11.1 K). As long as the cost of energy was cheap, this method was widely used.

Flow rate can easily be reduced by one-half of the 20°F (11.1 K) value by using a 40°F (22.2 K) temperature difference. The impact on pump flow rate is significant. The temperature difference selected depends upon the ability of the system to function with lower return water temperatures. Certain types of boilers can function with the low return water temperatures, while others cannot. Care must be taken in selecting the boiler type, coupled with supply and return water design temperatures. In specific instances, a low return water temperature could damage the boiler due to the condensation of combustion gases.

Lower flow rates could allow smaller pipe sizes, and pipe size, along with flow, affects pumping energy. A goal should be established for the pump horsepower to be selected. A small increase in some or all of the pipe distribution sizes could reduce the pump energy (hp or kW) horsepower needed for the system. When this goal is established and attained in the finished design, the concept and energy usage will be achieved. A reasonable goal can be expressed using the water transport factor equation adjusted to reflect kW (multiply hp by 0.746). Measurements of efficient designs indicate a performance of 0.026 kW/ton (0.007 kw/kWR[1]) being served as a reasonable goal for 10°F (5.6 K) delta-T systems. Adjusting the flow rate and delta-P variables in this formula will quickly show the benefits of larger pipes or lower flow rates (greater delta-T).

For instance, let us calculate the energy (kW) required with the modified water transport factor equation:

$$\text{pump kW} = \frac{(Q)(\Delta P)(0.746)}{(3960)(nP)(nM)} \quad \text{or} \quad \text{mkW} = \frac{(\text{L/s})(\Delta P \text{ kPa})(sg)}{(1000)E_{pump}E_{motordrive}}$$

where $nP$ ($E_{pump}$) is pump efficiency and $nM$ ($E_{motordrive}$) is electric motor efficiency.

We can create a performance index by solving the equation for 1 ton of cooling. That would produce an answer in kW/ton (kW/kWR). A typical condenser system might use 3 gpm per ton (0.054 L/s per kWR) cooling and a delta-P of 100 ft (300 kPa) with an 82% efficient pump and a 92% efficient motor. If we insert these variables into the above equation, the derived kW answer for 1 ton (3.5 kWR) of cooling would be 0.075 kW or 0.075 kW/ton (0.021 kW/kWR).

Now let us compare this index to the performance of an efficient design that increases piping and fitting size and therefore reduces delta-P to, say, 30 ft head (90 kPa) TDH, keeps the gpm the same, and uses an 85% efficient pump with a 92% efficient motor. (Since the most efficient motors cost the same as less efficient motors, almost everyone is now using premium efficiency motors.) If we solve again for kW, we get 0.022 kW/ton (0.006 kW/kWR).

This shows us we can reduce pumping energy by 71% by lowering the TDH from 100 to 30 ft (300 to 90 kPa) and selecting a more efficient pump. One author of this Guide has personally measured systems with the characteristics indicated above. If the average cost per kWh of electricity is $0.08 and we are pumping for a 1000 ton (3517 kWR) chiller, the annual operating cost difference would be over $37,000 per year.

If the system lasts 20 years, the improved system would save $740,000 in electricity costs. Of course, the obvious question is: "How much does it cost to increase pipe and fitting size and pump efficiency?" The answer will vary by project, but keep

---

1. kWR = Refrigeration cooling capacity in kWR.

in mind that efficient pumps cost no more than inefficient ones. Larger pipes and fittings cost more than smaller pipes and fittings and, in one author's experience, the cost of increasing the pipe size one size (going from a 10 in. [250 mm] pipe to a 12 in. [300 mm] pipe) is recovered in electrical cost in less than 18 months.

Using the formula above and carefully measuring existing projects, we can establish performance goals for our designs such as the ones in Table 10-1.

At times, reductions in pipe sizes to reduce first cost are suggested as *value engineering*. However, energy usage of the building may be greatly impacted: pump size and horsepower could well be increased. In order to be truly valid, *value engineering* should also include refiguring the life-cycle cost of owning and operating the building. These factors can also be applied to chilled-water systems, except that chilled-water systems have a smaller range of temperatures within which to work.

## System Volume

In small buildings, water system volume may relate closely to boiler or water-chiller operation. When pipe distribution systems are short and of small water volume, both boilers and water chillers may experience detrimental operating effects. Manufacturers of water chillers state that system water volume should be a minimum of 3 to 10 gal per installed ton of cooling (0.054 to 0.179 L/s per kWR). In a system less than this and under light cooling load conditions, thermal inertia coupled with the reaction time of chiller controls may cause the units to short-cycle or shut down on low-temperature safety control.

Similar detrimental effects may occur with small modular boilers in small systems. Under light load conditions, boilers may experience frequent short cycles of operation. An increase in system volume may eliminate this condition.

### Table 10-1    Example Performance Goals

| Component | Typical kW/ton (kW/kWR) | Efficient kW/ton (kW/kWR) | Delta | % Savings |
|---|---|---|---|---|
| Chiller | 0.62 (0.1763) | 0.485 (0.1379) | 0.135 (0.0384) | 22% |
| Cooling tower | 0.045 (0.0128) | 0.012 (0.0034) | 0.033 (0.0094) | 73% |
| Condenser-water pump | 0.0589 (0.0167) | 0.022 (0.0063) | 0.0369 (0.0104) | 63% |
| Chilled-water pump | 0.0765 (0.0218) | 0.026 (0.0074) | 0.0505 (0.0144) | 66% |
| Total water-side system | 0.8004 (0.2276) | 0.545 (0.155) | 0.2554 (0.0726) | 32% |

### Energy Usage

There are many opportunities to reduce energy usage in the design of hydronic systems:

- Reduction of flow rates by using larger supply-to-return temperature differences. For example, consider 30°F (16.7 K) delta-T on hot water systems and 14°F (7.8 K) or greater delta-T on chilled water systems. Reduction of pipe sizes could be considered to reduce first costs. However, to minimize energy usage, consider sizing systems at 3 fps (0.915 m/s).
- Reduction in pumping horsepower based on flow reduction and responsible sizing of piping. In today's electrical rate environment, cost of increases in pipe size can be offset with lower energy costs, providing investment rate of returns of over 25%.
- Utilization of reverse return concept to minimize pumping energy.
- Minimizing water-pressure drop across coils.
- Utilization of two-way valves, variable-speed pumping, and variable/low-flow heat generating and/or heat rejection equipment.
- Minimize the use of glycol and three-way valve applications except where necessary for freeze protection or other design and operating considerations.

### Sources of Further Information

Manufacturers of equipment—pumps, boilers, water chillers, control valves, and other devices—are valuable sources. There are also multiple Web sites that contain system design information. (Perform a search on the Internet for *hydronic piping design*.) Hydronic systems are also discussed in Chapters 12, 13, and 14 of the *ASHRAE Handbook—HVAC Systems and Equipment.*

## AIR

Using air as a means of energy distribution is almost universal in buildings, especially as a means of providing distributed cooling to spaces that need it. A key characteristic that makes air so widely used, however, is its importance in maintaining good IAQ. Thus, air distribution systems are not only a means for energy distribution, they serve the essential role of providing fresh or uncontaminated air to occupied spaces.

That being said, it is important to understand that properly distributing air to heat and cool spaces in a building is less efficient, from an energy usage perspective, than using hydronic or steam distribution. Air-distribution systems are often challenging to design because, for the energy carried per cross-sectional area, they take up the most space in a ceiling cavity and are frequent causes of space "conflicts" between disciplines (structural members, plumbing lines, heating/cooling pipes, light fixtures, etc.). Many approaches have been tried to better coordinate duct runs with other services—or even to integrate them in some cases.

Another tricky aspect of air system design is that there are temperature limitations on supply air due to the fact that air is the energy medium that directly impacts space occupants. While care must always be taken in how air is introduced into an occupied space, it is especially critical the colder the supply temperature gets. Low-temperature air supply systems offer many advantages in terms of green design, but an especially critical design aspect is avoiding occupant discomfort at the supply air/occupant interface.

Most of the same principles that were discussed under hydronic energy distribution systems apply to air systems with respect to temperature differences, carrier size, and driver power and energy. Thus, there are plenty of opportunities for applying green design techniques to such systems and for seeking innovative solutions.

### Energy Usage

There are many opportunities to reduce energy usage in the design of air-distribution systems:

* Size duct to minimize pressure drop
* Require SMACNA Seal Class "A" for all duct systems.
* VAV concepts can be used for many constant volume type systems.
* Minimize pressure drops of AHU components and duct-mounted accessories, such as sound attenuators, dampers, diffusers, etc.

### Sources of Further Information

Manufacturers of equipment—fans, ductwork, AHUs, dampers, and other air devices—are valuable sources. There are also multiple Web sites that contain system design information. Air systems are also discussed in Chapters 31 and 32 of the *ASHRAE Handbook—Fundamentals*, and in Chapters 16 through 25 of the *ASHRAE Handbook—HVAC Systems and Equipment.* Another source for information includes Sheet Metal and Air Conditioning Contractors' National Association, www.smacna.org.

## ELECTRIC

From the standpoint of space consumed, distribution of energy by electric means (wire, cables, etc.) is the most efficient. This advantage has often overcome the usual higher cost of electricity (per energy unit) as an energy form and has been one of the major reasons for electing to design all-electric buildings. The relative inefficiencies of electric resistance heating no longer make it a viable choice from an environmental or life-cycle cost perspective. However, when used in conjunction with a heat pump system, electrical energy as a source is efficient and less costly from an operating perspective.

## ASHRAE GreenTip #15:
## Variable-Flow/Variable-Speed Pumping Systems

### GENERAL DESCRIPTION

In most hydronic systems, variable flow with variable-speed pumping can be a significant source of energy savings. Variable flow is produced in chilled- and hot-water systems by using two-way control valves and in condenser-water cooling systems by using automatic two-position isolation valves interlocked with the chiller machinery's compressors. In most cases, variable flow alone can provide energy savings at a reduced first cost since two-way control valves cost significantly less to purchase and install than three-way valves. In condenser-water systems, even though two-way control valves may be an added first cost, they are still typically cost-effective, even for small (1 to 2 ton [3.5 to 7 kWR]) heat pump and air-conditioning units. (ASHRAE Standard 90.1-1999 requires isolation valves on water-loop heat pumps and some amount of variable flow on all hot-water and chilled-water systems.)

Variable-speed pumping can dramatically increase energy savings, particularly when it is combined with demand-based pressure reset controls. Variable-speed pumps are typically controlled to maintain the system pressure required to keep the most hydraulically remote valve completely open at design conditions. The key to getting the most savings is placement of the differential pressure transducer as close to that remote load as possible. If the system serves multiple hydraulic loops, multiple transducers can be placed at the end of each loop, with a high-signal selector used to transmit the signal to the pumps. With direct digital control (DDC) control systems, the pressure signal can be reset by demand and controlled to keep at least one valve at or near 100% open. If valve position is not available from the control system, a "trim-and-respond" algorithm can be employed.

Even with constant-speed pumping, variable-flow designs save some energy, as the fixed-speed pumps ride up on their impeller curves, using less energy at reduced flows. For hot-water systems, this is often the best life-cycle cost alternative, as the added pump heat will provide some beneficial value. For chilled-water systems, it is typically cost-effective to control pumps with variable-speed drives. It is very important to "right size" the pump and motor before applying a variable-speed drive in order to keep drive cost down and performance up.

### WHEN/WHERE IT'S APPLICABLE

Variable-flow design is applicable to chilled-water, hot-water, and condenser-water loops that serve water-cooled air-conditioning and heat-pump units. The limitations on each of these loop types are as follows:

• Chillers require a minimum flow through the evaporators. (Chiller manufacturers can specify flow ranges if requested.) Flow minimums on the evaporator side can be achieved via hydronic distribution system design using either a primary/secondary arrangement or primary-only variable flow with a bypass line and valve for minimum flow.

- Some boilers require minimum flows to protect the tubes. These vary greatly by boiler type. Flexible bent-water-tube and straight-water-tube boilers can take huge ranges of turn-down (close to zero flow). Fire-tube and copper-tube boilers, on the other hand, require a constant flow primary pump.

Variable-speed drives on pumps can be used on any variable-flow system. As described above, they should be controlled to maintain a minimum system pressure. That system pressure can be reset by valve demand on hot-water and chilled-water systems that have DDC control of the hydronic valves.

## PROS AND CONS

### Pro

1. Both variable-flow and variable-speed control save significant energy.
2. Variable-speed drives on pumps provide a "soft" start, extending equipment life.
3. Variable-speed drives and two-way valves are self-balancing.
4. Application of demand-based pressure reset significantly reduces pump energy and decreases the occurrence of system overpressurization, causing valves near the pumps to lift.
5. Variable-speed systems are quieter than constant-speed systems.

### Con

1. Variable-speed drives add cost to the system. (They may not be cost-effective on hot-water systems.)
2. Demand-based supply pressure reset can only be achieved with DDC of the heating/cooling valves.
3. Variable flow on condenser-water systems with open towers requires that supplementary measures be taken to keep the fill wet on the cooling towers. Cooling towers with rotating spray heads or wands can accept a wide variation in flow rates without causing dry spots in fill. Fitting the cooling tower with variable-speed fans can take advantage of lower flow rates (more "free area") to reduce fan energy while providing the same temperature of condenser water.

## KEY ELEMENTS OF COST

The following provides a possible breakdown of the various cost elements that might differentiate a variable-flow/variable-speed system from a conventional one and an indication of whether the net cost for the hybrid option is likely to be lower

(L), higher (H), or the same (S). This assessment is only a perception of what might be likely, but it obviously may not be correct in all situations. **There is no substitute for a detailed cost analysis as part of the design process.** The listings below may also provide some assistance in identifying the cost elements involved.

## First Cost

- Hydronic system terminal valves: two-way vs. three-way (applicable to hot-water and chilled-water systems)                L
- Bypass line with two-way valve or alternative means (if minimum chiller flow is required)                H
- Hydronic system isolation valves: two-position vs. none (applicable to condenser-water systems)                H
- Cooling tower wet-fill modifications (condenser-water systems)                H
- Variable-speed drives and associated controls                H
- DDC system (may need to allow demand-based reset) or pressure transducers                H
- Design fees                H

## Recurring Cost

- Pumping energy                L
- Testing and balancing (TAB) of hydronic system                L
- Maintenance                H
- Commissioning                H

## SOURCES OF FURTHER INFORMATION

CEC. 2002. *Part II: Measure Analysis and Life-Cycle Cost 2005, California Building Energy Standards*, P400-02-012. California Energy Commission, Sacramento, CA, May 16, 2002.

Taylor, S., P. Dupont, M. Hydeman, B. Jones, and T. Hartman. 1999. *The Cool-Tools Chilled Water Plant Design and Performance Specification Guide.* San Francisco, CA: PG&E Pacific Energy Center.

Taylor, S.T. 2002. Primary-only vs. primary-secondary variable flow chilled water systems. *ASHRAE Journal* 44(2):25–29.

Taylor, S.T., and J. Stein. 2002. Balancing variable flow hydronic systems. *ASHRAE Journal* 44(10):17–24.

# 11

# ENERGY CONVERSION SYSTEMS

## HEAT GENERATORS (HEATING PLANTS)

Considerable improvements in the seasonal efficiency of conventional heating plant equipment, such as boilers and furnaces, were made over the last several decades. Designers should verify claims of equipment manufacturers by reviewing documented data of this equipment demonstrating the efficiency ratings.

Some unconventional equipment and techniques to achieve greater efficiency or other possible green building design goals are described in the GreenTips at the end of this chapter.

## COOLING GENERATORS (CHILLED-WATER PLANTS)

Chilled-water plants are most often used in large facilities. Their benefits include higher efficiency and reduced maintenance costs in comparison to decentralized plants.

Generally, a chilled-water plant consists of:

- Chillers
- Chilled-water pumps
- Condenser-water pumps (for water-cooled systems)
- Cooling towers (for water-cooled systems) or air-cooled condensers
- Associated piping, connections, and valves

Because chilled-water temperature can be closely controlled, chilled-water plants have an advantage over direct-expansion systems because they allow air temperatures to be closely controlled also.

Often chilled-water plant equipment will be in a single central location, allowing system control, maintenance, and problem diagnostics to be performed efficiently. Chilled-water plants also allow redundancy to be easily designed into the system by adding one (or more) extra chillers, chilled-water pumps, condenser-water pumps, and cooling tower fans.

## Chiller Types

Electric chillers used within chiller plants employ either a scroll-, reciprocating-, screw-, or centrifugal-type compressor (in order of increasing size). Models can be offered with both air-cooled or water-cooled condensers, with the exception that centrifugal compressors today are water-cooled only. Absorption chillers, powered by steam, hot water, natural gas, or other hot gases, are used in some plants to offset high electric demand or consumption charges.

The various chiller types are used most often, though not exclusively, in the following situations.

*Electric chillers:*
• Where low to moderate electric consumption and demand charges prevail
• Where air-cooled heat dissipation is preferred
• Where condenser heat recovery is desired

*Absorption chillers:*
• Where low fuel (e.g., natural gas) costs prevail
• Where high electric demand charges prevail
• Where there is a plentiful source of heat available, its main use usually being for other functions

**Heat Pumps.** A heat pump is another means of generating cooling as well as heating using the same piece of equipment. There is usually an array of them used for a project, and they are generally distributed throughout the building (i.e., not part of a central plant). See the GreenTips at the end of this chapter on various heat-pump system types.

**Thermal Energy Storage (TES).** TES is a technique that has been encouraged by electricity pricing schedules where the off-peak rate is considerably lower than the on-peak rate. Cooling, in the form of chilled water or ice, is generated during off-peak hours and stored for use on-peak. Although not refrigeration equipment per se, the technique can usually reduce the size of refrigeration equipment—or obviate the need for adding a chiller to an existing plant.

The characteristics, merits, and cost factors of TES for cooling, as well as numerous reference sources, are presented in GreenTip #22.

## System Design Considerations

When a designer puts together a chilled-water plant, there are many design parameters to optimize. They include fluid flow rates and temperatures, pumping options, plant configuration, and control methods. For each specific application, the design professional should understand the client's needs and desires and implement the chiller plant options that best satisfy them.

**Fluid Flow Rates.** Flow rates were discussed in the "Media Movers" section of Chapter 8.

**Fluid Temperatures.** To allow the aforementioned lower flow rates, the chiller must be able to supply colder chilled-water temperatures and to tolerate higher condenser leaving temperatures.

**Pumping Options.** Pumps may be selected to operate with a specific chiller or they may be manifolded.

Pump-per-chiller arrangement advantages include:
*   Hydronic simplicity
*   Chiller and pump are controlled together
*   Pumps and chillers may be sized for one another

Manifolded-pump arrangement advantages include:
*   Simpler redundancy
*   Pumps may be centrally located

**Plant Configuration (Multiple Chillers).** Today, the most prevalent chiller plant configuration is the *primary-secondary* (decoupled) system. This system allows the flow rate through each chiller to be constant, yet it accommodates a reduction in pumping energy since the system water flow rate varies with the load.

Becoming more common are *variable primary flow* systems that also vary the flow through the chiller evaporators. New chiller controls allow this. Often these systems can be installed at a reduced cost when compared to primary-secondary systems since fewer pumps (and their attendant piping, connections, valves, fittings, and electrical draws) are required. These systems also save energy in comparison to the primary-secondary configuration.

(The subjects of plant configuration, pumping options, and control methods are discussed in detail in ASHRAE's *Fundamentals of Water System Design*, Trane's *Multiple-Chiller-System Design and Control Manual*, and Pacific Gas and Electric's *Chilled Water Plant Design Guide*.)

The designer should always review the overall use of energy within a facility and employ systems (including heat recovery systems) that interact with one another so as to minimize the overall energy consumption of the entire chilled-water plant.

## DISTRIBUTED ELECTRICITY GENERATION

One of the major opportunities for energy conservation in buildings is the use of on-site generation systems to provide both distributed electric power and thermal energy (otherwise wasted heat from the generation process), which can be used to meet the thermal loads of the building.

Distributed generation (DG) provides electricity directly to the building's electrical systems to offset loads that would otherwise have to be met by the utility grid (see Figure 11-1). The waste heat from that generation process goes through a heat recovery mechanism where it may be provided as heat to meet loads for conventional heating (such as space heating, reheat, domestic water heating) or for specialized processes. Alternatively, that heat energy, if at a sufficiently high temperature, can be used to power an absorption chiller to produce chilled water to meet either space cooling or process cooling loads. This is shown in Figure 11-2.

Any timing differences between the generation of the waste heat from the DG system and the thermal needs of the building can be handled utilizing a chilled-water TES system. This concept may also permit downsizing the absorption cooling system (so it does not need to be sized for the peak cooling load).

The overall usable energy value from the fuel input to the generation process is only about 30% (or less) if there is no waste heat recovery, but it can be over 70% if most or all the waste heat is able to be utilized. This increased system efficiency can have a radical impact on the economics of the energy systems because almost two-and-one-half times the useful value is obtained from the fossil fuel purchased. System sizing is generally done by evaluating the relative electrical and thermal

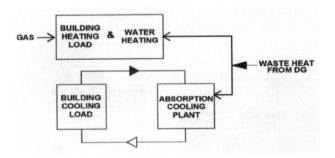

**Figure 11-1 Thermal energy storage and waste heat usage.**

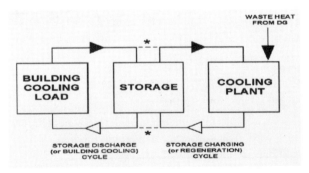

**Figure 11-2 Thermal uses of waste heat.**

loads over the course of the typical operating cycle and then selecting the system capacity to meet the *lesser* of the thermal or electrical loads. See Figure 11-3 showing comparative thermal and electrical energy for typical office buildings.

If it is sized for the greater of the two, then there will result a net waste of energy produced since it is seldom economical to sell electricity back to the grid. A key design issue that arises here is whether or not the system is being designed to improve efficiency, as discussed above, or as a baseload on-site generation system for purposes of improving the reliability of the electric and/or thermal energy supply. Either of these is a legitimate design criterion, although the goal, from a green design standpoint, almost always focuses on the energy-efficient strategy.

## DG Technologies

Technologies that can be used for this type of generation include engine-driven generators, micro-turbine-driven generators, or fuel cells. Typically, each uses natural gas as the input fuel. There are advantages and disadvantages associated with these technologies, which are summarized briefly in Table 11-1.

**Engine-Driven Generator (EDG).** This technology has been around the longest of the three and is in many ways the least expensive option. It produces a relatively high temperature of waste heat that can be more effectively utilized by the heat recovery systems. Disadvantages, however, include air pollution, acoustical impacts, and noise/vibration from the engines. EDG sets are available in sizes ranging from approximately 40 kW to several thousand kW.

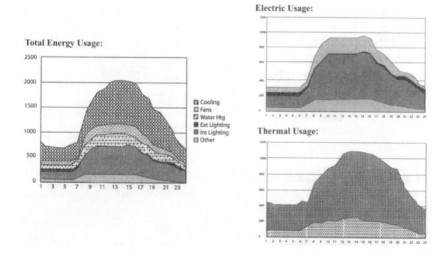

**Figure 11-3 Relationship of electric and thermal energy.**

### Table 11-1    Comparison of Power Generation Options

| Generation Option | Efficiency (%) | Typical Size | Installed Costs ($/kW) | O&M Costs ($/kWh) | Life-Cycle Costs (LCC assumes 20-year life cycle) ($/kWh) |
|---|---|---|---|---|---|
| Wind turbine | NA | 5 kW– 600 kW | $750– $1,250 | $0.001– $0.007 | $0.036 |
| Hybrid fuel cell— Gas turbine with heat recovery* | 56%–80% | 5 kW– 3 kW | $1,300– $1,500 | $0.005– $0.010 | $0.048 |
| Natural gas cogen on-site with heat recovery | 40%–48% | 30 kW– 300 kW | $600– $1,000 | $0.003– $0.010 | $0.052 |
| Natural gas power plant (combustion turbine) | 28%–45% | 500 kW– 150 kW | $600–$900 | $0.003– $0.008 | $0.052 |
| Fuel cell without heat recovery | 30%–60% | 5 kW– 3 kW | $1,900– $3,500 | $0.005– $0.010 | $0.072 |
| Natural gas cogen on-site without heat recovery | 20%–28% | 30 kW– 300 kW | $600– $1,000 | $0.003– $0.010 | $0.072 |
| Photovoltaic (PV) | 8%–13% | 1 kW– 50 kW | $3,500– $6,000 | $0.003– $0.005 | $0.166 |

Note: Life-cycle costs assume the same cost for gas as for all technologies. Central technologies will likely have lower rates. * Hybrid fuel cell microturbine plant is still in testing stage of development, not commercially available, and costs are projected costs for 2004. Does not include balance of system components, such as the inverter (alternating current to direct current power), wiring, and controls.

**Micro-Turbine Generator.** At the moment micro-turbine generators are somewhat more expensive in first cost than EDGs, but they have less air pollution and less severe acoustic and vibration impacts than EDGs. They also have longer operating lives and a projected lower cost. However, at the moment they are only available in sizes less than 100 kW per unit as compared to EDGs, which can be up in the several-hundreds-per-unit range.

**Fuel Cells.** Fuel cells are the most advanced form of power generation in terms of being a "clean and green" technology. They generate virtually no air pollution, have minimal acoustic and vibration impacts, and are considered the "wave of the future." At this point (early in the 2000s decade), however, the cost of fuel cell equipment is still so high as to make it the least attractive of these options from an economic standpoint. It is anticipated that this will change as continued development of the technology evolves over the next several years.

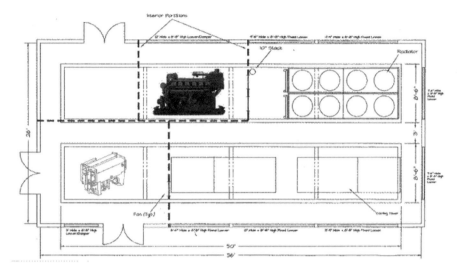

**Figure 11-4 Schematic of actual CHP system consisting of a reciprocating engine and an indirect-fired absorption chiller, along with other ancillary equipment (Patnaik 2004).**

## Combined Cooling, Heating, and Power (CCHP) Systems

Combined heating and power is also referred to as *cogeneration*, and with cooling, *trigeneration*. The larger, industrial-scale versions of such integrated energy systems have been in use through most of the 20th century. The design of such systems does not, however, follow an integrated approach, and the various pieces of equipment are brought together from standard manufacturer offerings. More recently, with the focus shifting to distributed power generation for the reasons cited above, as well as the California energy crisis of 2000, a new scale of CCHP has emerged, involving more integrated design of the components. These CCHP systems typically consist of one of the DG technologies described above, e.g., reciprocating engines or microturbines (possibly fuel cells, in the future) fired by natural gas, a heat exchanger to recover heat from the exhaust stream and/or jacket water, and, if there is a cooling demand, an absorption chiller. An intermediate medium such as hot water can be used to transfer heat between the exhaust stream of the prime-mover and the chiller (Figure 11-4), but lately a lot of development has gone into a dedicated heat recovery unit that also forms the generator of the absorption chiller (Rosfjord et al. 2004). Such integrated energy systems are admittedly in their infancy in the US, but there is a significant drive by organizations such as the DOE toward their commercialization as packaged systems.

The traditional application of CCHP has been to directly utilize the waste heat for space heating and/or domestic hot-water heating. More recently, with the maturation of absorption technology and advancement of related control systems, space or process cooling has been added to the mix by contractors and manufacturers. The former is put together "off-the-shelf" equipment, while the latter follows an integrated design approach with a better opportunity at optimization.

To illustrate the benefits of the DG-based CCHP systems, the following elementary view of the energy efficiencies, both at the component and system level, is offered. A typical reciprocating gas-fired engine has a thermal efficiency of about 35%; 65% of the energy in the fuel is ordinarily not utilized and wasted through the stack. The exhaust temperatures leaving the engine ($>500°F$) are typically high enough to drive, at the very least, a single-effect absorption chiller with a COP in the vicinity of 0.7. Factoring in the actual amount of available heat in the exhaust stream (temperature/enthalpy and flow rate), this translates to overall fuel utilization rates/efficiencies as high as 80%. The ratio of electrical to thermal load carrying capability shifts a bit when the prime-mover is a microturbine or a fuel cell, typical generation efficiencies being 25% (based on higher heating value) and 40%, respectively. However, the overall fuel utilization rate is relatively unchanged. The waste heat can also be directly utilized for desiccant dehumidification (regeneration).

What the integrated energy systems have done then is

- brought the power generation closer to the point of application/load ("distributed generation"), eliminating transmission and distribution losses, etc.;
- removed or reduced the normal electric and primary fuel consumption by independent pieces of equipment providing cooling, heating, and/or dehumidification (e.g., separate electric chiller and boiler), thereby improving substantially overall fuel utilization rates inclusive of the power generation process; and
- removed or reduced emissions of $CO_2$ and other combustion by-products associated with the operation of the cooling, heating, and/or dehumidification equipment.

Challenges include the matching of electrical and thermal loads, given the diversity of energy usage patterns in buildings. The consultant/contractor can determine if the heating/cooling components of the CCHP system will play a primary or complementary role (the latter involving other, more conventional equipment to fill in the missing thermal load). Numerous studies were done in this regard, noteworthy among which is "CHP: The Concept," by William A. Ryan of the Midwest CHP Application Center, Energy Resources Center, University of Illinois, Chicago (www.chpcentermw.org/presentations/WI-Focus-on-Energy-Presentation-05212003.pdf).

Further resources are given in GreenTip #16.

## REFERENCES

Patnaik, V. 2004. Experimental verification of an absorption chiller for BCHP applications. *ASHRAE Transactions* 110(1):503–507.

Rosfjord, T., T. Wagner, and B. Knight. 2004. UTC microturbine CHP product development and launch. Presented at the 2004 DOE/CETC Annual Workshop on Microturbine Applications.

## SUMMARY

All the pros and cons of these generation technologies must be taken into account in determining the most appropriate for a specific project. The key point, from a green design perspective, is that any of them are substantially "greener" than utilizing utility grid power for the simple reason that they afford the opportunity to use the waste heat from the electrical generation process.

## ASHRAE GreenTip #16:
## CHP Systems

### GENERAL DESCRIPTION

*CHP* stands for combined heating and power. Other abbreviations that have been used to describe such integrated energy systems are CCHP (includes *cooling*) and BCHP (building cooling, heating, and power). The goal, regardless of the abbreviation, is to improve system efficiencies or source fuel utilization by availing of the low-grade heat that is a by-product of the power generation process for heating and/or cooling duty. Fuel utilization efficiencies as high as 80% were reported (Adamson 2002). The resulting savings in operating cost, relative to a conventional system, are then viewed against the first cost, and simple paybacks under four years have been anticipated (LeMar 2002). This is particularly important from a marketing perspective, for both the distributed-generation and the thermal equipment provider. This is because, by themselves, a microturbine manufacturer and an absorption chiller manufacturer, for example, would find it difficult to compete with a utility and an electric chiller manufacturer, respectively, as the provider of low-cost power and cooling. Last, but by no means least, the higher (fossil-)fuel utilization rates result in reduced emissions of $CO_2$, the greenhouse gas with over 55% contribution to global warming (Houghton et al. 1990).

Gas engines, microturbines, and fuel cells have been at the center of CHP activity as the need for reliable power and/or grid independence has recently become evident. These devices are also being promoted to reduce the need for additional central-station peaking power plants. As would be expected, however, they come at a first cost premium, which can range from $1000–$4000/kW (Ellis and Gunes 2002). At the same time, operating (thermal) efficiencies have remained in the vicinity of those of the large, centralized power plants, even after the transmission and distribution losses are taken into account. This is particularly true of engines and microturbines (25%–35% thermal efficiency), while fuel cells promise higher efficiencies (~50%), albeit at the higher cost premiums ($3000–$4000/kW).

On the thermal side, standard gas-to-liquid or gas-to-gas heat exchanger equipment can be used for the heating component of the CHP system. This transfers the heat from the exhaust gases to the process/hydronic fluid or air, respectively. For the cooling component, the size ranges of distributed power generators offer a unique advantage in terms of flexibility in the selection of the chiller equipment. These can be smaller-end (relative to commercial) water-lithium bromide single- or double-effect absorption chillers or larger-end (relative to residential) single-effect or GAX ammonia-water absorption chillers (Erickson and Rane 1994). Such chillers have a typical COP of 0.7 and, as a rule of thumb, for thermal-to-electrical load matching, for every 4 kW of power generated, 1 ton of cooling may be achieved (Patnaik 2004).

Figure 11-5 illustrates typical operating conditions that an absorption chiller would see with a reciprocating engine.

## WHEN/WHERE IT'S APPLICABLE

CHP is particularly suited for applications involving distributed power generation. Buildings requiring their own power generation, either due to a stringent power reliability and/or quality requirement or remoteness of location, must also satisfy various thermal loads (space heating or cooling, water heating, dehumidification). A conventional fossil-fuel-fired boiler and/or electric chiller can be displaced, to some extent if not entirely, by a heat recovery device (standard heat exchanger) and/or an absorption chiller driven by the waste heat from the power generator. Since the source of heating and/or cooling is waste heat that would ordinarily have been rejected to the surroundings, the operating cost of meeting the thermal demand of the building is significantly mitigated if not eliminated.

Economic analyses suggest that CHP systems are ideally suited for base-loaded distributed power generation and steady thermal (heating and/or cooling) loading. This is also the desirable mode of operation for the absorption chiller. Peak-loading is then met by utility power. Alternatively, if utility power is used for base-loading

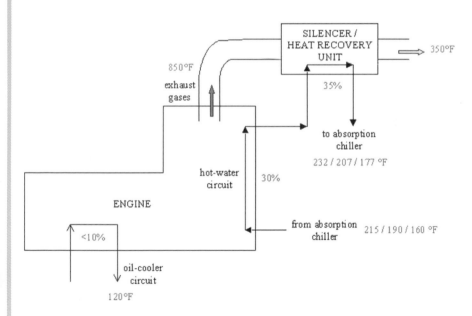

**Figure 11-5 Schematic of CHP system consisting of a gas-fired reciprocating engine showing typical operating temperatures (Patnaik 2004).**

and the DG meets the peaking demand, the thermal availability may be intermittent and require frequent cycling of the primary thermal equipment (boiler and/or chiller).

## PROS AND CONS

### Pro

- One of the primary advantages of CHP systems is the reduction in centralized (utility) peak-load generating capacity. This is especially true since one of the biggest contributors to summer peak loads is the demand for air conditioning. If some of this air-conditioning demand can be met by chillers fired by essentially free energy (waste heat), there is a double benefit.
- Additionally, DG-based CHP systems enabled the following:
  - Brought the power generation closer to the point of application/load ("distributed generation"), eliminating transmission and distribution losses, etc.
  - Removed or reduced the normal electric and primary fuel consumption by independent pieces of equipment providing cooling, heating, and/or dehumidification (e.g., separate electric chiller and boiler), thereby improving substantially overall fuel utilization rates inclusive of the power generation process.
  - Removed or reduced emissions of $CO_2$ and other combustion by-products associated with the operation of the cooling, heating, and/or dehumidification equipment.

### Con

- If the CHP system is to entirely replace a conventional boiler/electric chiller system, the ratio of electrical to thermal load of the building must match closely the relative performance (efficiencies) of the respective equipment.
- Start-up times for absorption chillers are relatively longer than those for vapor-compression chillers, particularly when coupled to microturbines, which themselves have large time constants. Such systems would require robust and sophisticated controls that take these transients into account.

## KEY ELEMENTS OF COST

The following provides a possible breakdown of the various cost elements that might differentiate a CHP system from a conventional one and an indication of whether the net cost for the alternative system is likely to be lower (L), higher (H), or the same (S). This assessment is only a perception of what might be likely, but it obviously may not be correct in all situations. **There is no substitute for a detailed cost analysis as part of the design process.** The listings below may also provide some assistance in identifying the cost elements involved.

First Cost

- Distributed power generator                              H
- Heat recovery device/heat exchanger                      L
- Absorption chiller                                       H
- Integrating control system                               H

Recurring Cost

- Distributed power generator (engine/microturbine/fuel cell)   S/H/L
- Heat recovery device/heat exchanger                      None
- Absorption chiller                                       None
- Integrating control system                               H

## SOURCES OF FURTHER INFORMATION

In keeping with the spirit of cross-cutting themes being promoted by ASHRAE, a number of technical sessions in recent meetings have been devoted to CHP, generally sponsored by technical committees on cogeneration systems (TC 1.10) and absorption/sorption heat pumps and refrigeration systems (TC 8.3). Presentations from these should be available on the ASHRAE Web site. The following is a link to recently sponsored programs by TC 8.3, including a couple of viewable presentations: http://tc83.ashraetcs.org/programs.html.

## REFERENCES

Adamson, R. 2002. Mariah Heat Plus Power[TM] Packaged CHP, Applications and Economics. Presented at the 2nd Annual DOE/CETC/CANDRA Workshop on Microturbine Applications at the University of Maryland, College Park, January.

Ellis, M.W., and M.B. Gunes. 2002. Status of fuel cell systems for combined heat and power applications in buildings. *ASHRAE Transactions* 108(1):1032–44.

Erickson, D.C., and M. Rane. 1994. Advanced absorption cycle: Vapor exchange GAX. *Proceedings of the ASME International Absorption Heat Pump Conference, New Orleans, LA, Jan. 19–21.*

Houghton, J.T., G.J. Jenkins, and J.J. Ephraums, eds. 1990. *Climate Change: The IPCC Scientific Assessment.* Intergovernmental Panel on Climate Change. New York: Cambridge UP.

LeMar, P. 2002. Integrated energy systems (IES) for buildings: A market assessment. Final report by Resource Dynamics Corporation for Oak Ridge National Laboratory, Contract No. DE-AC05-00OR22725, September 2002.

Patnaik, V. 2004. Experimental verification of an absorption chiller for BCHP applications. *ASHRAE Transactions* 110(1):503–507.

## ASHRAE GreenTip #17:
## Low-NO$_x$ Burners

### GENERAL DESCRIPTION

Low-NO$_x$ burners are natural gas burners with improved energy efficiency and lower emissions of nitrous oxides (NO$_x$).

When fossil fuels are burned, nitric oxide and nitrogen dioxide are produced. These pollutants initiate reactions that result in the production of ozone and acid rain. The NO$_x$ come from two sources: high-temperature combustion (thermal NO$_x$) and nitrogen bound to the fuel (fuel NO$_x$). For clean-burning fuels such as natural gas, fuel NO$_x$ generation is insignificant.

In most cases, NO$_x$ levels are reduced by lowering flame temperature. This can be accomplished by modifying the burner to create a larger (and therefore lower temperature) flame, injecting water or steam into the flame, recirculating flue gases, or limiting the excess air in the combustion process. In many cases a combination of these approaches is used. In general, reducing the flame temperature will reduce the overall efficiency of the boiler. However, recirculating flue gases and controlling the air-fuel mixture can improve boiler efficiency, so that a combination of techniques may improve total boiler efficiency.

Natural-gas-fired burners with lowered NO$_x$ emissions are available for commercial and residential heating applications. One commercial/residential boiler has a burner with inserts above the individual burners; this design reduces NO$_x$ emissions by 30%. The boiler also has a "wet base" heat exchanger to capture more of the burner heat and reduce heat loss to flooring.

NO$_x$ production is of special concern in industrial high-temperature processes because thermal NO$_x$ production increases with temperature. These processes include metal processing, glass manufacturing, pulp and paper mills, and cement kilns. Although natural gas is the cleanest-burning fossil fuel, natural gas can produce NO$_x$ emissions as high as 100 ppm or more.

A burner developed by MIT and the Gas Research Institute combines staged introduction combustion air, flue gas recirculation, and integral reburning to control NO$_x$ emissions. These improvements in burner design result in a low-temperature, fuel-rich primary zone, followed by a low-temperature, lean secondary zone; these low temperatures result in lower NO$_x$ formation. In addition, any NO$_x$ emission present in the recirculated flue gas is reburned, further reducing emissions. A jet pump recirculates a large volume of flue gas to the burner; this reduces NO$_x$ emissions and improves heat transfer.

The low-NO$_x$ burner used for commercial and residential space heating is larger in size than conventional burners, although it is designed for ease of installation.

## WHEN/WHERE IT'S APPLICABLE

Low $NO_x$ burners are best applied in regions where air quality is affected by high ground-level ozone and where required by law.

## PROS AND CONS

### Pro

1. Lowers $NO_x$ and CO emissions, where that is an issue.
2. Increases energy efficiency.

### Con

1. High cost.
2. Higher maintenance.

## KEY ELEMENTS OF COST

The following provides a possible breakdown of the various cost elements that might differentiate a low-$NO_x$ system from a conventional one and an indication of whether the net cost for the alternative system is likely to be lower (L), higher (H), or the same (S). This assessment is only a perception of what might be likely, but it obviously may not be correct in all situations. **There is no substitute for a detailed cost analysis as part of the design process.** The listings below may also provide some assistance in identifying the cost elements involved.

### First Cost

| | |
|---|---|
| • Conventional burner | L |
| • Low $NO_x$ burner | H |

### Recurring Cost

| | |
|---|---|
| • Maintenance | H |
| • Possible avoidance of pollution fines | L |

## SOURCE OF FURTHER INFORMATION

American Gas Association, www.aga.org.

## ASHRAE GreenTip #18:
## Combustion Air Preheating

### GENERAL DESCRIPTION

For fuel-fired heating equipment, one of the most potent ways to improve efficiency and productivity is to preheat the combustion air going to the burners. The source of this heat energy is the exhaust gas stream, which leaves the process at elevated temperatures. A heat exchanger, placed in the exhaust stack or ductwork, can extract a large portion of the thermal energy in the flue gases and transfer it to the incoming combustion air.

With natural gas, it is estimated that for each 50°F the combustion air is preheated, overall boiler efficiency increases by approximately 1%. This provides a high leverage boiler plant efficiency measure because increasing boiler efficiency also decreases boiler fuel usage. And, since combustion airflow decreases along with fuel flow, there is a reduction in fan-power usage as well.

There are two types of air preheaters: recuperators and regenerators. Recuperators are gas-to-gas heat exchangers placed on the furnace stack. Internal tubes or plates transfer heat from the outgoing exhaust gas to the incoming combustion air while keeping the two streams from mixing. Regenerators include two or more separate heat storage sections. Flue gases and combustion air take turns flowing through each regenerator, alternatively heating the storage medium and then withdrawing heat from it. For uninterrupted operation, at least two regenerators and their associated burners are required: one regenerator is needed to fire the furnace while the other is recharging.

### WHEN/WHERE IT'S APPLICABLE

While theoretically any boiler can use combustion preheating, flue temperature is customarily used as a rough indication of when it will be cost-effective. However, boilers or processes with low flue temperatures but a high exhaust gas flow may still be good candidates and must be evaluated on a case-by-case basis. Financial justification is based on energy saved rather than on temperature differential. Some processes produce dirty or corrosive exhaust gases that can plug or attack an exchanger, so material selection is critical.

### PROS AND CONS

#### Pro

1.  Lowers energy costs.
2.  Increasing thermal efficiency lowers $CO_2$ emissions.

Con

1. There are additional material and equipment costs.
2. Corrosion and condensation can add to maintenance costs.
3. Low specific heat of air results in relatively low U-factors and less economical heat exchangers.
4. Increasing combustion temperature also increases $NO_x$ emissions.

KEY ELEMENTS OF COST

The following provides a possible breakdown of the various cost elements that might differentiate a building with a combustion preheat system from one without and an indication of whether the net cost is likely to be lower (L), higher (H), or the same (S). This assessment is only a perception of what might be likely, but it obviously may not be correct in all situations. **There is no substitute for a detailed cost analysis as part of the design process.** The listings below may also provide some assistance in identifying the cost elements involved.

First Cost

| | |
|---|---|
| • Equipment costs | H |
| • Controls | S |
| • Design fees | H |

Recurring Cost

| | |
|---|---|
| • Overall energy cost | L |
| • Maintenance of system | H |
| • Training of building operators | H |

SOURCES OF FURTHER INFORMATION

DOE. 2002. *Energy Tip Sheet #1*, May. Office of Industrial Technologies, Energy Efficiency and Renewable Energy, US Department of Energy.

Fiorino, D.P. 2000. Six conservation and efficiency measures reducing steam costs. *ASHRAE Journal* 42(2):31–39.

## ASHRAE GreenTip #19:
## Combination Space/Water Heaters

### GENERAL DESCRIPTION

Combination space and water heating systems consist of a storage water heater, a heat delivery system (for example, a fan coil or hydronic baseboards), and associated pumps and controls. Typically gas-fired, they provide both space and domestic water heating. The water heater is installed and operated as a conventional water heater. When there is a demand for domestic hot water, cold city water enters the bottom of the tank, and hot water from the top of the tank is delivered to the load. When there is a demand for space heating, a pump circulates water from the top of the tank through fan coils or hydronic baseboards.

The storage tank is maintained at the desired temperature for domestic hot water (e.g., 140°F [60°C]). Because this temperature is cooler than conventional hydronic systems, the space heating delivery system needs to be slightly larger than typical. Alternatively, the storage tank can be operated at a higher water temperature; this requires tempering valves to prevent scalding at the taps.

The water heater can be either a conventional storage type water heater (either naturally venting or power vented) or a recuperative (condensing) gas boiler. Conventional water heaters have an efficiency of approximately 60%. By adding the space heating load, the energy factor increases because of longer runtimes and reduced standby losses on a percentage basis. Recuperative boilers can have efficiencies approaching 90%.

### WHEN/WHERE IT'S APPLICABLE

These units are best suited to buildings that have similar space and water heating loads, including dormitories, apartments, and condos. They are suited to all climate types.

### PROS AND CONS

Pro

1.   Reduces floor space requirements.
2.   Lowers capital cost.
3.   Improves energy efficiency.
4.   Increases tank life.

Con

1.   They are only available in small sizes.
2.   All space heating piping has to be designed for potable water.

3. No ferrous metals or lead-based solder can be used.
4. All components must be able to withstand prevailing city water pressures.

## KEY ELEMENTS OF COST

The following provides a possible breakdown of the various cost elements that might differentiate a combination space and water heating system from a conventional one and an indication of whether the net cost for the hybrid option is likely to be lower (L), higher (H), or the same (S). This assessment is only a perception of what might be likely, but it obviously may not be correct in all situations. **There is no substitute for a detailed cost analysis as part of the design process.** The listings below may also provide some assistance in identifying the cost elements involved.

### First Cost

- Conventional heating equipment                                L
- Combination space/domestic water heater                        H
- Sanitizing/inspecting space heating system                     H
- Piping and components able to withstand higher pressures        H
- Floor space used                                               L

### Recurring Cost

- Heating energy                                                 L
- Maintenance                                                    L

## SOURCES OF FURTHER INFORMATION

EERE. *A Consumer's Guide to Energy Efficiency and Renewable Energy*, EERE, US DOE, www.eere.energy.gov/consumerinfo/refbriefs/ad6.html.

Sustainable Sources, www.greenbuilder.com.

UG. *Combo Heating Systems: A Design Guide*. Union Gas, Chatham ON, CAN N7M5M1.

A ground-source heat pump (GSHP) extracts solar heat stored in the upper layers of the earth; the heat is then delivered to a building. Conversely, in the summer season, the heat pump rejects heat removed from the building into the ground rather than into the atmosphere or a body of water.

GSHPs can reduce the energy required for space heating, cooling, and service-water heating in commercial/institutional buildings by as much as 50%. GSHPs replace the need for a boiler in winter by utilizing heat stored in the ground; this heat is upgraded by a vapor-compressor refrigeration cycle. In summer, heat from a building is rejected to the ground. This eliminates the need for a cooling tower or heat rejector and also lowers operating costs because the ground is cooler than the outdoor air. (See Figure 11-6 for an example GSHP system.)

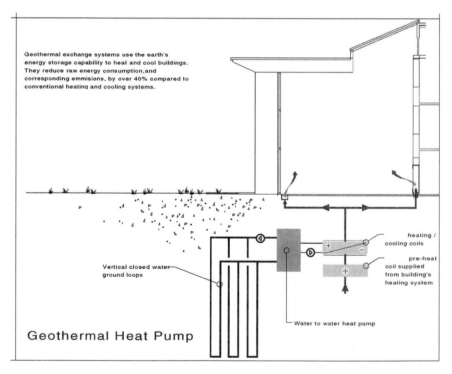

**Figure 11-6 Schematic example of GSHP closed-loop system.**

There are numerous types of GSHP loop systems. Each has its advantages and disadvantages. Visit the Geoexchange Geothermal Heat Pump Consortium Web site (www.geoexchange.org/about/how.htm) for a more detailed description of the loop options.

Water-to-air heat pumps are typically installed throughout a building with duct-work serving only the immediate zone; a two-pipe water distribution system conveys water to and from the ground-source heat exchanger. The heat exchanger field consists of a grid of vertical boreholes with plastic U-tube heat exchangers connected in parallel.

Simultaneous heating and cooling can occur throughout the building, as individual heat pumps, controlled by zone thermostats, can operate in heating or cooling mode as required.

Unlike conventional boiler/cooling tower-type water-loop heat pumps, the heat pumps used in GSHP applications are generally designed to operate at lower inlet-water temperature. GSHP are also more efficient than conventional heat pumps, with higher COPs and EERs. Because there are lower water temperatures in the two-pipe loop, piping needs to be insulated to prevent sweating; in addition, a larger circulation pump is needed because the units are slightly larger in the perimeter zones requiring larger flows.

GSHPs reduce energy use and, hence, atmospheric emissions. Conventional boilers and their associated emissions are eliminated since no supplementary form of energy is usually required. Typically, single packaged heat pump units have no field refrigerant connections and, thus, have significantly lower refrigerant leakage compared to central chiller systems.

GSHP units have life spans of 20 years or more. The two-pipe water-loop system typically used allows for unit placement changes to accommodate new tenants or changes in building use. The plastic piping used in the heat exchanger should last as long as the building itself.

When the system is disassembled, attention must be given to the removal and recycling of the HCFC or HFC refrigerants used in the heat pumps themselves and the antifreeze solution typically used in the ground heat exchanger.

## WHEN/WHERE IT'S APPLICABLE

The most economical application of GSHPs is in buildings that require significant space/water heating and cooling over extended hours of operation. Examples are retirement communities, multi-family complexes, large office buildings, retail shopping malls, and schools. Building types not well suited to the technology are buildings where space and water heating loads are relatively small or where hours of use are limited.

## PROS AND CONS

### Pro

1. Requires less mechanical room space.
2. Requires less outdoor equipment.
3. Does not require roof penetrations, maintenance decks, or architectural blends.
4. Has quiet operation.
5. Reduces operation and maintenance costs.
6. Requires simple controls only.
7. Requires less space in ceilings.
8. Loop piping, carrying low-temperature water, does not have to be insulated.
9. Installation costs are lower than for many central HVAC systems.

### Con

1. Requires surface area for heat exchanger field.
2. Higher initial cost overall.
3. Requires additional site coordination and supervision.

## KEY ELEMENTS OF COST

The following provides a possible breakdown of the various cost elements that might differentiate a GSHP system from a conventional one and an indication of whether the net cost for it is likely to be lower (L), higher (H), or the same (S). This assessment is only a perception of what might be likely, but it obviously may not be correct in all situations. **There is no substitute for a detailed cost analysis as part of the design process.** The listings below may also provide some assistance in identifying the cost elements involved.

### First Cost

- Conventional heating/cooling generators    L
- Heat pumps                                  H
- Outside piping system                       H
- Heat exchanger field                        H
- Operator training                           H
- Design fees                                 H

### Recurring Cost

- Energy cost (fossil fuel for conventional)  L
- Energy cost (electricity for heat pumps)    H
- Maintenance                                 L

SOURCES OF FURTHER INFORMATION

ASHRAE. 1995. *Commercial/Institutional Ground-Source Heat Pump Engineering Manual.* Atlanta: American Society of Heating, Refrigerating and Air-Conditioning Engineers, Inc.

Canadian Earth Energy Association, Ottawa ON, CAN K1P 6E2, www.earthenergy.org.

Caneta. GS-2000TM (a computer program for designing and sizing ground heat exchangers for these systems). Caneta Research Inc., Mississauga, ON, CAN L5N 6J7.

Kavanaugh, S.P., and K. Rafferty. 1997. *Ground-Source Heat Pumps: Design of Geothermal Systems for Commercial and Institutional Buildings.* Atlanta: American Society of Heating, Refrigerating and Air-Conditioning Engineers, Inc.

RETScreen (software for renewable energy analysis), Natural Resources Canada, www.retscreen.net.

## ASHRAE GreenTip #21:
## Water-Loop Heat Pump Systems

### GENERAL DESCRIPTION

A water-loop heat pump system consists of multiple water-source heat pumps serving local areas within a building and tied into a neutral-temperature (usually 60°F–90°F [15.5°C–30°C]) water loop that serves as both heat source and heat sink. The loop is connected to a central heat source (e.g., small boiler) and a central heat dissipation device (e.g., closed-circuit evaporative condenser or open-circuit cooling tower isolated from building loop via heat exchanger). These operate to keep the temperature of the loop water within range.

The water-source heat pump itself is an electric-driven, self-contained, water-cooled heating and cooling unit with a reversible refrigerant cycle (i.e., a water-cooled air-conditioning unit that can run in reverse). Its components include heat exchanger, heating/cooling coil, compressor, fan, and reversing controls, all in a common casing. The heat exchanger and coil are designed to accept hot and cold refrigerant liquid or gas. The units can be located either within the space (e.g., low, along outside wall) or remotely (e.g., in a ceiling plenum or in a separate nearby mechanical room).

Piping all of the water-to-refrigerant heat exchangers together in a common loop yields what is essentially an internal source heat recovery system. In effect, the system is capable of recovering heat energy (through the cooling process) and redistributing it where it is needed.

During the cooling mode, heat energy is extracted from room air circulated across the coil (just like a room air conditioner) and rejected to the water loop. In this mode, the unit's heat exchanger acts as a condenser and the coil as an evaporator. In the heating mode, the process is reversed: specifically, a reversing valve allows the heat exchanger to function as the evaporator and the coil as the condenser so that heat extracted from the water loop is "rejected" to the air being delivered to the occupied space, thus heating the space.

In addition to the components mentioned above, the system includes equipment and specialties normally associated with a closed hydronic system (e.g., pumps, filters, air separator, expansion tank, makeup system, etc.)

### WHEN/WHERE IT'S APPLICABLE

A water-source heat pump system is well qualified for applications where simultaneous heating and cooling needs/opportunities exist. (An example might be a building where, in certain seasons, south-side or interior rooms need cooling at the same time north-side rooms require heating.) Appropriate applications may include office buildings, hotels, schools, apartments, extended care facilities, and retail stores.

The system's characteristics may make it particularly suitable when a building is to be air conditioned in stages, perhaps due to cost constraints; once the basic system is in, additional heat pumps can be added as needed and tied into the loop. Further, since it uses low-temperature water, this system is an ideal candidate for mating with a hydronic solar collection system (since such solar systems are more efficient the lower the water temperature they generate).

## PROS AND CONS

### Pro

1. It can make use of energy that would otherwise be rejected to atmosphere.
2. Loop piping, carrying low-temperature water, does not have to be insulated.
3. When applied correctly, the system can save energy. (*Note:* Some factors tend to decrease energy cost, and some tend to increase it; which prevails will determine whether savings result.)
4. It is quieter than a system utilizing air-cooled condensers (i.e., through-the-wall room air conditioners).
5. Failure of one heat pump unit does not affect others.
6. It can condition (heat or cool) local areas of a building without having to run the entire system.

### Con

1. Multiple compressors located throughout a building can be a maintenance concern because of their being noncentralized and sometimes difficult to access (e.g., above the ceiling).
2. Effective water filtration is critical to proper operation of heat exchangers.
3. There is an increased potential for noise within the conditioned space from heat pump units.
4. Some of the energy used in the heating cycle is derived from electricity (used to drive the heat pump compressors), which may be more expensive than energy derived from fossil fuel.

## KEY ELEMENTS OF COST

The following provides a possible breakdown of the various cost elements that might differentiate a water-loop heat pump system from a conventional one and an indication of whether the net incremental cost for the alternative option is likely to be lower (L), higher (H), or the same (S). This assessment is only a perception of what might be likely, but it obviously may not be correct in all situations. **There is no substitute for a detailed cost analysis as part of the design process.** The listings below may also provide some assistance in identifying the cost elements involved.

First Cost

- Equipment costs (will vary depending on what type of conventional system would otherwise be used)     S/L
- Controls                                                S
- Design fees                                             S

Recurring Cost

- Overall energy cost                            L
- Maintenance of system                          H
- Training of building operators                 H

## SOURCES OF FURTHER INFORMATION

Geoexchange Geothermal Heat Pump Consortium, www.geoexchange.org.

Tri-State. Closed water-loop heat pump. Tri-State Generation and Transmission Association, Inc. http://tristate.apogee.net/cool/cchc.asp.

Trane. 1994. *Trane Water-Source Heat Pump System Design Application Engineering Manual,* SYS-AM-7. Lacrosse, WI: Trane Co.

There are several suitable media for storage of cooling energy, including:

1. Chilled water
2. Ice
3. Calcium chloride solutions (brine)
4. Glycol solutions
5. Concentrated desiccant solutions

Active thermal storage systems utilize a building's cooling equipment to remove heat, usually at night, from an energy storage medium for later use as a source of cooling. The most common energy storage media are ice and chilled water. These systems decouple the production of cooling from the demand for cooling, i.e., plant output does not have to match the instantaneous building cooling load. This decoupling increases flexibility in design and operations, thereby providing an opportunity for a more efficient air-conditioning system than with a non-storage alternative. Before applying active thermal storage, however, the design cooling load should be minimized.

Although many operating strategies are possible, the basic principle of a TES system is to reduce peak building cooling loads by shifting a portion of peak cooling production to times when the building cooling load is lower. Energy is typically charged, stored, and discharged on a daily or weekly cycle. The net result is an opportunity to run a chiller plant at peak efficiency during the majority of its operating period. A non-storage system, on the other hand, has to follow the building cooling load, and the majority of its operation is at part-load conditions. Part-load operation of chiller plants comes at the expense of efficiency.

Several buildings have demonstrated site energy reductions with the application of thermal energy storage (TES) as discussed in both the "Pro" and "Sources of Further Information" sections following.

In addition to the potential of site energy reduction, operation of TES systems can reduce energy resource consumption. This reduction is due to a shift toward using energy during periods of low aggregate electric utility demand. As a result, transmission and distribution losses are lower and power plant generating efficiencies can be higher because the load is served by base-load plants. Thermal storage can also have beneficial effects on CHP systems by flattening thermal and electric load profiles.

The ASHRAE *Design Guide for Cool Thermal Storage* (Dorgan and Elleson 1993) covers cool storage application issues and design parameters in some detail.

TES systems tend to perform well in situations where there is variability in loads. Successful applications of TES systems have included commercial office buildings, schools, worship facilities, convention centers, hotels, health care facilities, industrial processes, and turbine inlet air cooling.

1. *Capital cost savings.* Because TES allows downsizing the refrigeration system, the resulting cost savings (which may include *avoiding* having to add such equipment on an existing project) may substantially or entirely cover the added incremental cost of the storage system proper (see also Con 1 below). However, if the first cost is more than another design option, there are still life-cycle cost benefits due to a significant reduction in utility costs.

2. *Reduced size of refrigerating equipment.* The addition of aTES system allows the size of refrigerating equipment to be reduced since it will have to meet an average cooling load rather than the peak cooling load. Reduced refrigeration equipment size means less on-site refrigerant usage and lower probability of environmental impacts due to direct effects.

3. *Factors increasing energy efficiency.* Because TES allows operation of the refrigeration system at or near peak efficiency during all operating hours, the annual energy usage may be lower than non-storage systems that must operate at lower part-load ratios to meet instantaneous loads. In addition, since off-peak hours are usually at night when lower ambient temperatures prevail, lower condensing temperatures required for heat rejection would tend to increase refrigeration efficiency. A number of carefully documented examples of energy savings can be found in the literature, including Bahnfleth and Joyce (1994), Fiorino (1994), and Goss et al. (1996).

4. *Reduced environmental impacts.* Because TES systems shift the consumption of site energy from on-peak to off-peak periods, the total energy resources required to deliver cooling to the facility will be lower (Reindl et al. 1995; Gansler et al. 2001). In addition, in some electric grids, the last generation plants to be used to meet peak loads may be the most polluting per kW of energy produced (Gupta 2002); in such cases, emissions would be further reduced by the use of TES.

5. *Related high-efficiency technologies.* TES enables the practical incorporation of other high-efficiency technologies such as cold-air distribution systems and nighttime heat recovery.

6. *Electric power infrastructure.* TES can be effective at preventing or delaying the need to construct additional power generation and transmission equipment.

7. *Reduced first cost.* Liquid desiccant can be circulated in plastic pipes and does not need insulation.

### Con

1. *Capital cost increases.* Compared to a conventional system, the thermal storage element proper (water tank or ice tank) and any associated pumping, piping accessories, and controls add to the incremental capital cost. If the system's refrigeration equipment can be reduced in size sufficiently (see Pro 1), this burden may be mitigated substantially or balanced out.

2. *Factors decreasing energy efficiency.* The need to generate cooling at evaporator temperatures lower than conventional ones tends to decrease refrigeration efficiency. This reduction may be overcome, however, by factors that increase efficiency (see Pro 3 above).

3. *Engineering.* Successful TES systems require additional efforts in the design phase of a project.

4. *Space.* TES systems will require increased site space usage. The impact of site space usage can be mitigated by considering ice storage technologies.

5. *Operations.* Because a thermal storage system departs from the norm of system operation, continued training of facility operations staff is required as well as procedures for propagating system knowledge through a succession of facilities personnel.

6. *Controls.* Ice requires special control of melt rate to prevent uneven melting and to maximize performance.

7. *Treatment.* Calcium chloride brine needs management to prevent corrosion. Glycol needs management to prevent corrosion and toxicity.

8. *Added components.* Liquid desiccant needs small resistant heating to be above 77°F (25°C) to prevent crystallization (similar to compressor sumps to prevent condensation).

### KEY ELEMENTS OF COST

The following provides a possible breakdown of the various cost elements that might differentiate a TES system above from a conventional one and an indication of whether the net cost for the alternative option is likely to be lower (L), higher (H), or the same (S). This assessment is only a perception of what might be likely, but it obviously may not be correct in all situations. **There is no substitute for a detailed cost analysis as part of the design process.** The listings below may also provide some assistance in identifying the cost elements involved.

- Storage element (CHW, ice, glycol, and brine tanks)
  (Desiccant cost is higher than CHW and brine but similar to glycol)   H
- Additional pumping/piping re storage element   H
- Chiller/heat rejection system   L
- Controls   H
- Electrical (regarding chiller/heat rejection system)   S/L
- Design fees   H
- Operator training   H
- Commissioning   S/H
- Site space   H

- Electric energy   L
- Gas supply with low electrical demand   H
- Operator training (ongoing)   H
- Maintenance training   L

## SOURCES OF FURTHER INFORMATION

Bahnfleth, W.P., and W.S. Joyce. 1994. Energy use in a district cooling system with stratified chilled water storage. *ASHRAE Transactions* 100(1):1767–78.

California Energy Commission. 1996. Source Energy and Environmental Impacts of Thermal Energy Storage. Tabors, Caramanis & Assoc.

Dorgan, C., and J.S. Elleson. 1993. *Cool Storage Design Guide*. Atlanta: American Society of Heating, Refrigerating and Air-Conditioning Engineers, Inc.

Duffy, G. 1992. Thermal storage shifts to saving energy. *Eng. Sys.*

Elleson, J.S. 1996. *Successful Cool Storage Projects: From Planning to Operation*. Atlanta: American Society of Heating, Refrigerating and Air-Conditioning Engineers, Inc.

Fiorino, D.P. 1994. Energy conservation with stratified chilled water storage. *ASHRAE Transactions* 100(1):1754–66.

Galuska, E.J. 1994. Thermal storage system reduces costs of manufacturing facility (Technology Award case study). *ASHRAE Journal*, March.

Gansler, R.A., D.T. Reindl, T.B. Jekel. 2001. Simulation of source energy utilization and emissions for HVAC systems. *ASHRAE Transactions* 107(1):39–51.

Goss, J.O., L. Hyman, and J. Corbett. 1996. Integrated heating, cooling and thermal energy storage with heat pump provides economic and environmental solutions at California State University, Fullerton. *EPRI International Conference on Sustainable Thermal Energy Storage*, pp. 163–67.

Gupta, A. 2002. Director of Energy, NRDC. *New York Times*, March 17.

Lawson, S.H. 1988. Computer facility keeps cool with ice storage. *HPAC* 60(88):35–44.

Mathaudhu, S.S. 1999. Energy conservation showcase. *ASHRAE Journal* 41(4):44–46. Atlanta: American Society of Heating, Refrigerating and Air-Conditioning Engineers, Inc.

O'Neal, E.J. 1996. Thermal storage system achieves operating and first-cost savings (Technology Award case study). *ASHRAE Journal* 38(4).

Reindl, D.T., D.E. Knebel, and R.A. Gansler. 1995. Characterising the marginal basis source energy and emissions associated with comfort cooling systems. *ASHRAE Transactions* 101(1):1353–63.

### Links to Other Efficient Buildings Utilizing TES

Centex—Most efficient building in U.S. in 1999, www.energystar.gov/index.cfm?fuseaction=labeled_buildings.showProfile&profile_id=1306.

LEED Gold Building Hewlett Foundation, www.usgbc.org/Docs/Certified_Projects/Cert_Reg67.pdf.

## ASHRAE GreenTip #23:
## Double-Effect Absorption Chillers

### GENERAL DESCRIPTION

Chilled-water systems that use fuel types other than electricity can help offset high electric prices, whether those high prices are caused by consumption or demand charges. Absorption chillers use thermal energy (rather than electricity) to produce chilled water. A double-effect absorption chiller using high-pressure steam (115 psig) has a COP of approximately 1.20. Some double-effect absorption chillers use low-pressure steam (60 psig [414 kPa]) or 350°F–370°F (177°C–188°C) hot water, but with lower efficiency or higher cost.

Double-effect absorption chillers are available from several manufacturers. Most are limited to chilled-water temperatures of 40°F (4.3°C) or above, since water is the refrigerant. The interior of the chiller experiences corrosive conditions; therefore, the manufacturer's material selection is directly related to the chiller life. The more robust the materials, the longer the life.

### WHEN/WHERE IT'S APPLICABLE

Double-effect absorption chillers can be used in the following applications:

- When natural gas prices (used to produce steam) are significantly lower than electric prices.
- When the design team and building owner wish to have fuel flexibility to hedge against changes in future utility prices.
- When there is steam available from an on-site process; an example is steam from a turbine.
- When a steam plant is available but lightly loaded during the cooling season. Many hospitals have large steam plants that run at extremely low loads and low efficiency during the cooling season. By installing an absorption chiller, the steam plant efficiency can be increased significantly during the cooling season.
- At sites that have limited electric power available.
- In locations where district steam is available at a reasonable price (e.g., New York City).

### PROS AND CONS

#### Pro

1. Reduces electric charges.
2. Allows fuel flexibility, since natural gas, No. 2 fuel oil, propane, or waste steam may be used to supply thermal energy for the absorption chiller.
3. Uses water as the refrigerant, making it environmentally friendly.
4. Allows system expansion even at sites with limited electric power.
5. When the system is designed and controlled properly, it allows versatile use of various power sources.

## Con

1. Cost of an absorption chiller will be roughly double that of an electric chiller of the same capacity as opposed to 25% more for a single-effect absorption machine.
2. Size of an absorption chiller is larger than an electric chiller of the same capacity.
3. Although absorption chiller efficiency has increased in the past decade, the amount of heat rejected is significantly higher than that of an electric chiller of similar capacity. This requires larger cooling towers, condenser pipes, and cooling tower pumps.
4. Few plant operators are familiar with absorption technology.

## KEY ELEMENTS OF COST

The following provides a possible breakdown of the various cost elements that might differentiate an absorption chiller system from a conventional one and an indication of whether the net cost for the hybrid option is likely to be lower (L), higher (H), or the same (S). This assessment is only a perception of what might be likely, but it obviously may not be correct in all situations. **There is no substitute for a detailed cost analysis as part of the design process.** The listings below may also provide some assistance in identifying the cost elements involved.

### First Cost

| | |
|---|---|
| • Absorption chiller | H |
| • Cooling tower and associated equipment | H |
| • Electricity feed | L |
| • Design fees | H |
| • System controls | H |

### Recurring Cost

| | |
|---|---|
| • Electric costs | L |
| • Chiller maintenance | S |
| • Training of building operators | H |

## SOURCES OF FURTHER INFORMATION

ASHRAE. 2000. *2000 ASHRAE Handbook—HVAC Systems and Equipment*, p. 4.1. Atlanta: American Society of Heating, Refrigerating and Air-Conditioning Engineers, Inc.

ASHRAE. 2002. *2002 ASHRAE Handbook—Refrigeration*, Chapter 41. Atlanta: American Society of Heating, Refrigerating and Air-Conditioning Engineers, Inc.

Trane Co. 1999. *Trane Applications Engineering Manual, Absorption Chiller System Design*, SYS-AM-13. Lacrosse, WI: Trane Co.

ASHRAE GreenTip #24:
Gas-Engine-Driven Chillers

GENERAL DESCRIPTION

Chilled-water systems that use fuel types other than electricity can help offset high electric prices, whether those high prices are caused by consumption or demand charges. Gas engines can be used in conjunction with electric chillers to produce chilled water.

Depending on chiller efficiency, a gas-engine-driven chiller may have a cooling COP of 1.6 to 2.3.

Some gas engines are directly coupled to a chiller's shaft. Another option is to use a gas engine and switchgear. In such cases, the chiller may either be operated using electricity from the engine or from the electric utility.

WHEN/WHERE IT'S APPLICABLE

A gas engine is applicable in the following circumstances:

- When natural gas prices are significantly lower than electric prices.
- When the design team and building owner wish to have fuel flexibility to hedge against changes in future utility prices.
- At sites that have limited electric power available.

PROS AND CONS

Pro

1. Reduces electric charges.
2. Allows fuel flexibility if installed as a hybrid system (part gas engine and part electric chiller, so the plant may use either gas engine or electricity from utility).
3. Allows system expansion even at sites with limited electric power.
4. When the system is designed and controlled properly, allows versatile use of various fuel sources.
5. May be used in conjunction with an emergency generator if switchgear provided.

Con

1. Added cost of gas engine.
2. Additional space required for engine.
3. Due to amount of heat rejected being significantly higher than for similar capacity electric chiller, larger cooling towers, condenser pipes, and cooling tower pumps may be required.
4. Site emissions are increased.
5. Noise from engine may need to be attenuated, both inside and outside.
6. Significant engine maintenance costs.

The following provides a possible breakdown of the various cost elements that might differentiate a gas-engine-driven chiller from a conventional one and an indication of whether the net cost is likely to be lower (L), higher (H), or the same (S). This assessment is only a perception of what might be likely, but it obviously may not be correct in all situations. **There is no substitute for a detailed cost analysis as part of the design process.** The listings below may also provide some assistance in identifying the cost elements involved.

### First Cost

- Gas engine                                      H
- Cooling tower and associated equipment          H
- Electricity feed                                L
- Site emissions                                  H
- Site acoustics                                  H
- Design fees                                     H
- System controls                                 H

### Recurring Cost

- Electric costs                                  L
- Engine maintenance                              H
- Training of building operators                  H
- Emissions costs                                 H

### SOURCE OF FURTHER INFORMATION

NBI. 1998. *Gas Engine Driven Chillers Guideline*. Fair Oaks, CA: New Buildings Institute and Southern California Gas Company. www.newbuildings.org/downloads/guidelines/GasEngine.pdf.

## ASHRAE GreenTip #25:
## Gas-Fired Chiller/Heaters

### GENERAL DESCRIPTION

Chilled-water systems that use fuel types other than electricity can help offset high electricity prices, whether those high prices are caused by consumption or demand charges. Absorption chillers use thermal energy (rather than electricity) to produce chilled water. Some gas-fired absorption chillers can provide not only chilled water but also hot water. They are referred to as *chiller-heaters*.

A gas-fired absorption chiller has a cooling COP of approximately 1.0 and heating efficiency in the range of about 80%.

Gas-fired chiller-heaters are available from several manufacturers. Most are limited to chilled water supply temperatures of 40°F (4.3°C) or above, since water is the refrigerant. Some manufacturers offer dual-fuel capability (natural gas or No. 2 fuel oil).

### WHEN/WHERE IT'S APPLICABLE

Gas-fired chiller-heaters are applicable in the following circumstances:

- When natural gas prices are significantly lower than electric prices.
- At sites where a boiler can be eliminated by using the chiller-heater.
- When the design team and building owner wish to have fuel flexibility to hedge against changes in future utility prices.
- At sites that have limited electric power available.

### PROS AND CONS

#### Pro

1. Reduces electric charges.
2. Allows fuel flexibility, since either natural gas or No. 2 fuel oil may be used to supply thermal energy for the absorption chiller.
3. May allow a boiler to be eliminated.
4. Uses water as the refrigerant, making it environmentally friendly.
5. Allows system expansion even at sites with limited electric power.
6. When the system is designed and controlled properly, allows versatile use of various fuel sources.

#### Con

1. Cost will be roughly double that of the same capacity electric chiller.
2. Size of absorption chiller will be larger than the same capacity electric chiller and added space is required.

3.  The amount of heat rejected is significantly higher than from an electric chiller of similar capacity, approximately double that of a single-stage absorption machine, 50% greater for a two-stage unit.
4.  Larger cooling towers, condenser pipes, and cooling tower pumps are required compared with electric-drive machines.
5.  Few plant operators are familiar with absorption technology.

## KEY ELEMENTS OF COST

The following provides a possible breakdown of the various cost elements that might differentiate the gas-fired chiller/heater system from a conventional system and an indication of whether the net cost for the hybrid option is likely to be lower (L), higher (H), or the same (S). This assessment is only a perception of what might be likely, but it obviously may not be correct in all situations. **There is no substitute for a detailed cost analysis as part of the design process.** The listings below may also provide some assistance in identifying the cost elements involved.

### First Cost

| | |
|---|---|
| • Absorption chiller | H |
| • Possible boiler elimination | L |
| • Cooling tower and associated equipment | H |
| • Electricity feed | L |
| • Design fees | H |
| • System controls | H |

### Recurring Cost

| | |
|---|---|
| • Electric costs | L |
| • Chiller maintenance | S |
| • Training of building operators | H |

## SOURCES OF FURTHER INFORMATION

AGCC. 1994. *Applications Engineering Manual for Direct-Fired Absorption*. American Gas Cooling Center.

ASHRAE. 2000. *2000 ASHRAE Handbook—HVAC Systems and Equipment*, p. 4.1. Atlanta: American Society of Heating, Refrigerating and Air-Conditioning Engineers, Inc.

ASHRAE. 2002. *2002 ASHRAE Handbook—Refrigeration*, Chapter 41. Atlanta: American Society of Heating, Refrigerating and Air-Conditioning Engineers, Inc.

Trane. 1999. *Trane Applications Engineering Manual, Absorption Chiller System Design*, SYS-AM-13. Lacrosse, WI: Trane Co.

# ASHRAE GreenTip #26:
## Desiccant Cooling and Dehumidification

### GENERAL DESCRIPTION

There are two basic types of open-cycle desiccant process: solid and liquid desiccant. Each of these processes has several forms, and these should be investigated to determine the most appropriate for the particular application.

All systems have in common the need to have air contact the desiccant, during which moisture is absorbed from the air and the temperatures of both the air and desiccant are coincidently raised.

The moisture absorption process is caused by desiccant having a lower surface vapor pressure than the air. As the temperature of the desiccant rises, its vapor pressure rises and its useful absorption capability lessens. Some systems, particularly liquid types, have cooling of air and desiccant coincident with dehumidification. This can allow the need for less space and equipment.

The dehumidified air has then to be cooled by other means. Two supply air arrangements are available. One has the dehumidified air that is supplied to the building being the mixture of recycled and ventilation air that contacts the desiccant. Moisture as well as contaminants such as VOCs can be absorbed by the desiccant and recycled; particles of solid or liquid desiccant may also be carried over into the ducts and to building occupants.

The other arrangement combines energy recovery from building exhaust air that is typically much cooler and less humid than outdoor ventilation air. By dehumidifying the exhaust air to a sufficiently low humidity ratio (moisture content), it can be used to indirectly cool outdoor air for supply to the building that has not contacted desiccant. Using the recovered energy, this arrangement can be used either for processing only the ventilation air needed or the total supply air being 100% from outside.

So that desiccant can be reused, it has to be re-dried by a heating process generally called *reactivation* or *regeneration* in the case of solid types and reconcentration if a liquid. The re-drying can be either direct by contact of heated outdoor air with the desiccant or indirect. Indirect may be preferred, particularly in high humid climates due to the higher temperature needed to maximize the vapor difference and drying potential.

The energy storage benefit for liquid desiccant has been discussed previously in GreenTip #22.

Rotary desiccant dehumidifiers use solid desiccants such as silica gel to attract water vapor from the moist air. Humid air, generally referred to as *process air*, is dehumidified in one part of the desiccant bed while a different part of the bed is dried for reuse by a second airstream known as reactivation air. The desiccant rotates slowly between these two airstreams so that dry, high-capacity desiccant leaving the reactivation air is available to remove moisture from the moist process air.

Process air that passes through the bed more slowly is dried more deeply, so for air requiring a lower dew point, a larger unit (slower velocity) is required. The reactivation air inlet temperature changes the outlet moisture content of the process air. In turn, if the designer needs dry air, it is generally more economical to use high reactivation temperatures. On the other hand, if the leaving humidity need not be especially low, inexpensive low-grade heat sources such as waste heat or rejected cogeneration heat can be used.

The process air outlet temperature is higher than the inlet temperature primarily because the heat of sorption of the moisture removed is converted to sensible heat. The outlet temperature rises roughly in proportion to the amount of moisture that is removed. In most comfort applications, provisions must be made to remove excess sensible heat from the process air following reactivation. Cooling is accomplished with cooling coils, and the source of this cooling affects the operating economics of the system.

## WHEN/WHERE IT'S APPLICABLE

In general, applications that require a dew point at or below 40°F (4.3°C) may be candidates for active desiccant dehumidification. Examples include facilities handling hygroscopic materials; film drying; the manufacture of candy, chocolate, or chewing gum; the manufacture of drugs and chemicals; the manufacture of plastic materials; packaging of moisture-sensitive products; and the manufacture of electronics. Supermarkets often use desiccant dehumidification to avoid condensation on refrigerated casework. And when there is a need for a lower dew point and a convenient source of low-grade heat for reactivation is available, rotary desiccant dehumidifiers can be especially economical.

## PROS AND CONS

### Pro

1. Desiccant equipment tends to be very durable.
2. Often this is the most economical means to dehumidify below a 40°F dew point.
3. It eliminates condensate in the airstream, in turn, limiting the opportunity for mold growth.

### Con

1. Desiccant usually must be replaced, replenished, or reconditioned every five to ten years.
2. In comfort applications, simultaneous heating and cooling may be required.
3. The process is not especially intuitive and the controls are relatively complicated.

## KEY ELEMENTS OF COST

The following provides a possible breakdown of the various cost elements that might differentiate a building with a rotary desiccant dehumidification system from one without and an indication of whether the net cost is likely to be lower (L), higher (H), or the same (S). This assessment is only a perception of what might be likely, but it obviously may not be correct in all situations. **There is no substitute for a detailed cost analysis as part of the design process.** The listings below may also provide some assistance in identifying the cost elements involved.

### First Cost

- Equipment costs                                H
- Regeneration (heat source and supply)          H
- Ductwork                                        S
- Controls                                        H
- Design fees                                     S

### Recurring Cost

- Overall energy cost                            S/H
- Maintenance of system                           H
- Training of building operators                  H
- Filters                                         H

### SOURCE OF FURTHER INFORMATION

ASHRAE. 2000. *2000 ASHRAE Handbook—HVAC Systems and Equipment.* Atlanta: American Society of Heating, Refrigerating and Air-Conditioning Engineers, Inc.

# ASHRAE GreenTip #27:
# Indirect Evaporative Cooling

## GENERAL DESCRIPTION

Evaporative cooling of supply air can be used to reduce the amount of energy consumed by mechanical cooling equipment. Two general types of evaporative cooling—direct and indirect—are available. The effectiveness of either of these methods is directly dependent on the extent that dry-bulb temperature exceeds wet-bulb temperature in the supply airstream.

*Direct evaporative cooling* introduces water directly into the supply airstream, usually with a spray or wetted media. As the water absorbs heat from the air, it evaporates. While this process lowers the dry-bulb temperature of the supply airstream, it also increases the air moisture content.

Two typical forms of *indirect evaporative cooling* (IEC) are described below.

1. *Coil/cooling tower type of IEC.* This type uses an additional water-side coil to lower supply air temperature. The added coil is placed ahead of the conventional cooling coil in the supply airstream and is piped to a cooling tower where the evaporative process occurs. Because evaporation occurs elsewhere, this method of "precooling" does not add moisture to the supply air, but it is somewhat less effective than direct evaporative cooling. A conventional cooling coil provides any additional cooling required.

2. *Plate heat exchanger (PHE) type of IEC.* This comprises sets of parallel plates arranged into two sets of passages separated from each other.

    In a typical arrangement, exhaust air from a building is passed through one set of passages, during which it is wetted by water sprays. A stream of outdoor air is coincidently passed through the other set of passages and is cooled by heat transfer through the plates by the wetted exhaust air before being supplied to the building. Alternatively, the exhaust air may be replaced by a second stream of outdoor air.

    The wetted air is reduced in dry-bulb temperature to be close to its wet-bulb temperature. The stream of dry air is cooled to be close to the dry-bulb temperature of the wetted exhaust air.

    In some applications, the cooled stream of outdoor air is passed through the coil of a direct expansion refrigeration unit, where it is further cooled before being supplied to the building. This system is an efficient way for an all outdoor air supply system.

    The plates in the heat exchanger can be formed from various metals and polymers. Consideration needs to be given to preventing the plate material from corroding.

## WHEN/WHERE IT'S APPLICABLE

In climates with low wet-bulb temperatures, significant amounts of cooling are available. In such climates, the size of the conventional cooling system can be reduced as well.

In more humid climates, indirect evaporative cooling can be applied during non-peak seasons. It is especially applicable for loads that operate 24 hours a day for many days of the year.

## PROS AND CONS

### Pro

1. Indirect evaporative cooling can reduce the size of the conventional cooling system.
2. It reduces cooling costs during periods of low wet-bulb temperature.
3. It does not add moisture to the supply airstream (in contrast, direct evaporative cooling does add moisture).
4. It may be designed into equipment such as self-contained units.
5. There is no cooling tower or condenser piping in the PHE type of IEC described.

### Con

1. Air-side pressure drop (typically 0.2 to 0.4 in. w.c. [50 to 100 kPa]) increases due to an additional coil in the airstream.
2. To make water cooler in the coil/cooling tower type of IEC, the cooling tower fans operate for longer periods of time and consume more energy.
3. For the coil/cooling tower type of IEC, condenser piping and controls must be accounted for during design process.

## KEY ELEMENTS OF COST

The following provides a possible breakdown of the various cost elements that might differentiate an indirect evaporative cooling system from a conventional one and an indication of whether the net cost for the hybrid option is likely to be lower (L), higher (H), or the same (S). This assessment is only a perception of what might be likely, but it obviously may not be correct in all situations. **There is no substitute for a detailed cost analysis as part of the design process.** The listings below may also provide some assistance in identifying the cost elements involved.

### First Cost

- Indirect cooling coil      H
- Decreased conventional cooling system capacity      L
- Condenser piping, valves, and control      H

**Recurring Cost**

| | |
|---|---|
| • Cooling system operating cost | L |
| • Supply fan operating cost | H |
| • Tower fan operating cost | H |
| • Maintenance of indirect coil | S |

**SOURCES OF FURTHER INFORMATION**

ASHRAE. 1999. *1999 ASHRAE Handbook—HVAC Applications*, pp. 50.1–3. Atlanta: American Society of Heating, Refrigerating and Air-Conditioning Engineers, Inc.

ASHRAE. 2000. *2000 ASHRAE Handbook—Systems and Equipment*, p. 19.3–4; 44.7. Atlanta: American Society of Heating, Refrigerating and Air-Conditioning Engineers, Inc.

# 12

## ENERGY/WATER SOURCES

### RENEWABLE/NONRENEWABLE ENERGY SOURCES

There are often discussions about using renewable energy sources as a way to power the world, but little is done to actually implement it. This chapter focuses on ways to use renewable energy to offset nonrenewable energy sources. By definition, a renewable energy source (RES) is a fuel source that can be replenished in a short amount of time. The common renewable sources are solar, wind, hydro, and biomass. This is in contrast to the common nonrenewable energy sources such as coal, oil, natural gas, and nuclear.

The use of renewable energy is separated between utilizing the energy source on site versus paying for the renewable resource. Today, many utility companies offer the ability to purchase renewable energy mixes from their generation portfolio. These utilities generate their own renewable energy with either large-scale wind or solar facilities. Note that quite often, large-scale hydroelectric plants are not considered a renewable resource because of the size of the environmental impact of such facilities. *Green energy*, as it is sometimes called, can also be purchased by third-party resellers of energy. The concept is that renewable energy is put into the utility grid and you, as the end user, can purchase that power somewhere else on the grid. The bottom line is that you can, in most areas, purchase green power and you are not limited by the default utility offering. More than 600 utilities across the US now offer green power programs. According to the DOE's NREL, in 2005 total utility green power sales reached 2.7 billion kilowatt-hours (kWh), a 36% increase over 2004. An estimated 430,000 customers are participating in utility programs nationwide, up 20% from 2004. The US Environmental Protection Agency (EPA) has a publication, *Guide to Purchasing Green Power*, which is a good overview of the green power market and steps to take if you wish to participate.

The European Union (EU) Renewables Directive has been in place since 2001, at a time when oil and gas were still at a relatively low price. The directive aims to increase the EU's share of electricity produced from RES to 21% (up from 15.2% in 2001), thus contributing to reach the overall target of 12% of energy consumption from renewables by 2010. More than four million European consumers have already switched to a green power. The European Green Electricity Network (Eugene) provides contact information for green energy suppliers throughout Europe (www.eugenestandard.org).

Another consideration, also usually with little choice of options for the designer, is the nature of the default utility offering for energy resources from which the site energy source is derived. Whether electricity is generated from coal, imported oil, natural gas, or uranium, for example, may have broad implications for national or industry interests, the environment, or economics; however, there is little a designer can do about it—at least in choosing between conventional nonrenewable energy sources. This subject is addressed in more detail in the "Energy Resources" chapter of the *ASHRAE Handbook*, and the designer is referred to that source for more specific data.

The most important choice that designers do have is creating a building that is designed to minimize energy consumption. This is a focus of most of this book. Passive solar techniques—that is, those with minimal moving parts—can be incorporated into the design of a building. Daylighting, trombe walls, passive cooling, and natural ventilation are all considered ways to use the natural environment to help heat, light, and cool the building.

Other techniques can be used, such as solar water heating, solar ventilation preheat, photovoltaic (PV) systems, and wind systems, although these tend to be more complicated to integrate.

In many cases, renewable energy can be considered "free." The issue is that this free energy source usually needs some capital equipment to concentrate the diffuse nature of renewable energy to a useful form for the building. One way to illustrate this characteristic is to think of a gallon jug of, say, fuel oil compared to an array of hydronic solar collectors. The fuel oil could provide hot water, on demand, for hours in a simple water heater occupying a corner of a boiler room but will consume fossil fuels and produce emissions. The equivalent job done by solar collectors would require an array of collectors on a roof plus a tank, piping, and some controls (a simple thermosiphon type solar collector can operate without controls) but will produce no emissions and operate with the free solar energy.

While consideration of renewables is a highly touted element of green design, the design team should be well aware of the key characteristics of a particular renewable considered and develop creative strategies to make the most effective use of this free energy source.

Following is some additional discussion on the two main renewable energy sources (other than hydropower) in the world today.

## Sources of Further Information

CADDET—Centre for Analysis and Dissemination of Demonstrated Energy Technologies, International Energy Agency (IEA), a source of global information on proven, commercial applications covering the full range of renewable energy technologies, www.caddet.org.

EPA. *Guide to Purchasing Green Power.* www.epa.gov/grnpower/buygreenpower/guide.htm. Eugene. Green Energy Standard in Europe, www.eugenestandard.org.

*Green Energy In Europe,* www.greenprices.com.

## SOLAR

Solar energy is the primary energy source that fuels the growth of the earth's natural capital and drives wind and ocean currents that also can provide alternative energy sources. Solar energy has been successfully harnessed for human use since the beginning of time. Early civilizations and some modern ones used solar energy for many purposes: food and clothes drying, heating water for baths, heating adobe and stone dwellings, etc. Solar energy is free and available to anyone who wishes to utilize it.

Solar thermal heating for domestic hot-water and space heating has grown considerably over the years and is well established in several countries. Global installed capacity is estimated at 75 $GW_{th}$ (gigawatts of thermal energy) for flat plate and evacuated tube collectors and 24 $GW_{th}$ for unglazed plastic collectors (Weiss et al. 2006). In North America (USA and Canada) swimming pool heating is dominant with an installed capacity of about 19 $GW_{th}$ of unglazed plastic collectors, while in China (45 $GW_{th}$), Europe (11 $GW_{th}$), and Japan (6 $GW_{th}$), flat plate and evacuated tube collectors are mainly used to generate hot water for sanitary use and space heating. According to the European Solar Thermal Industry (ESTIF), at the end of 2005 the total capacity in operation in the EU reached 11.2 MWth (about 16 million m$^2$ of glazed collector area). Germany is the leader in terms of market volume, with 47% of the European market, followed by Greece (14%), Austria (12%), and Spain (6%). In terms of capacity in operation per capita, Cyprus, where more than 90% of all buildings are equipped with solar collectors, leads Europe with 480 $kW_{th}$/1.000 capita, followed by Austria and Greece at about 200 $kW_{th}$/1.000 capita. China dominates the world market (44%), with annual domestic sales of well over 10 millions m$^2$ (7 $GW_{th}$).

For many, a key impediment to increased solar use is economics. The cost of some solar technologies is perceived to be high compared to the fossil-based energy source it is offsetting. For example, while the simple payback of solar PV systems tends to still be rather long (although much improvement has occurred over the past couple of decades), the recent increase in the cost of energy and advances in solar energy justify a fresh look at the applications and the engineering behind those applications. In addition, public policy in many areas encourages solar and other renewable energy applications through tax incentives and encouraging or requiring repurchase of excess electrical energy generated by the utility provider.

The applicability and, consequently, the economics and public policy incentives available of different solar energy system types and applications depend much on the location. Besides economic considerations, there are other factors that should be studied.

## Solar Thermal Applications

Solar energy thermal applications range from low-temperature applications, such as domestic water or swimming pool heating, to medium- to high-temperature applications, such as absorption cooling or steam production for electrical generation. (See Figure 12-1 for examples of hardware.)

The capital cost of solar thermal systems generally increases with higher working fluid temperatures. The higher the delivery temperature, the lower the efficiency and the more solar collector area is generally required to deliver the same net energy. This is due to parasitic thermal losses inherent in solar collector design. Different solar collector types provide advantages and disadvantages, depending on the application, and there are significant cost differences between each solar collector type. Some common collector types are discussed below.

*Flat plate solar collectors* are best suited for processes requiring low-temperature working fluids (80°F to 160°F or 25°C to 70°C) and can deliver 80°F (25°C) fluid temperatures even during overcast conditions. The term *flat plate collector* generally refers to a hydronic coil-covered absorber housed in an insulated box with a single- or double-glass cover that allows solar energy to heat the absorber. Heat is removed by a fluid running through the hydronic coils. Its design makes it more susceptible to parasitic losses than an evacuated tube collector but more efficient in solar energy capture because flat plate collectors convert both direct and indirect solar radiation into thermal energy. This makes flat plate collectors the preferred choice for domestic hot-water and other low-temperature heating applications. Coupling a water-source heat pump (low fluid temperatures) with solar collectors

(a)  (b)  (c)

**Figure 12-1 Examples of solar hardware: (a) flat-plate collector, (b) evacuated tube collector, and (c) concentrating collector.**

provides heating efficiencies higher than GSHP applications and standard natural gas furnaces.

*Evacuated tube collectors* generally refer to a series of small absorbers consisting of small diameter (approximately 3/8 in. or 10 mm) copper tubing encased in a clear, cylindrical evacuated "thermos bottle" that minimizes parasitic losses even at elevated temperatures. Because of the relatively small absorber area, significantly more collector area is required than with the flat plate or concentrating collector.

Typical flat plate solar collector performance curves are illustrated in Figure 12-2. The collector's efficiency ($\eta$) is expressed as a function of the solar collector's working fluid inlet temperature ($T_{in}$), the ambient temperature ($T_o$), and the total solar radiation incident on the collector's surface ($G_T$).

*Concentrating collectors* generally refer to the use of a parabolic reflector that focuses the solar radiation falling within the reflector area onto a centrally located absorber. Concentrating collectors are best suited for processes requiring high-temperature working fluids (300°F to 750°F or 150°C to 400°C) and do not operate under overcast conditions. This type of collector converts only direct solar radiation, which varies dramatically with sky clearness and air quality (such as smog). Concentrating collectors rotate on one or two axes in order to track the sun and collect the available direct solar radiation.

Pros and cons of different solar collector types and other limitations of active solar systems can be found in a number of sources. Many of the key contributions came during the initial energy concerns on the 1970s, and a few are listed at the resources section at the end of this chapter.

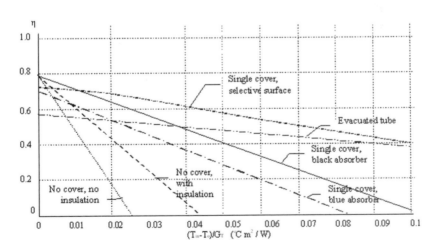

**Figure 12-2 Typical flat plate solar collector performance curves.**

The percentage of energy a solar system can provide is known as the solar fraction. The f-chart method developed by Sanford Klein at the University of Wisconsin provides an accurate assessment of the amount of energy a solar thermal system will provide. This modeling provides the designer the ability to vary system parameters, such as collector area, storage volume, operating temperature, and load, to optimize system design. The method was originally developed to aid engineers who did not have computer resources available to do the complex analyses required for solar thermal calculations. This method has since been adapted for automated computer analysis using modern tools.

The cost-effectiveness of thermal solar systems is also dependent on having a constant load for the energy the solar system provides. Since space heating requirements are generally seasonal in most climates, it is advisable that energy from the solar system have more uses than space heating alone. Domestic water heating is usually a much steadier year-round load, though typically often not very substantial. Solar cooling can extend utilization of solar collectors during the summer. The fact that peak cooling demand in summer is associated with high solar radiation availability offers an excellent opportunity to exploit solar energy with heat-driven cooling machines.

Solar hot water can be used to be the sole source of domestic hot water or to preheat incoming supply water. This can be as simple as having an uninsulated tank in a hot attic in a southern climate to having a batch water heater on the roof of a building. Pool heating is the reason for the vast majority of current sales of hot-water-based systems in most US markets. Water is taken from the pool and pumped through an unglazed collector and then returned back into the pool. In many climates, most of the energy for pool heating can be offset with this technique. Solar energy is a major source for domestic hot-water heating in several countries, such as China, Israel, Australia, and Greece, and is currently widely used in states such as Hawaii and Florida. Although it is an economically viable option for practically all areas of the US, a barrier to increased market acceptance is the current lack of trained installation contractors and limited information available in most of those regions.

Besides preheating water, solar energy can be used to also preheat incoming air before introduction to a building using systems integrated with the overall building design or collectors. This is a simple technological approach that can be economically viable and is well suited for northern climates or areas with high building heating load. The US DOE has published a bulletin on these transpired solar collectors, which is available for free downloading (see references below).

The main heat-driven cooling technologies that can be used for solar cooling include:

- Closed-cycle systems, for example, LiBr/$H_2O$ and $H_2O$/$NH_3$ absorption systems (see Chapter 41, *2002 ASHRAE Handbook—Refrigeration*) and adsorption (see Chapter 22, *2005 ASHRAE Handbook—Fundamentals*) cycles. They produce chilled water that can be used in combination with any air-conditioning equipment, such as an AHU, fan-coil systems, chilled ceilings, etc.

- Open cycles, for example, desiccant systems (see Chapter 22, *2005 ASHRAE Handbook—Fundamentals*). The term *open cycle* is used to indicate that the refrigerant is discarded from the system after providing the cooling effect and new refrigerant is supplied in its place in an open-ended loop.

Solar-assisted cooling systems employ solar thermal collectors connected to thermal-driven cooling devices. They consist of several main components (Figure 12-3), namely, the solar collectors, heat storage, heat distribution system, heat-driven cooling unit, optional cold storage, an air-conditioning system with appropriate cold distribution, and an auxiliary (backup) subsystem (integrated at different places in the overall system, used as an auxiliary heater parallel to the collector or the collector/storage or as an auxiliary cooling device or both).

Common flat plate solar collectors are used for reaching a driving temperature of 140°F–194°F (60°C–90°C) due to their low first cost (Balaras et al. 2006). With selective surface flat plate solar collectors, the driving temperature can be up to 248°F (120°C); however, the collector efficiency will be quite low at this temperature level. Stationary flat plate evacuated tube solar collectors are typically used for 175°F to 250°F (80°C–120°C) applications, and can reach higher temperatures but at a lower collector efficiency. Compound parabolic concentrators can reach 207°F–329°F (97°C–165°C).

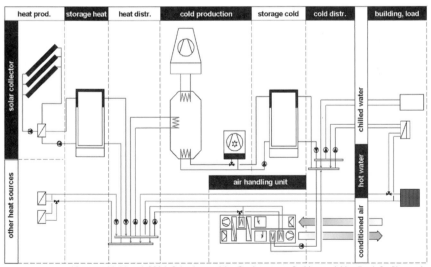

Henning, H.-M., ed. 2004. *Solar Assisted Air-Conditioning in Buildings—A Handbook for Planners)*.

**Figure 12-3 Schematic description of a solar air-conditioning system describing the integration of different component options.**

The average specific solar collector area averages 3.6 m/kW (136 ft$^2$/ton), ranging from 19 to 208 ft$^2$/ton (0.5 to 5.5 m/kW), depending on the employed technology. Adsorption and absorption systems typically use more than 76 ft$^2$/ton (2 m/kW) and usually lower than 189 ft$^2$/ton (5 m/kW). Overall, $H_2O/NH_3$ systems require larger specific collector areas than $LiBr/H_2O$ systems and, as a result, the installations are usually more expensive. The initial overall cost per installed cooling capacity in kW averages about 4000 Euro/kW, excluding the cost for distribution networks between the system and the application and the delivery units.

Each technology has specific characteristics that match the building's HVAC design, loads, and local climatic conditions. A good design must first exploit all available solar radiation and then cover the remaining loads from conventional sources. Proper calculations for collector and storage size depend on the employed solar cooling technology. Hot-water storage may be integrated between the solar collectors and the heat-driven chiller to dampen the fluctuations in the return temperature of the hot water from the chiller. The storage size depends on the application: if cooling loads mainly occur during the day, then a smaller storage will be necessary than when the loads peak in the evening. Heating the hot-water storage by the backup heat source should be strictly avoided. The storage's only function is to store excess heat of the solar system and to make it available when sufficient solar heat is not available. Specific guidelines and an overview of various applications are available in Balaras et al. (2006).

## Solar-Electric Systems

The direct conversion from sunlight to electricity is accomplished with PV systems. These systems continue to drop in price and many states (as well as the federal government) offer tax incentives for the installation of these systems. The most common application is a grid-tied system, where electricity is directly fed into the grid.

Other applications of PV may be attractive and provide additional value other than the cost of offsetting utility power. Whenever power lines must be extended, PV should be investigated. It is cost-effective to use stand-alone PV applications for signs, remote lights, and blinking traffic lights. Many times, if an existing small grid line is available and additional power is needed, it is less expensive to add the PV system rather than increase the power line. This may be true of remote guard shakes, restroom facilities, or other outbuildings. PV should also be considered as part of the uninterruptible power system for a building. Every building has battery backup for certain equipment. Batteries can be centralized and the PV fed directly into this system. The cost can be less than other on-site generation and fuel storage.

Technical and legal factors have generally been worked out that have hindered grid-tied solar PV systems from applications in the past. In temperate climates,

United Solar Ovonic, provided courtesy of DOE/NREL, Golden, CO

**Figure 12-4 Building-integrated PV roofing.**

grid-tied PV systems provide a good method to reduce peak summer electrical demand, since peak solar gains generally correspond to peak air-conditioning demand. This was a major factor that led to the passing of the California "Million Solar Roofs Initiative" in early 2006. The German "100.000 rooftops—HTDP" program, which began in early 1999, had a goal of 300 MW and was successfully completed in 2003 with the parallel introduction of the EEG (German renewable energy feed-in law) that came into effect in 2000. The combination between EEG and HTDP secured commercially oriented PV investors a full payback of their investment and the breakthrough of the market. In total, nearly 1800 MWp of PVs had been installed in the EU by 2005, exhibiting an annual growth in 2005 of 18.2% with respect to the 2004 market, reaching the limits of its supply capacities for the first time due to the present shortage of silicon. Other countries, such as Japan, are also developing grid-tied PV networks.

The economic viability of PV systems is dependent on many factors, and technology improvements have brought the cost per peak watt down to $10 or less. The integration of PV systems in with building materials is a new development that may, in the long run, help make PV systems viable in most areas. One of the key ways to incorporate solar energy technologies is to incorporate it directly into the architecture of the building. Trombe walls can replace spandrel panels and other glass facades; PV panels can become overhangs for the building or parking shading structures. Using these elements as part of the building doubles the value of the systems. An example of these technologies is the concept called *building-integrated photovoltaics* (BIPV). A picture of one BIPV concept is shown in Figure 12-4. These are being developed by the NREL, among others.

Finally, PV is very visible and can be used as marketing tool for a building project.

PV collectors are addressed in GreenTip #30.

## Sources of Further Information for Solar Energy

AGORES—A Global Overview of Renewable Energy Sources, the official European Commission renewable energy information centre and knowledge gateway, with a global overview of RES, www.agores.org.

ASES—American Solar Energy Society, www.ases.org.

Balaras, C.A., H.-M. Henning, E. Wiemken, G. Grossman, E. Podesser, and C.A. Infante Ferreira. 2006. Solar cooling—An overview of european applications and design guidelines. *ASHRAE Journal* 48(6):14–22.

Beckman, W., S.A. Klein, and J.A. Duffie. 1977. *Solar Heating Design*. New York: John Wiley and Sons.

BEER—Building Energy Efficiency Research Project at the Department of Architecture, The University of Hong Kong, www.arch.hku.hk/research/beer/.

EPIA—European Photovoltaic Industry Association, www.epia.org.

ESTIF—European Solar Thermal Industry, www.estif.org.

EUREC—European Renewable Energy Centres Agency, www.eurec.be.

EUROSOLAR—The European Association for Renewable Energies e.V., www.eurosolar.org.

Henning. H-M., ed. 2004. *Solar Assisted Air-Conditioning in Buildings—A Handbook For Planners*. Wien: Springer-Verlag.

International Energy Agency (IEA) Photovoltaic Power Systems Programme, www.iea-pvps.org.

International Energy Agency (IEA) Solar Heating And Cooling Programme, www.iea-shc.org.

ISES—International Solar Energy Society, www.ises.org.

Kreith, F. and J.F. Kreider. 1978. *Principles of Solar Engineering*. New York: McGraw Hill.

National Renewable Energy Laboratory, The Center for Buildings and Thermal Systems, Solar Energy Research, www.nrel.gov/buildings_thermal/solar.html.

SEIA—Solar Energy Industries Association, www.seia.org.

Solar Energy Industries Association (SEIA), Washington, DC www.seia.org/.

*Transpired Air Collectors*, Energy Efficiency and Renewable Energy, US Department of Energy, www.eere.energy.gov/de/transpired_air.html.

Weiss, W., I. Bergmann, and G. Faninger. 2006. *Solar Heat Worldwide*. Solar Heat & Cooling Programme, International Energy Agency, www.iea-shc.org/welcome/IEASHCSolarHeatingWorldwide2006.pdf.

WRDC—World Radiation Data Centre maintained for the World Meteorological Organization (WMO) provides solar radiation and radiation balance data (the world network), http://wrdc-mgo.nrel.gov/.

## WIND

Using prevailing breezes and wind energy is one of the most promising alternative technologies today. Wind turbine design and power generation have become more reliable over the last 30 years. Consistency and velocity of available breezes are essential to successful application of wind for natural ventilation and electrical generation. Information on velocities, durations, and direction of winds in a project area is generally available from National Oceanic and Atmospheric Administration.

From a natural ventilation perspective for passive cooling of buildings, other factors, such as temperature and relative humidity, must also be taken into consideration. Many areas can use outdoor air/natural breezes to provide passive cooling only during a limited period of the year, requiring designers wanting to use natural ventilation to carefully analyze both climatic conditions and cost to implement. The benefits of using natural ventilation must be weighed against the consistency of breezes, potential for higher relative humidity levels (which may affect IAQ), and the impact on occupant comfort and HVAC operation.

There are many excellent examples where natural ventilation provides the cooling needed throughout the year, but most are located in cool, arid climates. Even in moderate climates, such as that in Atlanta, Georgia, natural ventilation can provide passive cooling during some periods of the year but, in doing so, the designer must consider the impact on HVAC operation and building control. These elements may be very difficult to get to work effectively with operable windows. Additional programming and control points are required for successful operation. The potential of natural ventilation, its appropriate use, the design and dimensioning methodologies, the need for an integrated design approach, and how to overcome barriers are available in a handbook by Alvarez et al. (1998). (See also GreenTip #13 on hybrid ventilation in Chapter 8.)

Incorporating one or more wind turbine generators at a building site is another, more active approach. Initially, it should be recognized that there is a disadvantage of scale: one or two wind turbines alone are destined to be less cost-effective than a wind farm with hundreds. Wind turbines should first be evaluated by looking at the wind resource at the site. For most small-scale wind turbines, steady winds over 20 mph are preferred to maximize the cost effectiveness. Other issues include noise, vibration, building geometry (if on the building), wind pattern interrupters, wildlife effect, periodic maintenance, safety, visual impact, and community acceptance; these must also be taken into account.

## Sources of Further Information for Wind Energy

Alvarez, S., E. Dascalaki, G. Guarracino, E. Maldonado, S. Sciuto, and L. Vandaele. 1998. *Natural Ventilation of Buildings—A Design Handbook*. F. Allard, ed. London: James & James Science Publishers Ltd.

EWEA—The European Wind Energy Association, www.ewea.org.

NOAA Web site, www.noaa.gov.

## HYDRO

The ability to use hydro is limited on an individual building scale. On rare occasions, older building renovations can take advantage of existing river dams and generate hydroelectric on site. This is especially true in the Northeast, where the primary power supply in the 1800s was hydro. In some cases, small streams can be diverted and small turbines installed—this is called *low-impact hydro*, as a minimal dam is needed and water is returned to the same stream.

In some cases, pressure reduction values can be replaced with small turbines. Although not really free energy as the water was pressurized to bring it on site, it is better to recover the energy rather than throw it away through the pressure reduction valve.

## BIOMASS

Biomass is the conversion of plant and animal matter into useful energy. Several ideas should be investigated to be included in a building. Wood chip conversion to heat (and possibly then electricity) can replace boiler systems in buildings. Automatic feed systems can now take wood chips and make hot water as conveniently as oil and natural gas systems. In many locations wood chips are free for the cost of the transportation, saving further cost. Pellet burners are a similar concept, except for the additional cost of the wood pellets. Burners are also available for corn cobs. In some cases, manufacturers will lease equipment for the energy savings, resulting in no additional cost to the end-user.

The production and distribution of bio-oils and bio-diesel fuels derived from bio-based materials is growing in momentum. Heating systems are available to burn used vegetable oils. These systems are useful near the point of production of used vegetable oil—usually the food service industry. Grocery stores and restaurants may benefit from such technologies. Most diesel-fueled machines can run on bio-diesel without any (or minor) modifications. In addition, hauling and waste costs for the oil are eliminated, increasing the attractiveness of the technology.

## WATER

Since 70% of our planet is covered with water, it may seem that water is an inexhaustible source. However, 97% of the water on earth consists of seawater, and over 2% is ice. The remaining part is either located at great depths or is heavily polluted. This means that merely 0.003% can be immediately used for the supply of drinking water.

The importance of this natural base material is still increasing: the presence of fresh drinking water contributes to the social and cultural structure and the level of development of a country. Therefore, it seems logical that this natural resource would be valued, protected, and safeguarded for future generations. Still, this does not happen. Water is still being wasted, polluted, and misused. As a consequence, global water shortage is already a reality in many countries around the world. Recent prognoses indicate that by 2025 over 50 countries, with a joint population of 3 billion people, will face water shortage, as the demand for water will have increased by 650%.

Water is also a resource to the building. Energy is required to treat and pump water as it is brought to the site. As with renewable energy sources, conservation is the first step in minimizing environmental impact. Water is no different. The topic of water supply and usage is discussed in more detail in Chapter 12. We will focus here on how to supply water to a building system. Rainwater collection and storage can be used to meet irrigation needs. In some areas, this water can be filtered and used for other nonpotable uses, such as fountains and some limited washing. Other sources of water are the collection of cooling tower bleeds and collection of condenser water. These make good sources for irrigation systems and minimize the need for purchased water and waste water treatment.

## Sources of Further Information for Water

Green Venture, www.greenventure.on.ca.

North Carolina Cooperative Extension Service: Water quality and waste management, www.bae.ncsu.edu/programs/extension/publicat/wqwm.

The World's Water, www.worldwater.org.

# ASHRAE GreenTip #28:
## Passive Solar Thermal Energy Systems

### GENERAL DESCRIPTION

*Passive solar thermal energy systems* utilize solar energy, mainly for space heating, via little or no use of conventional energy or other mechanisms than the building design and orientation. All abovegrade buildings are "passive solar"; making buildings collect, store, and use solar energy wisely is then the challenge to building designers. A building that intentionally optimizes passive solar heating can visually be a "solar building," but many reasonably sized features that enhance energy collection and storage can be integrated into the design without dominating the overall architecture.

To be successful, a well-designed passive solar building needs (1) an *appropriate thermal load* such as space heating; (2) *aperture*, such as clear, glazed windows; (3) *thermal storage* to minimize overheating and to use the heat at night; (4) *control*, either manual or automatic, to address overheating; and (5) *night insulation* of the aperture so that there is not a net heat loss.

### HIGH-PERFORMANCE STRATEGIES

The following strategies are general in nature and are presented as a guideline to help maximize the performance of a passive thermal solar design.

Do conservation first. Minimizing the heating load will reduce conventional and renewable heating systems' sizes and yields the best economics. Insulate, including the foundation, and seal the building well. Use quality exterior windows and doors.

In the northern hemisphere, the aperture must face due south for optimal performance. If not possible, make it within ±10 degrees of due south. In the southern hemisphere, this solar aperture looks north.

Minimize use of east- and west-facing glazing. They admit solar energy at non-optimal times, at low angles that minimize storage, and are difficult to control by external shading. Also reduce north glazing in the colder regions of the northern hemisphere, due to high heat loss rates.

Use optimized and/or moveable external shading devices, such as overhangs, awnings, and side fins. Internal shading devices should not be relied upon for passive solar thermal control—they tend to cause overheating.

Use high-mass "direct gain" designs. The solar collector (windows) and storage (floors and potentially walls) are part of the occupied space and typically have the highest solar savings fraction, which is the percent of heating load met by solar. "Directly irradiated" thermal masses are much more effective than indirect, thus floors or trombe walls are often best.

Use vertical glazing. Horizontal or sloped windows and skylights are hard to control and insulate.

Calculate the optimal thermal mass—it is often around 8 in. thick (about 20 cm) high-density concrete for direct-gain floors over conditioned basements. The optimization should direct the designer to a concept that will capture and store the highest amount of solar energy without unnecessarily increasing cost or complexity. For all direct-gain surfaces, make sure they are of high-absorptivity (dark color) and are *not* covered by floor carpet, tile, much furniture, or other ways that prevent or slow solar energy absorption. Be sure the thermal mass is highly insulated from the outdoor air or ground.

Seeking a very high annual solar savings fraction ($f_s$) often leads to disappointment and poor economics, so keep expectations reasonable. Even a 15% annual fraction represents a substantial reduction in conventional heating energy use. A highly optimized passive solar thermal single-family house, in an appropriate climate, often only has about a 40% solar fraction. Combining passive with active solar, PVs, wind, and other renewable energy sources is often the most satisfactory way to achieve a very high annual solar savings fraction.

An old solar saying, of unknown origin, is "the more passive a building, the more active the owner." Operating a passive solar building to optimize collection has thus been called *solar sailing* and requires time and experience. Be sure that passive solar and the building operator are good matches for each other.

## KEY ELEMENTS OF COST

Passive solar energy systems must be engineered, otherwise poor performance is likely.

Window sizes are typically larger than for conventional design. Some operable windows and/or vents, placed high and low, are needed for overheat periods. Proper solar control must also be foreseen.

Concrete floors and walls are commonly thicker in the storage portion of a passive solar building. A structural engineer's services are likely required.

Conventional backup systems are still needed. They *will be used* during cloudy and/or cold weather, so select high-efficiency equipment with low-cost fuel sources.

Night insulation *must* be used consistently or else there will be a net heat loss. Making the nightly installation and removal automatic is recommended, but costly.

## SOURCES OF FURTHER INFORMATION

ASHRAE. 2003. *2003 ASHRAE Handbook—Applications*, Chapter 33, "Solar Energy Use." Atlanta: American Society of Heating, Refrigerating and Air-Conditioning Engineers, Inc.

Balcomb, J.D. 1984. *Passive Solar Heating Analysis: A Design Manual*. Atlanta: American Society of Heating, Refrigerating and Air-Conditioning Engineers, Inc.

Crosbie, M.J., ed. 1998. *The Passive Solar Design and Construction Handbook*, Steven Winter Associates. New York: John Wiley & Sons.

## GENERAL DESCRIPTION

Both passive and active solar thermal energy systems rely on capture and use of solar heat. *Active solar thermal energy systems* differ from *passive* systems in the way they utilize solar energy, mainly for space and water heating, by also using some conventional energy. This can allow for a greatly enhanced collection, storage, and use of the solar energy.

To be successful, a well-designed active solar system needs (1) an *appropriate thermal load* such as potable water, space air, or pool heating; (2) *collectors* such as flat-plate "solar panels;" (3) *thermal storage* to use the heat at a later time; and (4) *control*, typically automatic, to optimize energy collection and storage and for freeze and overheat protection.

The "working fluid" or "coolant" that moves heat from the collectors to the storage device is typically water, a water/glycol solution, or air. The heat storage medium is often water but could be a rock-bed or a high-mass building itself for air-coolant systems.

The solar energy collectors are most often of "fixed" orientation and nonconcentrating, but they could be "tracking" and/or concentrating. "Flat-plate" collectors are most common and are typically installed as fixed and nonconcentrating. Large surface areas of collectors, or mirrors and/or lens for concentration, are needed to gather heat and to achieve higher temperatures.

There are many different types of active solar thermal energy systems. For example, one type often used for potable water heating is flat-plate, pressurized water/glycol coolant, and two-tank storage. An internal double-wall heat exchanger is typically employed in one tank, known as the "preheat tank," and the other tank, plumbed in series, is a conventional water heater. These preheat tanks are now widely available due to nonsolar use as "indirect water heaters." One-tank systems typically have an electric-resistance heating element installed in the top of the special tank.

## HIGH-PERFORMANCE STRATEGIES

The following strategies are general in nature and are presented as a guideline to helping maximize the performance of an active thermal solar design.

Do conservation first. Minimizing the heating load will reduce conventional and renewable heating systems' sizes, and yields the best economics.

In the northern hemisphere, the solar collectors must face due south for optimal performance. If not possible, make them within ±10 degrees of due south. In the southern hemisphere, the solar collectors look north.

When using flat-plate collectors for space heating, mount them at an angle equal to the local latitude plus 15 to 20 degrees. For water heating, use the local latitude plus 10 degrees.

Calculate the optimal thermal storage—about one day's heat storage (or less) often yields the best economics. Place the thermal storage device within the heated space and be sure it is highly insulated, including under its base.

Seeking a very high annual solar savings fraction ($f_s$) often leads to disappointment and poor economics, so keep expectations reasonable. Even a 25% annual fraction represents a substantial reduction in conventional energy use. A highly optimized active solar thermal domestic water heating system, in an appropriate climate, often has about a 60% solar fraction. Combining active solar with passive, PVs, wind, and other renewable energy sources is often the most satisfactory way to achieve a very high annual solar savings fraction.

## KEY ELEMENTS OF COST

Active solar energy systems must be engineered; otherwise poor performance is likely. Fortunately, good design tools, such as F-Chart software, are readily available.

Well-designed, factory-assembled collectors are recommended, but are fairly expensive. Site-built collectors tend to have lower thermal performance and reliability.

Storage tanks must be of high quality and be durable. Water will eventually leak, so proper tank placement and floor drains are important. For rock storage, moisture control and air-entrance filtration are important for mold growth prevention.

Quality and appropriate pumps, fans, and controls can be somewhat expensive. Surge protection for all the electrical components is highly advised. Effective grounding, for lightning and shock mitigation, is normally required by building code. For liquid coolants in sealed loops, expansion tanks and pressure relief valves are needed. For domestic water heating, a temperature-limiting mixing valve is required for the final potable water to prevent scalding; even nonconcentrating systems can produce 180°F (80°C) or so water at times. All thermal components require insulation for safety and reduced heat loss.

Installation requires many trades: a contractor to build or install the major components, a plumber to do the piping and pumps (water systems), an HVAC contractor to install ducts (air systems) and/or space-heating heat exchangers, an electrician to provide power, and a controls specialist.

Conventional backup systems are still needed. They *will be used* during cloudy and/or cold weather, so select high-efficiency equipment with low-cost fuel sources.

## SOURCES OF FURTHER INFORMATION

ASHRAE. 1988. *Active Solar Heating Systems Design Manual*. Atlanta: American Society of Heating, Refrigerating and Air-Conditioning Engineers, Inc.

ASHRAE. 2003. *2003 ASHRAE Handbook—Applications*, Chapter 33, "Solar Energy Use." Atlanta: American Society of Heating, Refrigerating and Air-Conditioning Engineers, Inc.

Howell, J., R. Bannerot, and G. Vliet. 1982. *Solar-Thermal Energy Systems: Analysis and Design.* New York: McGraw-Hill.

Klein, S., and W. Beckman. 2001. *F-Chart Software.* Madison, WI: The F-Chart Software Company. Available from www.fchart.com/index.shtml.

## ASHRAE GreenTip #30:
## Solar Energy System—Photovoltaic

### GENERAL DESCRIPTION

Light shining on a PV cell, which is a solid-state semiconductor device, liberates electrons that are collected by a wire grid to produce direct current electricity.

The use of solar energy to produce electricity means that PV systems reduce greenhouse gas emissions, electricity cost, and resource consumption. Electrical consumption can be reduced. Because the peak generation of PV electricity coincides with peak air-conditioning loads (*if* the sun shines then), peak electricity demands (from the grid) may be reduced, though it is unlikely without substantial storage capacity.

PV can also reduce electrical power installation costs where the need for trenching and independent metering can be avoided. The public appeal of using solar energy to produce electricity results in a positive marketing image for PV-powered buildings and, thus, can enhance occupancy rates in commercial buildings.

While conventional PV design has focused on the use of independent applications in which excess electricity is stored in batteries, grid-connected systems are becoming more common. In these cases, electricity generated in excess of immediate demand is sent to the electrical grid, and the PV-powered building receives a utility credit. Grid-connected systems are often integrated into building elements. Increasingly, PV cells are being incorporated into sunshades on buildings for a doubly effective reduction in cooling and electricity loads.

PV power is being applied in innovative ways. Typical economically viable commercial installations include the lighting of parking lots, pathways, signs, emergency telephones, and small outbuildings.

A typical PV module consists of 33 to 40 cells, which is the basic block used in commercial applications. Typical components of a module are aluminum, glass, tedlar, and rubber; the cell is usually silicon with trace amounts of boron and phosphorus.

Because PV systems are made from a few relatively simple components and materials, the maintenance costs of PV systems are low. Manufacturers now provide 20-year warranties for PV cells.

PV systems are adaptable and can easily be removed and re-installed in other applications. Systems can also be enlarged for greater capacity through the addition of more PV modules.

### WHEN/WHERE IT'S APPLICABLE

PV is well suited for rural and urban off-grid applications and for grid-connected buildings with air-conditioning loads. The economic viability of PV depends on the distance from the grid, electrical load sizes, power line extension costs, and incentive programs offered by governmental entities or utilities.

PV applications include prime buildings, outbuildings, emergency telephones, irrigation pumps, fountains, lighting for parking lots, pathways, security, clearance, billboards, bus shelters or signs, and remote operation of gates, irrigation valves, traffic signals, radios, telemetry, or instrumentation.

Grid-connected PV systems are better suited for buildings with peak loads during summer cooling operation but are not as well suited for grid-connected buildings with peak wintertime loads.

Note that a portion of a PV electrical system is direct current, so appropriate fusing and breakers may not be readily available. A PV system is also not solely an electrical installation; other trades, such as roofing and light steel erectors, may be involved with a PV installation. When a PV system is installed on a roof or wall, it will likely result in envelope penetrations that will need to be sealed.

## PROS AND CONS

### Pro

1. Reduces greenhouse gas emissions.
2. Reduces nonrenewable energy demand, with the ability to help offset demand on the electrical grid during critical peak cooling hours.
3. Enhances green-image marketing.
4. Lowers electricity consumption costs and may reduce peak electrical demand charges.
5. Reduces utility infrastructure costs.
6. Increases electrical reliability for the building owner; may be used as part of an emergency power backup system.

### Con

1. Relatively high initial capital costs.
2. Requires energy storage in batteries or a connection to electrical utility grid.
3. May encounter regulatory barriers.
4. High-capacity systems require large-building envelope areas that are clear of protuberances and have uninterrupted access to sunshine.
5. Capacity to supply peak electrical demand can be limited, depending on sunshine during peak hours.

## KEY ELEMENTS OF COST

The following provides a possible breakdown of the various cost elements that might differentiate a PV system from a conventional one and an indication of whether the net cost for this system is likely to be lower (L), higher (H), or the same

(S). This assessment is only a perception of what might be likely, but it obviously may not be correct in all situations. **There is no substitute for a detailed cost analysis as part of the design process.** The listings below may also provide some assistance in identifying the cost elements involved.

First Cost

- PV modules                                    H
- Wiring and various electrical devices          H
- Battery bank                                   H
- Instrumentation                                H
- Connection cost (if grid-connected)            H

Recurring Cost

- Electricity                                    L

SOURCES OF FURTHER INFORMATION

California Energy Commission, Renewable Energy Program, www.energy.ca.gov/renewables/.

Canadian Renewable Energy Network, www.canren.gc.ca.

National Renewable Energy Laboratory (NREL), *PV for Buildings*, www.nrel.gov/buildings/pv/index.html.

NRC. *Photovoltaic Systems Design Manual*. Natural Resources Canada, Office of Coordination and Technical Information, Ottawa ON CAN K1A 0E4.

NRC. RETSCREEN (renewable energy analysis software), Natural Resources Canada, Energy Diversification Research Laboratory, Varennes PQ CAN J3X 1S6; tel 1 450 652 4621; www.retscreen.gc.ca.

*Photovoltaics*, Solar Energy Technologies Program, Energy Efficiency and Renewable Energy, US Department of Energy, www1.eere.energy.gov/solar/photovoltaics.html.

PV Power Resource Site, www.pvpower.com.

Renewable Energy Deployment Initiative (REDI) (a Canadian federal program that supports the deployment of renewable technologies; some technologies qualify for incentives), www.nrcan.gc.ca/redi.

School of Photovoltaic and Renewable Energy Engineering, University of New South Wales, www.pv.unsw.edu.au.

Solar Energy Industries Association (SEIA), Washington, DC, www.seia.org/.

Sustainable Sources, www.greenbuilder.com.

WATSUN-PV (simulation software), University of Waterloo, Waterloo ON CAN.

# ASHRAE GreenTip #31:
## Solar Protection

### GENERAL DESCRIPTION

Shading the building's transparent surfaces from solar radiation is mandatory during summer and sometimes even necessary during winter. This way, it is possible to prevent solar heat gains when they are not needed and to control daylighting for minimizing glare problems. Depending on the origin of solar radiation (direct, diffuse, reflected), it may be possible to select different shading elements that provide more effective solar control.

Depending on the specific application and type of problem, there may be different options for selecting the optimum shading device. The decision can be based on several criteria, from aesthetics to performance and effectiveness or cost. Different types of shading elements are suitable for a given application, result in varying levels of solar control effectiveness, and have a different impact on indoor daylight levels, natural ventilation, and overall indoor visual and thermal comfort conditions.

There are basically three main groups of solar control devices: (1) *External shading devices* can be fixed and/or movable elements. They have the most apparent impact on the aesthetics of the building. If properly designed and accounted for, they can become an integral part of the building's architecture, integrated into the building envelope. *Fixed types* are typically variations of a horizontal overhang and a vertical side fin, with different relative dimensions and geometry. Properly designed and sized, fixed external shading devices can be effective during summer, while during winter they allow the desirable direct solar gains through the openings. This is a direct positive outcome given the relative position of the sun and its daily movement in winter (low solar elevation) and summer (high solar elevation). *Movable types* are more flexible, since they can be adjusted and operated either manually or automatically for optimum results and typically include various types and shapes of awnings and louvers. (2) *Interpane shading devices*, are usually adjustable and retractable louvers, roller blinds, screens, or films, which are placed within the glazing. This type of a shading device is more suitable for solar control of scattered radiation or sky diffuse radiation. Given that the incident solar radiation is already absorbed by the glazing, thus increasing its temperature, one needs to take into account the heat transfer component to the indoor spaces. (3) *Internal shading devices* are very common because of indoor aesthetics, offering privacy control, and their easy installation, accessibility, and maintenance. Although on the interior, they are very practical and, most of the time, necessary; their overall thermal behavior needs to be carefully evaluated, since the incident solar radiation is trapped inside the space and will be absorbed and turn into heat if not properly controlled (i.e., reflecting solar radiation outward through the opening). Numerous types or combinations of the various shading devices are also possible, depending on the application.

## HIGH-PERFORMANCE STRATEGIES

*Natural shading:* Deciduous plants, trees, and vines offer effective natural shading. It is critical for their year-round effectiveness not to obstruct solar radiation during winter in order to increase passive solar gains. Plants also have a positive impact on the immediate environment surrounding the building (microclimate) by taking advantage of their evaporative cooling potential. However, the plants need some time to grow, may cause moisture problems if they are too close to opaque elements, and can suffer from plant disease. The view can be restricted and some plants, especially large leafless trees, can still obstruct solar radiation during winter and may reduce natural ventilation. In general, for deciduous plants, the shading effect is best for east and west orientations, along with southeast and southwest.

*Louvers:* Also referred to as *venetian blinds* and can be placed externally (preferable) or internally (easier maintenance and installation in existing buildings). The external louvers can be fixed in place with rotating or fixed tilt of the slats. The louvers can also be retractable. The slats can be flat or curved. Slats from semitransparent material allow for outdoor visibility. The louvers can be operated manually (slat tilt angle, up or down movement) or they can be electrically motor driven. Adjusting the tilt angle of the slats or raising/lowering the panel can change the conditions from maximum light and solar gains to complete shading. Louvers can also be used to properly control air movement during natural ventilation. Slat curvature can be utilized to redirect incident solar radiation before entering into the space. Slat material can have different reflective properties and can also be insulated. During winter, fully closed louvers with insulated slats can be used at night for providing additional thermal insulation at the openings.

*Awnings:* External or internal awnings can be fixed in place, operated manually, or driven electrically by a motor that can also be automated. A preference lies with light-colored materials for high solar surface reflectivity. Awnings are easily installed on any type and size of opening and may also be used for wind protection during winter to reduce infiltration and heat losses.

## KEY ELEMENTS OF COST

*Natural shading:* Natural shading is usually low cost, reduces glare, and, depending on the external building facade, can improve aesthetics. Plants should be carefully selected to match local climatic conditions in order to optimize watering needs.

*Louvers:* External electrically driven and automated units have a higher cost and need to account for maintenance costs of motors but are more flexible and effective. Louvers are difficult to clean on a regular basis. Nonretractable louvers somewhat obstruct outward vision.

*Awnings:* Awning fabric needs periodic replacement depending on local wind conditions. Electrically driven and automated units have a higher cost and need to account for maintenance costs of motors.

SOURCES OF FURTHER INFORMATION

Argiriou, A., A. Dimoudi, C.A. Balaras, D. Mantas, E. Dascalaki, and I. Tselepidaki. 1996. *Passive Cooling of Buildings.* M. Santamouris and A. Asimakopoulos, eds. London: James & James Science Publishers Ltd.

Baker, N. 1995. *Light and Shade: Optimising Daylight Design.* European Directory of sustainable and energy efficient building. London: James & James Science Publishers Ltd.

Givoni, B. 1994. *Passive and Low Energy Cooling of Buildings.* New York: Van Nostrand Reinhold.

Stack, A., J. Goulding, and J.O. Lewis. *Solar Shading.* European Commission, DG TREN, Brussels. http://erg.ucd.ie/down_thermie.html.

# 13

# LIGHTING SYSTEMS

Coverage of the subject of lighting in a guide for HVAC&R designers is not intended to make them into lighting experts but rather to familiarize them with the basic process of lighting design and the materials, allowing HVAC design engineers to interact effectively with lighting designers and architects in creating effective systems. Lighting has a significant impact on building loads and energy usage; therefore, it is important that the HVAC&R designer understand the role played by these other team members.

Proper lighting system design should always involve an experienced professional, preferably one who is a member of the green building design team.

## ELECTRIC LIGHTING

### Efficient Lighting Design

Lighting design as a practice and a profession has evolved considerably since the early days of interest in energy efficiency. Lower lighting levels (footcandles [Lux]) have become common and, due to significant advances in lighting equipment efficiency, power levels are much lower than 25 years ago.

For any lighting designer setting out to perform a lighting design, consulting the *Lighting Handbook* of the Illuminating Engineering Society of North America (IESNA) is highly recommended. However, assuming some familiarity with the process, the following guidance should yield good results.

Many buildings can be efficiently lighted using carefully selected standard lighting systems. Assuming that some basic rules of spacing and lamp or luminaire wattage are followed, successful lighting designs with lighting power densities of 0.7–1.0 W/ft$^2$ (8–11 W/m$^2$) can be easily applied to most building types.

Most energy codes have lighting power limits between 1.0 and 1.5 W/ft$^2$ (11 and 16 W/m$^2$). The primary exceptions are storage buildings (less), retail stores (more), and hospitality facilities (more). However, even in some stores and hospitality buildings, the same efficient lighting approaches suitable for schools, offices, and most other common building types can also be used.

### Efficient Lighting Systems

Tables 13-1, 13-2, and 13-3 contain a listing of one state's standard lighting systems and related criteria that will generally satisfy modern (IESNA) light level recommendations, as well as comply with energy codes and meet a modest project budget.

- Table 13-1 lists two common lighting systems that can be used for a wide variety of project types at very low cost.
- Table 13-2 lists lighting designs suitable for private and open office areas and similar spaces, such as exam rooms in clinics and hospitals.
- Table 13-3 lists a variety of common lighting systems that can be used in industrial and commercial applications

For most spaces, designers should employ lighting layouts that conform to these criteria. Spacing measurements are taken from the plan-view center of the luminaire. Luminaires should be mounted at least one-third of the indicated mounting distance away from any ceiling-high partition.

If more than one type of luminaire (excluding exit signs) is to be located within one space enclosed with ceiling high partitions, the spacing between different luminaires must be the larger of the required spacing for the two luminaries.

### Table 13-1    General Use Systems

| Primary Application | Luminaire Type | Lamps or Total Lamp Watts | Spacing between Luminaries (in Plan View) | Lamp Ballast System |
|---|---|---|---|---|
| General use spaces of all types | Nominal 4 ft (1.2 m) recessed or surface-mounted fluorescent troffer (parabolic, basket, lensed, etc.) with high-efficiency electronic ballast | (2) F32T-8 lamps | No less than 8 ft (2.4 m) OC | Maximum 56 W, 45–48 W, low-ballast-factor ballast preferred |
| General use spaces of all types | Nominal 2 ft (0.6 m) recessed or surface-mounted fluorescent troffer (parabolic, basket, lensed, etc.) with electronic ballast | (3) F17T-8 lamps | No less than 8 ft (2.4 m) OC | Maximum 52 W |

## Table 13-2  Lighting for Offices, Including Commercial, Academic, and Institutional

| Primary Application | Luminaire Type | Lamps or Total Lamp Watts | Spacing between Luminaires in Plan View | Lamp Ballast System |
|---|---|---|---|---|
| Open offices | Suspended linear fluorescent fixtures, consisting of nominal 4 ft (1.2 m) sections in continuous rows with electronic ballast(s)  Nominal 4 ft (1.2 m) recessed or surface-mounted fluorescent troffers | One F54T-5HO or two F32T-8 or two F28T-5  Two F32T-8 lamps | Continuous rows no closer than 15 ft (4.6 m) apart  Regular grid 8 ft (2.4 m) OC | Maximum 60 input W per 4 ft (1.2 m) unit, minimum ceiling height 10 ft (3.0 m)  Maximum 48 input W per luminaire |
| Very small private offices < 105 ft² (10 m²) | One recessed or suspended 4 ft (1.2 m) linear fluorescent fixture | Three F32T-8 lamps | One luminaire per office | Maximum 90 input W; minimum ceiling height 9 ft (2.7 m) for suspended fixtures |
| Small private offices 105–125 ft² (10–12 m²) | Two recessed or suspended 4 ft (1.2 m) linear fluorescent fixtures | Two F32T-8 lamps per fixture | No less than 6 ft (1.8 m) OC | Maximum 48 input W to each luminaire; minimum ceiling height 9 ft (2.7 m) for suspended fixtures |

**Table 13-2 *(Continued)* Lighting for Offices, Including Commercial, Academic, and Institutional**

| | | | | |
|---|---|---|---|---|
| Small private offices 125–160 ft$^2$ (10–15 m$^2$) | Two recessed or suspended 4 ft (1.2 m) linear fluorescent fixtures | Two F32T-8 lamps per fixture | No less than 6 ft (1.8 m) OC | Maximum 56 input W to each luminaire; minimum ceiling height 9 ft (2.7 m) for suspended fixtures |
| Medium private offices 160–200 ft$^2$ (15–19 m$^2$) | Two recessed or suspended 4 ft (1.2 m) linear fluorescent fixtures<br><br>Three recessed or suspended 4 ft (1.2 m) linear fluorescent fixtures<br><br>Four recessed 2 ft (0.6 m) linear fluorescent fixtures | Three F32T-8 lamps per fixture<br><br>Two F32T-8 lamps per fixture<br><br>Two F17T-8 lamps per fixture | No less than 6 ft (1.8 m) OC | Maximum 72 input W to each luminaire<br><br>Maximum 48 input W to each luminaire<br><br>Maximum 36 input W per fixture |
| Executive offices and conference rooms 200-250 ft$^2$ (19–23 m$^2$) | Four recessed or suspended 4 ft (1.2 m) linear fluorescent fixtures four recessed 2 ft (0.6 m) linear fluorescent fixtures | Two F32T-8 lamps per fixture<br><br>Two F32T-8U lamps per fixture | No less than 8 ft (2.4 m) OC | Maximum 48 input W to each luminaire |

Key to table abbreviations: OC = on center; < = less than.

### Table 13-3 Other Common Lighting Systems

| Primary Application | Luminaire Type | Lamps or Total Lamp Watts | Spacing between Luminaires in Plan View | Lamp Ballast System |
|---|---|---|---|---|
| Lobbies, atriums, etc.<br><br>Industrial space | Metal halide, induction, or multiple compact fluorescent lamps (of equivalent lamp watts with electronic ballasts), downlights, pendants, etc. | 100 W or less<br><br>150 W or less<br><br>250 W or less<br><br>400 W or less | No less than 12 ft (3.7 m) OC<br>No less than 15 ft (4.6 m) OC<br>No less than 18 ft (5.5 m) OC<br>Not less than 22 ft (6.7 m) OC | Mounting height at least 12 ft (3.7 m) AFF; only recommended for high bay spaces |
| Corridors, lobbies, meeting rooms, etc. | Compact fluorescent (including twin tube, quad tube, or triple tube) or metal halide downlights, wallwashers, monopoints, and similar directional luminaires<br><br>Wall sconces using any of the above light sources | 40 W or less<br><br>60 W or less<br><br>80 W or less<br><br>100 W or less | No less than 6 ft (1.8 m) OC<br>No less than 8 ft (2.4 m) OC<br>Not less than 10 ft (3.0 m) OC<br>No less than 12 ft (3.7 m) OC | Any space height |
| Undercabinet and undershelf task lighting | Hardwired undercabinet or undershelf fluorescent luminaires, nom. 2, 3, or 4 ft (0.6, 0.9, or 1.2 m) in length and employing an electronic ballast | No greater than 8.5 W/ft (27.09 W/m) of luminaire | When mounted underneath permanent overhead cabinets | Luminaires may be mounted end-to-end if needed to accommodate cabinet length |

## Table 13-3 *(Continued)* Other Common Lighting Systems

| | | | | |
|---|---|---|---|---|
| Lobby, executive office, and conference room accent lighting | Low-voltage downlights, accent lights, or monopoint lights having an integral transformer | Rated at 50 W or less | No less than 8 ft (2.4 m) OC | For accent lighting only; should not be used for general lighting |
| Copyroom, storeroom, etc. | Nominal 4 ft (1.2 m) recessed or surface-mounted fluorescent troffer, wraparound, strip lights, etc., with electronic ballast | One or two lamps totaling 64 W or less | No less than 8 ft (2.4 m) OC | Maximum 60 input W to each luminaire |
| Small utility, storage, and closet spaces | Single-lamp fluorescent with electronic ballast (strip, wrap, industrial, or other fixture) | 32 W | One luminaire in a closet, electric room, or other small space | Maximum 35 input W to each luminaire |
| Storage and utility spaces | Industrials, wraparounds, strip lights, etc., consisting of nominal 4 ft (1.2 m) sections | Two F32T-8 | Individual 12-lamp luminaires 8 ft (2.4 m) OC  Continuous single-lamp rows no closer than 8 ft (2.4 m) apart | 48 input W per two lamps |
| Bathroom vanities and stairwells  Exit signs | Two-lamp fluorescent with electronic ballast (wrap, cove, troffer, corridor, vanity, valence, or other fixture)  LED | Two F32T-8  No greater than 3.5 W per sign | One luminaire per vanity in a toilet or locker room or one luminaire per landing in a stairwell | Maximum 48 input W to each luminaire |

Key to table abbreviations: AFF = above finished floor; LED = light-emitting diode; OC = on center; < = less than.

*None* of the following luminaires should be employed:

- Luminaires employing Edison (standard screw-in) baseline voltage sockets or halogen lamps using any sockets rated over 150 W
- Luminaires designed for incandescent or halogen low-voltage lamps exceeding 75 W
- Track lighting systems of any kind or voltage of operation
- Line-voltage monopoints permitting the installation of track luminaires.

However, for every 20 luminaires meeting these requirements, a single hard-wired luminaire of any type (except track) rated not more than 150 lamp watts may be placed as desired. This permits architects, interior designers, and lighting designers the ability to add lighting for aesthetic effects or décor without an unreasonable energy burden. Note that if more than one such luminaire is permitted, any number of them may be located in any of the project's spaces. In other words, in a project with one hundred luminaires from Tables 13-1, 13-2, or 13-3, five decorative luminaires would be permitted, and they could all be installed over the receptionist's desk.

These lists are not intended to be comprehensive but rather straightforward and instructive. There are certainly other good (and efficient) designs not listed. Professional lighting design assistance may be needed to reach optimum performance for specific conditions.

In all cases, be certain to review the subsequent sections on other aspects of lighting for detailed information on product specifications and additional energy-saving ideas.

### When/Where Applicable

The above systems are generally good for most conventional space types with ordinary ceiling systems. These include:

- Typical private offices
- Typical open office areas
- Office area corridors
- Conference rooms and classrooms
- Meeting and seminar rooms
- Most laboratories
- Equipment, server, and cable rooms
- Building lobbies
- Elevator lobbies
- Building "core" and circulation areas
- Industrial areas, shops, and docks
- "Big box" and grocery stores

For commercial buildings, these recommendations assume standard acoustical tile ceilings. For industrial buildings, open bar joist construction is assumed. Ceiling heights are assumed to be standard as well.

## Efficient Lamps and Ballasts

The lighting industry has made significant improvements in technology over the last 25 years. Recent improvements, while not as widely publicized, continue to permit good lighting at decreasing power levels. Proper specification of lamps and ballasts are an important part of achieving these results.

**Specifications.** The following specifications are recommended to ensure the latest technology is being employed.

***Ballasts.*** Ballasts for all fluorescent lamps and for HID lamps rated 150 W and less should be electronic. Harmonic distortion should be less than 20%,

- *T-8 System Ballasts*

  Four-foot (1.2-meter) T-8 fluorescent lighting systems should employ "high-efficiency" electronic ballasts. Because instant-start ballasts are the most efficient and least costly, they should be used in all longer duty cycle applications where the lights are turned on and off infrequently. Fluorescent systems controlled by motion sensors in spaces where the lights will be turned on and off frequently should employ program-start ballasts.

  Designers are strongly encouraged to use low-ballast-factor ballast (BF < 0.80) whenever possible. T-8 low-ballast-factor and normal-ballast-factor ballasts should be "high efficiency" electronic, not exceeding 28 input W per 4 ft (1.2 m) lamp at BF > 0.85 and not exceeding 24 input W for BF > 0.70 (ANSI free air rating).

  In lieu of the above, electronic dimming ballasts may be used as needed.

- *Metal Halide Ballasts*

  Metal halide ballasts 150 W and less should be electronic.

  Metal halide systems greater than 150 W should use linear-reactor pulse-start type ballasts wherever 277-volt power is available; for other voltages, pulse-start lamps and ballasts shall be used.

***Lamps.*** The lamps listed in Table 13-4 represent the best common lamp types to employ. Note that this list is not comprehensive, and a better choice for a particular project may not be listed. However, for the majority of applications, this list is a good guide.

The lamps recommended for primary lighting systems are both energy efficient and have long lives, representing excellent cost benefits. Lamps for other applications are generally less efficient, have shorter lives, or both. Also, designers should use the minimum number of different types of lamps on a project to reduce maintenance costs and improve stocking.

## Table 13-4 Common Lamp Types

| Generic Lamp Types | Applications | Requirements |
|---|---|---|
| 4 ft (1.2 m) T-8 lamps F32T-8 | Primary lighting systems in commercial, institutional, and low bay industrial spaces | TCLP compliant (low mercury) lamps with barrier coat and high lumen phosphor (minimum 3,100 initial lumens) Premium long-life-rated lamp |
| Fluorescent T-5 and T-5HO lamps F14T-5, F21T-5, and F28T-5; F24T-5HO, F39T-5HO, and F54T-5HO | Primary lighting systems in commercial, institutional, and low and high bay industrial spaces | Standard T-5 and T-5HO lamps |
| Metal halide pulse start lamps over 250 W | Primary lighting systems in large spaces with very high ceilings and/or special lighting requirements | Pulse-start lamps only— Be certain to specify pulse-start ballasts Use linear reactor ballasts on 277-volt systems |
| Fluorescent T-8 lamps F17T-8, F25T-8 | Secondary and specialized applications in commercial, institutional, and low bay industrial spaces | TCLP compliant (low mercury) lamps with barrier coat and 800 series phosphor Premium long-life-rated lamps |
| Compact fluorescent long lamps F40TT-5, F50TT-5, and F55TT-5 | Specialized applications in commercial, institutional, and low bay industrial spaces | Standard long twin-tube lamps |
| Compact fluorescent 4-pin lamps CF13, CF18, CF26, CF32, CF42, CF57 and CF70 | Downlighting, wallwashing, sconces, and other common space and secondary lighting systems in commercial, institutional, and low bay industrial spaces | Standard twin-, quad-, triple-tube and four-tube lamps. |
| Halogen MR16 lamps | Accent lighting for art and displays only; do not use for general lighting | Halogen IR 12-volt compact reflector lamps |

## Table 13-4 (Continued) Common Lamp Types

| | | |
|---|---|---|
| Ceramic PAR and T HID Lamps PAR20, PAR30, PAR38, ED17, and T-6 | Downlighting, accent lighting, and other special, limited applications in commercial, institutional, and low bay industrial spaces; retail display lighting | 39-, 70-, 100-, and 150-W ceramic lamps |
| Halogen infrared reflecting PAR30 and PAR38 lamps | Applications requiring full range dimming; retail display lighting | Halogen IR 50-, 60-, 80-, and 100-W reflector lamps |

### Lighting Power Density Criteria

A complete, hardwired lighting system, to be efficient, should be installed with the following lighting power density (LPD) requirements:

- Private offices shall not exceed a connected power density of 0.9 W/ft$^2$ (10 W/m$^2$).
- Open office areas shall not exceed a connected power density of 0.8 W/ft$^2$ (9 W/m$^2$).
- Conference rooms and similar spaces shall not exceed a connected power density of 1.2 W/ft$^2$ (13 W/m$^2$).
- Core areas, including lobbies, elevator lobbies, mailrooms, lunchrooms, restrooms, copy rooms, locker rooms, and similar spaces, shall not exceed a connected LPD of 0.8 W/ft$^2$ (9 W/m$^2$).
- Hallways, corridors, storage rooms, mechanical and electrical rooms, and similar spaces shall not exceed a connected power density of 0.7 W/ft$^2$ (8 W/m$^2$).
- Any other space not listed shall not exceed a connected power density of 0.6 W/ft$^2$ (6 W/m$^2$).

Additional lighting, such as lighting within furniture systems, should not be installed in a space unless a more complete analysis and design are undertaken. These systems need to be carefully coordinated with the permanent lighting systems of the building. *Exception*: portable plug-in lamps and under-cabinet luminaires attached to the underside of modular furniture, overhead cabinets, bins, or shelves should be used where needed.

There are a few space types, such as video teleconferencing rooms, showrooms, retail space, and food service space, that usually require more lighting power than provided above. For these uncommon space types, an appropriate LPD requirement should be determined from ANSI/ASHRAE/IESNA Standard 90.1-2004 (or the latest edition).

## Application Notes

**Open Offices.** For general illumination in spaces with ceilings 9 ft, 6 in. (2.9 m) or higher, consider suspended linear fluorescent indirect, direct/indirect, or semi-indirect lighting systems, supplemented by task lights. General layouts should be between 0.6 and 0.8 W/ft$^2$ (6 and 9 W/m$^2$), using high-performance luminaires and T-8, T-5, or T-5HO lamps. Task lights can be used where needed. If troffers are preferred, consider using T-8 lamps as specified with low-ballast-factor ballasts. This will ensure appropriate energy use while maintaining recommended light levels. Task lighting should be added underneath shelves and bins as required.

**Private Offices.** Suspended linear fluorescent lighting should be a first consideration for private offices, although recessed troffers can also be used. Luminaires should use T-8 or T-5 lamps. LPD should be around 0.8–0.9 W/ft$^2$ (9–10 W/m$^2$). Task lights can be used where needed.

**Executive Offices, Board and Conference Rooms.** Executive offices can be designed similarly to private offices (above). If desired, a premium approach using compact fluorescent downlights, wallwashers, and/or halogen accent lights can be used, but the overall design should not exceed 1.2 W/ft$^2$ (13 W/m$^2$). If the number of executive offices is high, lighting power levels should be reduced to match the recommendations for private offices (above).

**Classrooms.** Classrooms should be lighted using direct/indirect classroom lighting systems, with about 0.9 W/ft$^2$ (10 W/m$^2$) of connected power using T-8 "super" lamps and efficient electronic ballasts.

**Corridors.** In general, corridors should be lighted using compact fluorescent sconces, downlights, ceiling-mounted or close-to-ceiling decorative diffuse fixtures, or similar equipment. Power density should be about 0.5–0.6 W/ft$^2$ (5–6 W/m$^2$) overall. Note that these luminaires may be equipped with emergency battery backup when needed as an alternative to less attractive "bug eye" type emergency lighting.

**High Bay Spaces.** Industrial, grocery, and retail space without ceilings or with very high ceilings (usually 15 ft [4.6 m] and above) need special lighting fixtures. For mounting heights up to 20–25 ft [6.1–7.6 m], first try to use fluorescent industrial luminaires employing T-8 lamps, keeping in mind that two "super" 4 ft (1.2 m) T-8 lamps and a high light output "overdrive" ballast at 77 W produce as much light (mean lumens) as a 100 W metal halide lamp that, with ballast, consumes 120 W. And four "super" T-8 lamps with overdrive ballasts at 154 W produce as much light as a 175 W pulse-start metal halide (195–205 W) or a standard 250 W metal halide lamp (286–295 W with ballast).

For mounting heights above 25 ft (7.6 m), consider T-5HO high-bay luminaires. Similar savings relative to metal halide are possible. High-wattage metal halide should be reserved for very high mounting (above 50 ft [15.2 m]) and for special applications such as sports lighting.

**Other Applications.** The following luminaire types are generally recommended for these areas:

- Artwork, bulletin/display surfaces, etc., use compact fluorescent wallwashers or low-voltage monopoint lights.
- Utility spaces, including cable and equipment rooms, use two-lamp strip lights, industrials, or surface luminaires.
- Lobby spaces, cafeterias, and other public spaces, as much as possible use appropriate selections from among these luminaires.

For commercial buildings, these recommendations assume standard acoustical tile ceilings. For industrial buildings, open bar joist construction is assumed. Ceiling heights are assumed to be standard as well.

### Sources of Further Information

California Energy Commission. Title 24 2005 Reports and Proceedings. Available from www.energy.ca.gov.

IESNA. *Lighting Handbook*. Illuminating Engineering Society of North America. Available from www.iesna.org.

## DAYLIGHT HARVESTING

### General Description

Most buildings have *some* type of natural light that is transmitted into the building through windows and/or skylights. The majority of commercial, industrial, and institutional buildings have windows and, in some cases, skylights, clerestories, and more extensive fenestration systems.

From an energy perspective, the optimal use of daylight is to control the dimming or extinguishing of the electric lighting system when space lighting is supplemented by natural light transferred through fenestration in the building envelope. This process, called *daylight harvesting* (see Figure 13-1), is discussed in this section because of its significant potential. The prediction of daylight harvesting savings is not easy. There are added capital costs for daylight harvesting elements, such as dimming ballasts and photoelectric controls. It is important to justify these costs by accurately predicting the potential energy savings of daylight harvesting techniques.

From a lighting design perspective, daylight can be treated as any other light source and used to compose lighting design solutions with illuminance, luminance, contrast, color, and other lighting design elements. However, the lighting designer is challenged to deal with the fact that the light source location is given and that, in most cases, the only means available to change its characteristics is through blinds, shades, or other mechanical forms of attenuation and shielding.

Health Science Center Research Building, St. Louis, MO

**Figure 13-1 Example of daylight harvesting rendering.**

It is possible to simulate the performance of natural lighting to determine the amount and, to a certain extent, the quality of available daylight under varying conditions of season, time of day, and weather. However, this is exhaustive analytical work of a highly specialized nature, and it is recommended that appropriate experts perform such studies. In the meantime, some buildings can benefit tremendously from some *simple* daylight harvesting considerations.

### Basic Toplighting

Basic toplighting involves using simple skylights in the roof. (This is not to say that other toplighting configurations, such as the clerestory, roof monitor, or sawtooth roof, are not workable; indeed, they often have advantages over horizontal skylights.) Needless to say, there are many architectural considerations including structure, waterproofing, and other details. However, when used in a manner similar to light fixtures, laid out to provide uniform illumination, toplights are an acceptable way to illuminate single-story, large spaces using daylight.

Toplighting is best when a number of smaller skylights are used, much the way lighting systems use many light fixtures rather than one big light source in the middle of the room. Skylights should be diffuse or prismatic, *not* clear. Skylights do not have to incorporate light control louvers, since the optimum size of the skylight is chosen for "passive" skylighting (i.e., no active or moving elements needed to regulate the amount of interior light).

To determine the optimum size of skylights, one can download a program called SkyCalc from www.energydesignresources.com/resource/129/. This program, which is optimized for California and the Northwest in the US, can be applied with some care anywhere in North America. It takes into account location, utility rates, and other basic data and yields recommended skylight area.

*Note*: The ideal amount of fenestrated roof is generally around 5%. Most architects design skylights that are too big. An HVAC&R designer's input here, especially when backed up by calculations, can save a lot of energy.

### When/Where It's Applicable

Daylight harvesting is most likely to be suitable for large volume, single-story or top-story space types with ordinary structures. These include:

- Gyms
- Industrial workspace
- "Big box" retail
- Grocery stores
- Exhibition halls
- Storage
- Warehousing

In each case, automatic lighting controls that dim or extinguish electric lights when there is adequate daylight are essential or the energy savings will not be realized. (See also subsequent section on "Lighting Controls.")

### Pros/Cons of Daylighting

### Pro

1. Daylight harvesting offers significantly reduced energy costs (can be more than 60%) and reduced HVAC load (as long as solar gains don't outweigh electric lighting reductions).
2. It extends the electric lighting maintenance cycle (lamps last two to three times as long in calendar years).
3. It ensures low power use.
4. There are improved human factors and enjoyment of space.

### Con

1. Daylight harvesting requires close architectural, structural, and lighting design coordination.
2. There is no assurance that the design will meet exact project lighting requirements.
3. There is increased building cost.
4. There is a risk of poor design or installation workmanship, resulting in roof leaks.
5. Daylighting may not be suitable for uncommon room shapes, sizes, and/or finishes.
6. There is net decreased roof insulation.

### Sources of Further Information

NBI. 2001. *Advanced Lighting Guidelines*. www.newbuildings.org/lighting.htm. New Buildings Institute.
SkyCalc. www.energydesignresources.com/resource/129/.

## THE LIGHT CONVEYOR

The light conveyor is a specialized technique whereby light from a source is transmitted some distance from the source to light spaces, either along its length or some distance away. The source can be either natural light or an artificial source. It is described in GreenTip #32.

## LIGHTING CONTROLS

### General

While all modern energy codes require automatic shutoff controls for commercial buildings, implementing automatic controls in all building projects is a sound money- and energy-saving idea. There are two ways to reduce lighting energy use through controls:

- Turn lights off when not needed (reduce hours).
- Reduce lighting power to minimum need (reduce kilowatts).

By code, each interior space enclosed by ceiling-high partitions must have separate local switching and some form of automated OFF control (occupancy sensing, time-based scheduling, or other). In addition, wherever possible, provide separate switching for lights in daylighted zones. In order to comply with code requirements and ensure maximum energy savings, specify the most appropriate lighting control option(s) as described below and outlined in Table 13-5.

### Control Options

1. Ceiling-mounted motion sensor with transformer/relay, auxiliary relay, and series switch. The sensor should be located to look down upon the work area in order to detect small hand motion as well as major movements. The sensor may be mounted to the upper wall if a ceiling location is not workable. More than one sensor can be used for a large room or a room with obstructions, such as a library or server room. The main transformer relay should control the overhead lighting system (usually 277 volt) and the auxiliary relay should control at least one-half of a receptacle to switch task lights and other applicably controlled plug loads. Note that the light switch is in series so that it can only turn lights off in an occupied room; it can not override the motion sensor's OFF control.

2. Similar to the above, but without auxiliary relay and connection to receptacle. Must control at least two luminaires or banks of lighting groups and be equipped with two manual override (OFF) switches for either high-low light level switching or alternate fixture switching. May also be used to control multiple dimmers or a multi-channel preset dimming controller.

**Table 13-5    Recommended and Optional Lighting Controls**

| Type of Space | Minimum Recommended Control | Optional Control(s) |
|---|---|---|
| Private office, exam room | 1 | 2    2+4   1+8   2+4+8   2+8 |
| Open office | 3 | 3+4   3+8        3+4+8 |
| Conference rooms, teleconference rooms, boardrooms, classrooms | 2 | 1    2+8        1+8 |
| Server rooms, computer rooms, and other clean work areas | 5 | |
| Toilet rooms, copy rooms, mail rooms, coffee rooms | 5 or 6 | |
| Individual toilets, janitor closets, electrical rooms, and other small spaces | 6 | |
| Public corridors | 3 | |
| Corridors, hallways, lobbies (private spaces only) | 3 | 3+8 |
| Public lobbies | 7 | 7+8 |
| Industrial work areas | 7 | 7+8 |
| Warehousing and storage | 9 (HID systems) 3 or 5 (fluorescent systems) | 3 or 5 +8 (fluorescent) |
| Stores, newsstands, food service | 7 | |
| Mechanical rooms | Manual switching only | |
| Stairs | None | Motion sensors can be used to reduce light levels to minimum egress lighting levels only |

3.  Ceiling-mounted motion sensors connected to programmable time controller. During programmed ON times, the lights remain on. During programmed OFF times, motion within the space initiates lights on for a time out period. Controller shall be programmable according to the day of the week and shall have an electronic calendar to permit programming holidays.

4.  Workstation motion sensor connected to a plug strip or task light with auxiliary receptacle.

5.  One or more ceiling-mounted motion sensors with transformer/relay, minimum two luminaires controlled.

6. Switchbox motion sensor; one or more luminaires controlled.
7. Programmable time controller with manual override switch(es) located in a protected or concealed location. Separate zones for retail and similar applications where displays can be controlled separately from general lighting. May also control dimmers.
8. In addition to any of the above, an automatic daylighting sensor connected to dimming ballast(s) in each luminaire in the daylighted zone.
9. A motion sensor connected to a high-low lighting system.

When using controls such as motion sensors or daylight sensors, be very thorough and carefully read the manufacturer's literature. Different sensors work for different applications, and their sensing systems are optimized. For instance, avoid wallbox motion sensors except in spaces where their sensing field is appropriate. For spaces with small-motion work, a look-down sensor (from the ceiling) generally works much better than a look-out sensor (from a wallbox).

### Applicability

The above controls are applicable to most commercial, institutional, and industrial buildings. Use common sense in special spaces, keeping in mind safety and security. Never switch path-of-egress lighting systems except with properly designed emergency transfer controls.

### Pros/Cons

### Pro

1. Low to moderate costs for most space types
2. Virtually no maintenance
3. Generally will lower energy use.

### Con

1. If controls are not properly commissioned, unacceptable results may occur until they are fixed.
2. There is no assurance that the controls meet exact project lighting requirements.
3. Substitutions and value engineering can easily cause bad results.

## COST CONSIDERATIONS

### Lighting Systems

The systems described above are generally low-to-moderate-cost lighting systems. On average, they also use low-maintenance lamps and ballasts. The combination of low first cost, low maintenance, and low energy use leads to lighting choices that are among the most economical available.

### Lamps and Ballasts

The costs of premium lamps and ballasts over conventional lamps and ballasts can be as much as 100% of the cost of the materials. This can increase the cost of a lighting system by 20%–30%.

However, premium lamps, especially the T-8 4 ft (1.2 m) lamps shown, offer the following specific benefits:

- Increased lamp life by as much as 50%. In a T-8 application, this can be 5,000 to 10,000 hours. The cost of replacing a lamp is about 75% labor and 25% material. Relamping cost savings alone pay the difference.
- Reduced lamp energy use when used with the correct ballast. In a typical T-8 application, this means achieving energy savings of around 6 W per lamp. At 3000 annual hours and 8.5 cents per kWh, the combination saves over $1.50 per year in energy costs, per lamp. The premium for a two-lamp ballast and lamps is about $12.00. The energy savings pay for the added costs in about four years. (Costs are as of year 2003.)

### Daylight Harvesting

Daylight harvesting is a potentially complex undertaking in which the first cost of lighting remains the same, the cost of lighting controls increases, and the added cost of skylights and/or structural changes/complications are incurred as well. To be cost-effective, this needs to be offset by a combination of HVAC energy savings, lighting energy savings, HVAC system first-cost reduction, and perhaps savings from utility incentives or tax credits. Expect daylight harvesting systems to yield a four-to-five-year simple payback *with* a utility incentive, six to eight years or more *without*.

### Controls

The lighting control systems described above are generally of low to moderate cost. However, using better quality sensors and separate transformer/relay packs with remote sensors costs much more than wallbox devices. Savings can range from modest to considerable, depending on the building and occupants.

### SOURCES OF FURTHER INFORMATION

*Advanced Lighting Guidelines*, www.newbuildings.org/lighting.htm.
Architectural Energy Corporation, www.archenergy.com.
BetterBricks, www.betterbricks.com.
California Energy Commission, Title 24 2005 Reports and Proceedings, available from www.energy.ca.gov.
The Collaborative for High Performance Schools, www.chps.net.

Energy Trust of Oregon, www.energytrust.org.

Heschong Mahone Group, www.h-m-g.com.

IESNA. *Lighting Handbook*. Illuminating Engineering Society of North America, available from www.iesna.org.

New Buildings Institute, www.newbuildings.org.

RealWinWin, www.realwinwin.com.

Rising Sun Enterprises, www.rselight.com.

Savings by Design, www.savingsbydesign.com.

## GENERAL DESCRIPTION

A light conveyor is large pipe or duct with reflective sides that transmits artificial or natural light along its length.

There are two types of such light-directing devices. The first is a square duct or round pipe made of plastic. By means of how the inside of the duct or pipe is cut and configured, light entering one end of the pipe is both reflected off these configurations (just as does light through a prism) and transmitted through. That reflected light continues to travel down the pipe, but the relatively small amount of light transmitted through the pipe provides continuous lighting along the pipe. Because some light is absorbed and escapes along the length of the pipe (i.e., is "lost"), the maximum distance that light can be "piped" into a building is about 90 ft (27.4 m).

There are a few installations where sun-tracking mirrors concentrate and direct natural light into a light pipe. In most applications, however, a high-intensity electric light is used as the light source. Having the electric light separate from the space where the light is delivered isolates the heat, noise, and electromagnetic field of the light source from building occupants. In addition, the placement of the light source in a maintenance room separate from building occupants simplifies replacement of the light source.

A second light-directing device is a straight tube with a highly reflective interior coating. The device is mounted on a building roof and has a clear plastic dome at the top end of the tube and a translucent plastic diffusing dome at the bottom end. The tube is typically 12 to 16 in. (300 to 400 mm) in diameter. Natural light enters the top dome, is reflected down the tube, and is then diffused throughout the building interior. The light output is limited by the amount of daylight falling on the exterior dome.

## WHEN/WHERE IT'S APPLICABLE

The first light conveyor system is best suited to building applications where there is a need to isolate electric lights from the interior space (for example, operating rooms or theaters) or where electric light replacement is difficult (for example, swimming pools or tunnels). For the reflective tube system, each device can light only a small area ($10 \text{ ft}^2 [1 \text{ m}^2]$) and is best suited to small interior spaces with access to the roof, such as interior bathrooms and hallways.

## PROS AND CONS

### Pro

1. A light conveyor transports natural light into building interiors.
2. The first type of light conveyor isolates the electric light source from the lighted space.

3. The first type of light conveyor reduces lighting glare.
4. It lowers lighting maintenance costs.

Con

1. A light conveyor may have greater capital costs than traditional electric lighting.
2. The tube type may increase roof heat loss.
3. The tube type runs the risk of poor installation, resulting in leaks.
4. The effectiveness may not be worth the additional cost.

## KEY ELEMENTS OF COST

Because of the specialized nature of these techniques, it is difficult to address specific cost elements. As an alternative to conventional electric lighting techniques, it could add to or reduce the overall cost of a lighting system—and the energy costs required—depending on specific project conditions. A designer should not incorporate any such system without thoroughly investigating its benefits and applicability and should preferably observe such a system in actual use.

## SOURCE OF FURTHER INFORMATION

Preliminary Evaluation of Cylindrical Skylights
McKurdy, Harrison and Cooke
23rd Annual SESCI Conference
Solar Energy Society of Canada, Inc.
116 Lisgar, Suite 702
Ottawa ON
Canada K2P 0C2
tel: 613-234-4151
fax: 613-234-2988
www.solarenergysociety.ca

# 14

# PLUMBING AND FIRE PROTECTION SYSTEMS

Plumbing and fire protection systems are normally not considered within the purview of the HVAC&R designer's expertise. Nevertheless, both subsets of designers, in practice, must work closely in putting together a functional building mechanical system. Indeed, frequently the designer of HVAC&R systems and plumbing systems is one and the same. For detailed design guidelines and information refer to the National Fire Protection Association (NFPA) and the American Society for Plumbing Engineers (ASPE).

It is important in green building design for the practitioners of each design discipline to be familiar with what the other disciplines may bring to an effective green design. This is especially so with plumbing design. The editors of this Guide have thus chosen to include discussion of some key aspects of plumbing design that can have an impact on green design—including several significant ASHRAE GreenTips. Several of these GreenTips may have an impact in other areas as well. For instance, point-of-use hot-water heaters would not only save heating energy and distribution energy, but they could also result in the use of less water.

## WATER SUPPLY

This basic resource is obviously essential at every building site, but what has changed over the last several decades is the realization that it is fast becoming a *precious* resource. While the total amount of water in its various forms on the planet is finite, the amount of fresh water, of a quality suitable for the purposes for which it may be used, is not uniformly distributed (e.g., 20% of the world's fresh water is in the United States' Great Lakes); elsewhere it is often nonexistent or in very meager supply. Nevertheless, water must be allocated somehow to the world's populated lands, many of which are undergoing rapid development. In short, it is becoming more and more difficult to provide for the adequate and equitable distribution of the world's water supply to those users for whom it is essential.

This trend has implications for not only how *prudently* we use the water we have, but what we do to avoid contaminating water supplies. While many of the measures to protect and preserve the world's freshwater supplies are beyond ASHRAE's purview, there are a number of simple things relating to building sites that *can* be done as part of a green design effort.

While it is obvious that water purity must meet the health and safety standards prescribed by the authorities (see below), there are other techniques that can contribute not only to using lesser quantities of the highest-quality water needed at a site but to requiring less energy to "process" that water (i.e., distribute it, heat it, and dispose of it once used).

Some key techniques of this nature are covered below.

Regarding the quality of the water provided for a given building/building site, the design team should:

- Supply local or municipal water that exceeds the requirements of the EPA or more stringent local requirements for potable and heated water.
- Meet the EPA's national primary drinking water regulations (NPDWR), including maximum contaminant level, by testing or by installing appropriate treatment systems. The quality of the municipal water supply shall be evaluated at applicable points, including restrooms/showers, kitchen/pantry areas, drinking fountains, architectural fountains, and/or indoor water features.
- Exceed NPDWR maximum contaminant level goals and secondary standards by testing or by installing appropriate treatment systems. The quality of the municipal water supply shall be evaluated at applicable points, including restrooms/showers, kitchen/pantry areas, drinking fountains, architectural fountains, and/or indoor water features.

Some of the techniques for reducing the demand for domestic water in a building are well known and in fairly widespread use. These include water restrictors (such as flow-control shower heads) and spring-closing or timed lavatory faucets, especially in public or semi-public washrooms. GreenTip #33 addresses other water-conserving fixtures that can be used.

GreenTip #34 would also impact the amount of pure fresh water used on a site by making use of "used" water (known as graywater) for purposes where potability is not a requirement.

## DOMESTIC WATER HEATING

One of the earliest techniques used for lowering the energy required for domestic water heating, going back to the mid-1970s, was reducing the temperature of the water supplied. The pre-Arab-oil-crisis norm for domestic hot water was 140°F (60°C), and the energy-saving recommendation thereafter was 105°F to 110°F (40°C to 43°C). While this can save heating energy, where hot/cold mixing valves are utilized (as with some shower controls), it may also cause more hot water to be used. In addition, the dangers of *legionellosis*, which thrives at the lower heated water ranges, have revised the prudent recommendation for the hot-water supply temperature to 105°F–120°F (40°C–49°C). Protection against possible *legionellosis* can be achieved by generating 140°F (60°C) centrally (in a well-insulated heater/storage tank) and then mixing to the lower temperature through a mixing valve.

Another energy-reducing technique (point-of-use water heating), which can also reduce the water quantity used, is covered in GreenTip #35.

Another technique to reduce water-heating cost is combining the function of domestic and space water heating, where allowed by codes. See Chapter 11, "Energy Conversion Systems," for GreenTip #19 on combination space and water heating systems.

Yet another water heating technique, with more limited and specialized application, is covered in GreenTip #36.

Solar heating utilizing PVs is often used for preheating or full heating of domestic hot-water systems.

Strategies should be considered to preheat and/or fully heat domestic hot water utilizing waste heat (i.e., flash steam, etc.) from the main building heating system.

## SANITARY WASTE

See GreenTip #34, which deals with sanitary waste water and a strategy to conserve potable water.

## STORM DRAINAGE

GreenTip #37 also deals with a strategy for conserving potable water, though the water "source" differs.

## FIRE PROTECTION

There are limited opportunities to integrate the components or subsystems of the fire protection systems as a green strategy on a project. It is imperative that all code issues be fully understood when considering this strategy.

## FIRE SUPPRESSION SYSTEMS

These systems are designed for life safety of the occupants. However, there may be times when there is an opportunity to utilize the water source, or the water in the sprinkler and standpipe piping, as a heat sink for heat pump systems.

## SMOKE MANAGEMENT SYSTEMS

These systems are designed for life safety of the occupants. However, there may be times when there is an opportunity to combine these systems with the central air-handling equipment in the building, thereby saving on materials and resources that would have been needed if the systems were not optimally integrated.

## ASHRAE GreenTip #33:
## Water-Conserving Plumbing Fixtures

### GENERAL DESCRIPTION

Water conservation strategies save building owners both consumption and demand charges.

Further, municipal water and wastewater treatment plants save operating and capital costs for new facilities. As a general rule, water conservation strategies are very cost-effective when properly applied.

The Energy Policy Act of 1992 set reasonable standards for the technologies then available. Now there are plumbing fixtures and equipment capable of significant reduction in water usage. For example, a rest stop in Minnesota that was equipped with ultra-low-flow toilets and waterless urinals has recorded a 62% reduction in water usage.

Tables 14-1, 14-2, and 14-3 list the maximum water usage standards established by the Energy Policy Act of 1992 for typical fixture types. Also listed is water usage for flush-type and flow-type fixtures. Listing of conventional fixture usage allows comparison to the low-flow and ultra-low-flow fixture usage.

### Table 14-1    EPCA Maximum Flows

| Fixture Type | Energy Policy Act of 1992 Maximum Water Usage |
|---|---|
| Water closets, gpf* (L/f) | 1.6 (6.1) |
| Urinals, gpf (L/f) | 1.0 (3.8) |
| Shower heads, gpm* (L/s) | 2.5 (0.16) |
| Faucets, gpm (L/s) | 2.5 (0.16) |
| Replacement aerators, gpm (L/s) | 2.5 (0.16) |
| Metering facets, gal/cycle (L/cycle) | 0.25 (0.95) |

\* Note: gpf = gallons per fixture (L/f = liters per fixture); gpm = gallons per minute (L/s = liters per second). The gpm (L/s) value is at flowing water pressure of 80 psi (552 kPa).

### Table 14-2    Flush-Fixture Flows

| Flush-Fixture Type | Water Use, gpf (L/f) |
|---|---|
| Conventional water closet | 1.6 (6.1) |
| Low-flow water closet | 1.1 (4.2) |
| Ultra-low-flow water closet | 0.8 (3.0) |
| Composting toilet | 0.0 |
| Conventional urinal | 1.0 (3.8) |
| Waterless urinal | 0.0 |

**Table 14-3    Flow-Fixture Flows**

| Flow-Fixture Type | Water Use, gpm (L/s) |
|---|---|
| Conventional lavatory | 2.5 (0.16) |
| Low-flow lavatory | 1.8 (0.11) |
| Kitchen sink | 2.5 (0.16) |
| Low-flow kitchen sink | 1.8 (0.11) |
| Shower | 2.5 (0.16) |
| Low-flow shower | 1.8 (0.11) |
| Janitor sink | 2.5 (0.16) |

## WHEN/WHERE IT'S APPLICABLE

Applicable state and local codes should be checked prior to design as some of them have "approved fixture" lists; some code officials have not approved the waterless urinal and low-flush toilet technologies. Waterless urinals and low-flow lavatory fixtures usually pay back immediately. Toilet technology continues to evolve rapidly, so be sure to obtain test data and references before specifying; some units work very well, while others perform marginally.

Options that should be considered in design of water-conserving systems include:

- Infrared faucet sensors
- Delayed action shutoff or automatic mechanical shutoff valves (metering faucets at 0.25 gal per cycle [0.95 L/cycle])
- Low-flow or ultra-low-flow toilets
- Lavatory faucets with flow restrictors
- Low-flow kitchen faucets
- Domestic dishwashers that use 10 gal (38 L) per cycle or less
- Commercial dishwashers (conveyor type) that use 120 gal (455 L) per hour
- Waterless urinals
- Closed cooling towers (to eliminate drift) and filters for cleaning the water.

## PROS AND CONS

### Pro

1. Water conservation reduces a building's potable water use, in turn reducing demand on municipal water supply and lowering costs and energy use associated with water.

2. It reduces a building's overall waste generation, thus putting fewer burdens on the existing sewage system.
3. It may save capital cost since some fixtures, such as waterless urinals and low-flow lavatories, may be less expensive to install initially.

## Con

1. Some states and municipalities have "approved fixture" lists that may not include certain newer and more efficient fixtures. However, the design engineer would likely have the option to go to a review process in order to get new fixture technologies put on the approved fixture list.
2. Maintenance of these fixtures is different and will require special training of staff.

## KEY ELEMENTS OF COST

The following provides a possible breakdown of the various cost elements that might differentiate a building utilizing water-conserving plumbing fixtures from one that does not and an indication of whether the net cost is likely to be lower (L), higher (H), or the same (S). This assessment is only a perception of what might be likely, but it obviously may not be correct in all situations. **There is no substitute for a detailed cost analysis as part of the design process.** The listing below may also provide some assistance in identifying the cost elements involved.

### First Cost

- Low-flow and ultra-low-flow flush water closets       S/H
- Waterless urinals                                     S/L
- Low-flow shower heads                                 S
- Metering faucets                                      S
- Electronic faucets                                    M
- Dual-flush water closets                              S/H
- Water-conserving dishwashers                          S/H

### Recurring Cost

- Potable water                                         L
- Sewer discharge                                       L
- Maintenance                                           L/S
- Training of building operators                        S/H
- Orientation of building occupants                     S
- Commissioning                                         S

SOURCES OF FURTHER INFORMATION

The American Society of Plumbing Engineers, www.aspe.org.

Del Porto, D., and C. Steinfeld. 1999. *The Composting Toilet System Book.* The Center for Ecological Pollution Prevention.

Public Technology, Inc., US Department of Energy and the US Green Building Council. 1996. *Sustainable Building Technical Manual—Green Building Design, Construction and Operations.* Public Technology, Inc.

US Environmental Protection Agency, *How to Conserve Water and Use It Effectively,* www.epa.gov/OW/you/chap3.html.

US Green Building Council, www.usgbc.org.

## ASHRAE GreenTip #34:
## Graywater Systems

### GENERAL DESCRIPTION

Graywater is generally wastewater from lavatories, showers, bathtubs, and sinks that is not used for food preparation. Graywater is further distinguished from blackwater, which is wastewater from toilets and sinks that contains organic or toxic matter. Local health code departments have regulations that specifically define the two kinds of waste streams in their respective jurisdictions.

Where allowed by local code, separate blackwater and graywater waste collection systems can be installed. The blackwater system would be treated as a typical waste stream and piped to the water treatment system or local sewer district. However, the graywater would be "recycled" by collecting, storing (optional), and then distributing it via a dedicated piping system to toilets, landscape irrigation, or any other function that does not require potable water.

Typically, for a commercial graywater system, such as for toilet flushing in a hotel, a means of short-term on-site storage, or, more appropriately, a surge tank, is required. Graywater can only be held for a short period of time before it naturally becomes blackwater. The surge tank would be provided with an overflow to the blackwater waste system and a potable makeup line for when the end-use need exceeds stored capacity.

Distribution would be accomplished via a pressurized piping system requiring pumps and some low level of filtration. Usually, there will be a requirement for the graywater system to be a supplemental system. Therefore, systems will still need to be connected to the municipal or localized well service.

### WHEN/WHERE IT'S APPLICABLE

Careful consideration should be given before pursuing a graywater system. While a graywater system can be applied in any facility that has a nonpotable water demand and a usable waste stream, the additional piping and energy required to provide and operate such a system may outweigh any benefits. Such a system is best applied where the ratio of demand for nonpotable water to potable water is relatively high and consistent, as in restaurants, laundries, and hotels.

Some facilities have a more reliable graywater volume than others. For example, a school would have substantially less graywater in the summer months. This may not be a problem if the graywater was being used for flushing since it can be assumed that toilet use would vary with occupancy. However, it would be detrimental if graywater were being used for landscape irrigation.

## PROS AND CONS

### Pro

1. A graywater system reduces a building's potable water use, in turn reducing demand on the municipal water supply and lowering costs associated with water.
2. It reduces a building's overall wastewater generation, thus putting less tax on the existing sewage systems.

### Con

1. There is an added first cost associated with the additional piping, pumping, filtration, and surge tank required.
2. There are additional materials and their associated embodied energy costs.
3. There is negative public perception of graywater and health concerns regarding ingestion of nonpotable water.
4. Costs include maintenance of the system, including the pumps, filters, and surge tank.
5. Local health code authority has jurisdiction, potentially making a particular site infeasible due to that authority's definition of blackwater versus graywater.

## KEY ELEMENTS OF COST

The following provides a possible breakdown of the various cost elements that might differentiate a building utilizing a graywater system from one that does not and an indication of whether the net incremental cost is likely to be lower (L), higher (H), or the same (S). This assessment is only a perception of what might be likely, but it obviously may not be correct in all situations. **There is no substitute for a detailed cost analysis as part of the design process.** The listings below may also provide some assistance in identifying the cost elements involved.

### First Cost

| | |
|---|---|
| • Collection systems | H |
| • Surge tank | H |
| • Water treatment | H |
| • Distribution system | H |
| • Design fees | H |

### Recurring Cost

| | |
|---|---|
| • Cost of potable water | L |
| • Cost related to sewer discharge | L |
| • Maintenance of system | H |

- Training of building operators        H
- Orientation of building occupants     S
- Commissioning cost                    H

## SOURCES OF FURTHER INFORMATION

Advanced Buildings Technologies and Practices, www.advancedbuildings.org.

The American Society of Plumbing Engineers, www.aspe.org.

Del Porto, D., and C. Steinfeld. 1999. *The Composting Toilet System Book*. The Center for Ecological Pollution Prevention.

Ludwig, A. 1997. Builder's Greywater Guide and Create an Oasis with Greywater. Oasis Design.

Public Technology Inc., US Department of Energy and the US Green Building Council. 1996. *Sustainable Building Technical Manual—Green Building Design, Construction and Operations*. Public Technology, Inc.

US Green Building Council. *LEED Reference Guide, Version 2.0*, June 2001.

## ASHRAE GreenTip #35:
## Point-of-Use Domestic Hot-Water Heaters

### GENERAL DESCRIPTION

As implied by the title, point-of-use domestic hot-water heaters provide small quantities of hot water at the point of use, without tie-in to a central hot water source. A cold water line from a central source must still be connected, as well as electricity, for heating the water.

There is some variation in types. Typically, such as for lavatories, the device may be truly instantaneous, or it may have a small amount of storage capacity. With the instantaneous type, the heating coil is sized such that it can heat a normal-use flow of water up to the desired hot-water temperature (120°F [49°C], say). When a small tank (usually three to ten gallons) is incorporated in the device, the electric heating coil is built into the tank and can be sized somewhat smaller because of the small amount of stored water available.

The device is usually installed under the counter of the sink or bank of sinks.

A similar type of device boosts the water supply (which is cold water) up to near boiling temperature (about 190°F [88°C]). This is usually used for purposes of quickly making a cup of coffee or tea without having to brew it separately in a coffeepot or teapot.

### WHEN/WHERE IT'S APPLICABLE

These devices are applicable wherever there is a need for a hot-water supply that is low in quantity and relatively infrequently used *and* it is excessively inconvenient or costly to run a hot-water line (with perhaps a recirculation line as well) from a central hot-water source. Typically, these are installed in lavatories or washrooms that are isolated or remote, or both. However, they can be used in any situation where there is a hot-water need but where it would be too inconvenient and costly to tie in to a central source. (There must, of course, be available a source of incoming water as well as a source of electricity.)

### PROS AND CONS

#### Pro

1. A point-of-use device is a simple and direct way to provide small amounts of domestic hot water per use.
2. Long pipe runs—and, in some cases, a central hot-water heating source—can be avoided.
3. Energy is saved by avoiding heat loss from hot-water pipes and, if not needed, from a central water heater.

4. In most cases where applicable, it has lower first cost.
5. It is convenient—especially as a source of 190°F–210°F (88°C–99°C) water supply.
6. When installed in multiple locations, central equipment failure does not knock out all user locations.
7. It may save floor space in the central equipment room if no central heater is required.
8. Water is saved by not having to run the faucet until the water warms up.

## Con

1. This is a more expensive source of heating energy (though cost may be trivial if usage is low and may be exceeded by heat losses saved from central heating method).
2. Water impurities can cause caking and premature failure of electrical heating coil.
3. It cannot handle changed demand for large hot-water quantities or too-frequent use.
4. Maintenance is less convenient (when required) since it is not centralized.
5. Temperature and pressure (T&P) relief valve and floor drain may be required by some code jurisdictions.

## KEY ELEMENTS OF COST

The following provides a possible breakdown of the various cost elements that might differentiate a point-of-use domestic hot-water heater from a conventional one and an indication of whether the net incremental cost for the system is likely to be lower (L), higher (H), or the same (S). This assessment is only a perception of what might be likely, but it obviously may not be correct in all situations. **There is no substitute for a detailed cost analysis as part of the design process.** The listings below may also provide some assistance in identifying the cost elements involved.

## First Cost

- Point-of-use water heater equipment         H
- Domestic hot-water piping to central source (including insulation thereof) L
- Central water heater (if not required) and associated fuel and flue gas connections         L
- Electrical connection         H
- T&P relief valve and floor drain (when required by code jurisdiction)         H

Recurring Cost

- Energy to heat water proper                                          H
- Energy lost from piping not installed (and perhaps central heater)   L
- Maintenance/repairs, including replacement                          H

## SOURCES OF FURTHER INFORMATION

The American Society of Plumbing Engineers, www.aspe.org.

*Domestic Water Heating Design Manual.* 1998. Chicago: American Society of Plumbing Engineers.

Fagan, D. 2001. A comparison of storage-type and instantaneous heaters for commercial use. *Heating/Piping/Air Conditioning Engineering*, April.

## ASHRAE GreenTip #36:
## Direct-Contact Water Heaters

### GENERAL DESCRIPTION

A direct-contact water heater consists of a heat exchanger in which flue gases are in direct contact with the water. It can heat large quantities of water for washing and/or industrial process purposes. Cold supply water enters the top of a heat exchanger column and flows down through stainless steel rings or other devices. Natural gas is burned in a combustion chamber, and the flue gases are directed up the heat exchanger column. As the gases move upward through the column, they transfer their sensible and latent heat to the water. A heat exchanger or water jacket on the combustion chamber captures any heat loss from the chamber. The gases exit only a few degrees warmer than the inlet water temperature. The heated water may be stored in a storage tank for "on-demand" use. Direct-contact water heaters can be 99% efficient when the inlet water temperature is below 59°F (15°C).

The low-temperature combustion process results in low emissions of $NO_x$ and CO; thus, the system is in effect a low-$NO_x$ burner. It is also a low-pressure process since heat transfer occurs at atmospheric pressure.

Although there is direct contact between the flue gases and the water, there is very little contamination of the water. Direct-contact systems are suitable for all water-heating applications, including food processing and dairy applications; the water used in these systems is considered bacteriologically safe for human consumption.

### WHEN/WHERE IT'S APPLICABLE

The high cost of direct-contact water heaters (due to stainless steel construction) restricts their use to where there is a large, almost continuous, demand for hot water. Appropriate applications include laundries, food processing, washing, and industrial processes. The system can also be used for closed-loop (or recirculating) applications such as space heating. However, efficiency—the primary benefit of direct-contact water heating—will be reduced because of the higher inlet water temperature resulting from recirculation.

### PROS AND CONS

#### Pro

1. Increases part-load and instantaneous efficiency.
2. Reduces $NO_x$ and CO emissions.
3. Increases safety.
4. Increases system response time.

## Con

1. High cost.
2. Less effective in higher-pressure or closed-loop applications or where inlet water temperatures must be relatively high.
3. Results in considerable water usage beyond that required for the process itself due to high evaporation rate.

## KEY ELEMENTS OF COST

The following provides a possible breakdown of the various cost elements that might differentiate a direct-contact water heater from a conventional one and an indication of whether the net cost for the alternative option is likely to be lower (L), higher (H), or the same (S). This assessment is only a perception of what might be likely, but it obviously may not be correct in all situations. **There is no substitute for a detailed cost analysis as part of the design process.** The listings below may also provide some assistance in identifying the cost elements involved.

## First Cost

- Water heater                              H
- Operator training (unfamiliarity)         H

## Recurring Cost

- Water heating energy                       L

Direct-contact boilers are two to three times the price of indirect or conventional boilers, primarily because of the stainless steel construction. In high and continuous water use applications, however, the payback period can be under two years.

## SOURCES OF FURTHER INFORMATION

NSF. 2000. *NSF/ANSI 5-2000e: Water Heaters, Hot Water Supply Boilers, and Heat Recovery Equipment.* Ann Arbor, MI: National Sanitation Foundation and American National Standards Institute.

QuikWater, High Efficiency Direct Contact Water Heaters, www.quikwater.com

## ASHRAE GreenTip #37:
## Rainwater Harvesting

### GENERAL DESCRIPTION

Rainwater harvesting has been around for thousands of years. Rainwater harvesting is a simple technology that can stand alone or augment other water sources. Systems can be as basic as a rain barrel under a downspout or as complex as a pumped and filtered graywater system providing landscape irrigation, cooling tower makeup, and/or building waste conveyance.

Systems are generally composed of five or less basic components: (1) a catchment area, (2) a means of conveyance from the catchment, (3) storage (optional), (4) water treatment (optional), and (5) a conveyance system to the end use.

The catchment area can be any impermeable area from which water can be harvested. Typically this is the roof, but paved areas such as patios, entries, and parking lots may also be considered. Roofing materials such as metal, clay, or concrete-based are preferable to asphalt or roofs with lead-containing materials. Similarly, care should be given when considering a parking lot for catchment due to oils and residues that can be present.

Conveyance to the storage will be gravity fed like any stormwater piping system. The only difference is that now the rainwater is being diverted for useful purposes instead of literally going down the drain.

Commercial systems will require a means of storage. Cisterns can be located outside the building (above grade or buried) or placed on the lower levels of the building. The storage tank should have an overflow device piped to the storm system and a potable water makeup if the end-use need is ever greater than the harvested volume.

Depending on the catchment source and the end use, the level of treatment will vary. For simple site irrigation, filtration can be achieved through a series of graded screens and paper filters. If the water is to be used for waste conveyance, then an additional sand filter may be appropriate. Parking lot catchments may require an oil separator. The local code authority will likely decide acceptable water standards, and, in turn, filtration and chemical polishing will be a dictated parameter, not a design choice.

Distribution can be via gravity or pump depending on the proximity of the storage tank and the end use.

### WHEN/WHERE IT'S APPLICABLE

If the building design is to include a graywater system or landscape irrigation—and space for storage can be found—rainwater harvesting is a simple addition to those systems.

When a desire exists to limit potable water demand and use, depending on the end-use requirement and the anticipated annual rainfall in a region, harvesting can be provided as a stand-alone system or to augment a conventional makeup water system.

Sites with significant precipitation volumes may determine that reuse of these volumes is more cost-effective than creating stormwater systems or on-site treatment facilities.

Rainwater harvesting is most attractive where municipal water supply is either nonexistent or unreliable, hence its popularity in rural regions and developing countries.

## PROS AND CONS

### Pro

1. Rainwater harvesting reduces a building's potable water use, in turn reducing demand on the municipal water supply and lowering costs associated with water.
2. Rainwater is soft and does not cause scale buildup in piping, equipment, and appliances. It could extend the life of systems.
3. It can reduce or eliminate the need for stormwater treatment or conveyance systems.

### Con

1. There is added first cost associated with the cisterns and the treatment system.
2. There are additional materials and their associated embodied energy costs.
3. The storage vessels must be accommodated. Small sites or projects with limited space allocated for utilities would be bad candidates.
4. Costs include maintenance of the system, including the catchments, conveyance, cisterns, and treatment systems.
5. There is no US guideline on rainwater harvesting. The local health code authority has jurisdiction, potentially making a particular site infeasible due to backflow prevention requirements, special separators, or additional treatment.

## KEY ELEMENTS OF COST

The following provides a possible breakdown of the various cost elements that might differentiate a building utilizing rainwater harvesting from one that does not and an indication of whether the net cost is likely to be lower (L), higher (H), or the same (S). This assessment is only a perception of what might be likely, but it obviously may not be correct in all situations. **There is no substitute for a detailed cost analysis as part of the design process.** The listings below may also provide some assistance in identifying the cost elements involved.

First Cost

- Catchment area      S
- Conveyance systems      S
- Storage tank      H
- Water treatment      S/H
- Distribution system      S/H
- Design fees      H

Recurring Cost

- Cost of potable water      L
- Maintenance of system      H
- Training of building operators      H
- Orientation of building occupants      S
- Commissioning cost      H

## SOURCES OF FURTHER INFORMATION

The American Society of Plumbing Engineers, www.aspe.org.

American Water Works Association, WaterWiser, The Water Efficiency Clearing-house, www.waterwiser.org.

Gerston, J. *Rainwater Harvesting: A New Water Source.*

Irrigation Association, www.irrigation.org.

Public Technology Inc., US Department of Energy and the US Green Building Council. 1996. *Sustainable Building Technical Manual—Green Building Design, Construction and Operations.* Public Technology, Inc.

US Green Building Council. 2001. *LEED Reference Guide, Version 2.0.*

Waterfall, P.H. 1998. *Harvesting Rainwater for Landscape Use,* http://ag.arizona.edu/pubs/water/az1052/.

# 15

# BUILDING CONTROL SYSTEMS

## INTRODUCTION

Control systems play an important part in the operation of a building and, thus, determine whether many of the green design aspects included in the original plan actually function as intended. Controls for HVAC and related systems have evolved over the years, but in general they can be described as either distributed or centralized. Distributed controls are generally packaged devices that are provided with the equipment. A building automation system (BAS) is a form of centralized control capable of controlling HVAC and other systems such as life safety, lighting, water distribution, and security.

Control systems are at the core of building performance. When they work well, the indoor environment promotes productivity with the light, comfort, and ventilation people need to carry out their tasks. When they break down, the results are higher utility bills, loss of productivity, and discomfort. In modern buildings, microprocessors in direct digital BASs control lights, ventilation, space temperature and humidity, plumbing systems, electrical systems, life-safety systems, and other building systems. The control system can assist in conserving resources through the scheduling, staging, and modulation of equipment to meet the needs of the occupants and systems that they are designed to serve. The control system can assist with operation and maintenance through accumulation of equipment runtimes, display of trend logs, and use of automated alarms. Finally, the control system interface can be used as a repository for building maintenance information where operation and maintenance manuals or equipment ratings, such as pump curves, are stored as electronic documents available through a hyperlink on the control system graphic for the appropriate system. This chapter presents the key issues to designing, commissioning, and maintaining control systems for optimal performance.

This chapter is divided into seven sections as follows:

- *Control system role in delivering energy efficiency.* Through scheduling, unloading, and fault detection, controls have the capability of reducing building energy usage by up to 20% in a typical commercial building.

- *Control system role in delivering water efficiency.* Used primarily in landscape irrigation and leak detection, controls can reduce water usage.
- *Control system role in delivering IEQ.* In most commercial buildings, controls play a crucial role in providing IEQ. Controls regulate the quantities of outdoor air brought into the building, zone ventilation, zone temperature, and relative humidity and can monitor the loading of air filters.
- *Control system commissioning process.* Of all the building systems, controls are the most susceptible to problems in installation. These can be addressed by a thorough process of performance verification
- *Control system role in attaining LEED certification.* This section describes the elements of LEED certification that can be addressed by control system design and implementation.
- *Designing for sustained efficiency.* Control systems help ensure continued efficient building operation by enabling M&V of building performance and serving as a repository of maintenance procedures.
- *References to other sources of information.* Useful resources for design and commissioning of successful building controls.

## CONTROL SYSTEM ROLE IN DELIVERING ENERGY EFFICIENCY

Data from the report, "Energy Impact of Commercial Building Controls and Performance Diagnostics: Market Characterization, Energy Impact of Building Faults and Energy Savings Potential," written for the US DOE in November 2005, reports the following:

- 17 quads/year of energy are used in commercial buildings
- Only 10% of commercial buildings (33% of the floor space) have a building management system
- Less than 10% of the commercial building space has automated lighting control
- A potential for nearly 1 quad/year of energy savings (11% of total building energy) exists through installing BASs in all buildings and fully commissioning them.

Thus, the potential for energy savings by installing building automation is huge. The study indicated that commercial building energy usage could be reduced by 2% to 11% (0.34 to 1.8 quads/year) with proper management and control.

Building automation can save energy through a variety of methods, including:

- *Reduction of equipment runtime.* Examples include scheduled control of lighting and air-conditioning systems inside of buildings and photoelectric controls for site lighting.
- *Efficient unloading of equipment.* Examples include daylight control using dimming or stepped lighting in spaces with access to natural light and variable-speed control of pumps, fans, and compressors for cooling systems.

- *Automated fault detection and diagnostic systems.* Examples include controls that report when dampers or valves are stuck open or closed.

The GreenTips included in this chapter give two detailed examples of how controls can be used to deliver energy efficiency in building air distribution systems.

## CONTROL SYSTEM ROLE IN DELIVERING WATER EFFICIENCY

Control systems can deliver building water efficiency in two main areas: landscape irrigation and leak detection. Additionally, controls can be used to regulate and monitor on-site wastewater treatment plants where those systems are in place.

Landscape irrigation can make a significant contribution to the annual water use of a building. The level of contribution depends on the amount and type of landscaping, as well as on geographical region. The simplest irrigation controls are based on time clocks that open valves for a set duration for a set number of periods per week. Unless the duration and frequency of watering is adjusted throughout the irrigation season, the use of time clock controllers often results in excessive water use. Some time clock controllers allow for 365-day programming in order to automatically address seasonally varying irrigation requirements.

An improvement on the time clock controller is to add moisture sensors, which can enable the system to bypass a watering period if ground moisture levels are above a setpoint. Still more sophisticated controllers can gather data about local weather conditions, either directly via sensors or indirectly via a remote weather station, and use that data to adjust the amount of water delivered to the landscape.

Integrating landscape irrigation controls into the BAS provides a number of advantages. Among these are the ability to adjust schedules and setpoints from a single location, the ability to perform remote diagnostics, and the ability to track system performance and water use. Integrating water meters into the BAS enables continuous measurement of water consumption. Water consumption data can be analyzed during unoccupied periods to determine whether leaks are present in the water distribution system. The judicious placement of submeters can allow building maintenance staff to find the system or location in which the leak or leaks are present. Continuous water meter data can also be used to identify processes and areas of high water use and guide post-construction water conservation efforts.

Finally, some advanced green buildings feature on-site wastewater treatment plants. Such plants generally include pumps, fans, and sensors to monitor characteristics of the treated water, such as dissolved oxygen levels and total suspended solids. Depending on the goals and complexity of the wastewater treatment plant, an industrial supervisory control and data acquisition (SCADA) system may be necessary to ensure maximization of throughput, efficient energy use, and code compliance. The SCADA system may be stand-alone or it may be integrated into the BAS.

## CONTROL SYSTEM ROLE IN DELIVERING IEQ

Factors regulated by building control systems that impact IEQ include operative temperature, relative humidity, outdoor airflow rates, and light levels. The first two factors are addressed in *ANSI/ASHRAE Standard 55-2004, Thermal Environmental Conditions for Human Occupancy*, the third factor is addressed in *ANSI/ASHRAE Standard 62.1-2004, Ventilation for Acceptable Indoor Air Quality*, and the fourth factor is addressed in Chapter 10 of the IESNA *Lighting Handbook*.

### Thermal Comfort

The operative temperature is approximately equal to the average of the room air temperature and the mean radiant temperature (MRT) for most building interior environments. Radiant heating systems raise the MRT and thus lower the room air temperature required to provide a comfortable operative temperature. Radiant cooling systems lower the MRT and thus raise the room air temperature required to provide a comfortable operative temperature. Stand-alone thermostats and BASs typically measure and control room air temperature. When designing radiant systems, it is important to keep the distinction between operative temperature and air temperature in mind. For a discussion of these two quantities, see Chapter 53 of the *2003 ASHRAE Handbook—HVAC Applications*. ASHRAE Standard 55-2004 describes methods for determining acceptable operative temperatures that depend on occupant clothing insulation levels and metabolic rates.

Although less common than thermostats and temperature sensors, stand-alone humidistats and humidity sensors can be used to control room humidity levels. ASHRAE Standard 55-2004 specifies a maximum humidity ratio of 0.012. The standard does not specify a minimum humidity ratio for thermal comfort but notes that nonthermal comfort factors (e.g., skin drying) may be used to establish such a minimum.

### Air Quality and Ventilation

Procedures for determining minimum outdoor airflow rates are described in ASHRAE Standard 62.1-2004. Building controls play a role in ensuring that minimum outdoor airflow rates are achieved in three areas: VAV system control, mixed-mode ventilation systems, and dynamic reset of outdoor air intake flows.

It is difficult to ensure that minimum outdoor airflow rates are met by VAV systems over the entire range of operating conditions in the absence of controls designed, installed, and maintained specifically for that purpose. One means of achieving such control is to measure the supply airflow rate and $CO_2$ concentrations in the return air duct, mixed air plenum, and outdoor air duct on a continuous basis. The BAS can determine if the outdoor airflow rate is sufficient based on this information and adjust the mixed air dampers accordingly. If airflows are also measured at each VAV box, this control routine can be improved upon further. As described in ASHRAE Standard 62.1-2004, zone air distribution effectiveness can change when the temperature of the supply air changes. Therefore, it may be necessary for the control system to reset the minimum outdoor airflow after a seasonal switchover of supply air temperature.

Mixed-mode ventilation refers to the combination of mechanical ventilation and operable windows providing natural ventilation. If the windows are sized properly, it is not necessary to provide mechanical ventilation when the windows are open. Perhaps the most straightforward way to control mixed-mode ventilation systems is to use the output of a window switch to shut down the terminal unit (e.g., VAV box) when the window is open. When the occupant decides that it is time to shut the window, the mechanical HVAC system is brought back online. It may be necessary to install an alarm or override based on space temperature in order to provide freeze protection.

It may be desirable to reset the minimum outdoor airflow based on changes in occupancy or changes in zone air distribution effectiveness. Occupancy can be estimated directly by a card reader system, for example. Occupancy can be estimated indirectly by measuring the $CO_2$ concentration in the occupied space. If $CO_2$ levels are used to estimate occupancy, it is important to provide design minimum outdoor airflow rates for a set period in advance of occupancy to compensate for the lag between increased occupancy and increased $CO_2$ levels.

### Lighting Levels

The BAS can maintain desired light levels by either adjusting electric light output or controlling the amount of daylight entering the building. One or more photocells may be used to measure the light level in the occupied space and the output used to brighten or dim electric lights. Alternatively, photocell output may be used to switch between multiple light levels. Photocell output may also be used to raise or lower window blinds or adjust louvers to keep daylighting at comfortable levels and eliminate glare.

Of course, human perceptions of thermal and visual comfort vary considerably between individuals. For this reason, it is a principle of good design that building occupants should be given as much control over their thermal and visual environments as possible. Room thermostats, operable windows, dimming switches, and adjustable blinds are all means of giving people this control. Integrating these manually operable controls into the BAS can contribute significantly toward optimizing both IEQ and energy efficiency.

### CONTROL SYSTEM COMMISSIONING PROCESS

The commissioning process is defined in *ASHRAE Guideline 0-2005, The Commissioning Process*, as:

> A quality focused process for enhancing the delivery of a project. The process focuses upon verifying and documenting that the facility and all of its systems

and assemblies are planned, designed, installed, tested, operated, and maintained to meet the Owner's Project Requirements.

The full process from project planning through occupancy and operations is explained in Guideline 0 and in the *2003 ASHRAE Handbook—HVAC Applications*, Chapter 42, "New Building Commissioning." Guidelines for applying the commissioning process to green buildings in general are found in Chapter 3 of this Guide. Additional information relative to design reviews and commissioning for designers and commissioning providers can be found on the Energy Design Resources Web site at www.energydesignresources.com/publication/gd/.

This section will cover salient elements of applying the commissioning process to controls in green buildings and will focus on what the design engineer and commissioning provider can do during the design phase to facilitate a successful commissioning program.

## Include Commissioning Engagement in Design Fees

The owners and designers should include in the design fees sufficient time to fully engage in the commissioning process during design, including responding to design review comments and incorporating commissioning requirements into the project specifications.

## Conduct and Participate in Design Reviews

During the design reviews, the commissioning provider should see that the following controls-related elements are included in the contract documents. This list of items is not comprehensive, but it provides an idea of the type of issues that should be addressed.

**Provide Detailed Control Descriptions.** One of the most prevalent reasons why control systems fail to perform as intended is that insufficient forethought is given to the sequence prior to the contractor programming the sequence in the field. The designers and, later, the control contractor's programmer often do not think the sequence out and consider how it will (or will not) function during all possible modes and scenarios of weather, loads, staging, and interactions. This issue can be mitigated by the following:

1. Ensure that the designer provides one-line flow diagrams of the major controlled systems, showing interfaces and control authorities between packaged and central control.
2. Ensure that the designer includes detailed sequences of operation that include brief system narrative, alarms, what initially starts equipment, staging, failure and standby functions, power outage response and reset requirements, interlocks to other systems, control authorities with packaged controls, and energy efficiency strategies with setpoints given.

**Match Control Strategies to Operator Capabilities.** If the operators do not understand the features or sequences sufficiently and there is not a qualified controls contractor maintaining the system, the advanced features or sequences that have problems will likely be overridden or disabled. Designers and design reviewers should make sure the complexity of control schemes matches the expected level of technical expertise of the operators according to those identified in the OPR document.

**Strategies Relying on Drift-Prone Sensors.** Control sensor and loop recalibrations are a necessary function for maintaining high-performing systems. The OPR document will define the operator training and skill sets needed to maintain the system functioning at design efficiency. Major control strategies that depend on sensors that are known to drift should be avoided or, if called for, then the necessary training and recalibration programs must be institutionalized into the building maintenance culture. For example, consider the case of a chiller staging sequence utilizing supply and return temperatures and a flowmeter(s). The sensors may well drift over time and typical accepted errors in these types of sensors will yield load calculations that may disrupt proper staging. This strategy can result in high overall efficiency, but it requires a regular calibration and maintenance check.

**Requirements for System Architecture Rationale.** Ensure that in the requirements for the controls submittal, the controls contractor is required to provide calculations and rationale for the number and layout of the primary (peer-to-peer) and secondary (application-specific) controllers in relation to the total number of points and other network traffic. Require that the contractor describe how many points can be reasonably trended without appreciably affecting point value refresh rates and describe the impacts on network speed that alternative layouts would have.

The BAS performance requirements should be defined in the contract documents (i.e., specifications). The performance requirements are defined during the predesign phase and contained in the OPRs in accordance with ASHRAE Guideline 0-2005.

**Requirements for Clear Control Sequences.** Ensure that the requirements for the control drawing submittals in the specifications include statements requiring the following:

- A brief overview narrative of the system, generally describing its purpose, components, and function.
- All interactions and interlocks with other systems.
- Detailed delineation of control between any packaged controls and the BAS, listing what points only the BAS monitors and what BAS points are control points and are adjustable.
- Start-up, warm-up, normal, and unoccupied operating modes and power failure and alarm sequences.
- Capacity control sequences and equipment staging.
- Initial and recommended values for all adjustable settings, setpoints, and parameters that are typically set or adjusted by operating staff and any other

control settings or fixed values, delays, etc., that will be useful during testing and operating the equipment.

- To facilitate review and referencing in testing procedures, all sequences shall be written in short statements, each with a number for reference.

**Requirements for Clear Control Drawings.** Ensure that the control drawing submittal requirements include at least the following:

- The control drawings shall contain graphic schematic depictions of all systems showing each component (valves, dampers, actuators, coils, filters, fans, pumps, speed controllers, piping, ducting, etc.), each monitored or control point and sensor, and all interlocks to other equipment. Drawings should include fan and pump flow rates as well as horsepower.
- The schematics will include the system and component layout of any equipment that the control system monitors, enables, or controls, even if the equipment is primarily controlled by packaged controls.
- Provide a full points list, including point abbreviation key, point type, system point with which it is associated, point description, units, panel ID, and field device.
- Network architecture drawing showing all controllers, workstations, printers, and other devices in a riser format and including protocols and speeds for all trunks.
- Sketches of all graphics screens for review and approval.

## Specify a Systems Manual

Ensure that the commissioning scope for the commissioning provider, contractor, and designer includes a systems manual that, among other things, includes narratives explaining all energy efficiency features and strategies, a setpoint and parameter table indicating the impacts of changing the values, a recalibration and retesting frequency table, suggested "smart alerts" in the control system to send alerts on malfunctioning sensors and actuators, and a list of standard trend logs to view to verify proper performance. The building operators need to be trained on the systems manual and its contents in order to properly maintain the system.

## CONTROL SYSTEM ROLE IN ATTAINING LEED CERTIFICATION

In Chapter 6, the USGBC's LEED products are discussed. This section explicitly discusses how controls can be used in various sections of the LEED for New Construction (LEED-NC) Rating System, specifically in reference to the latest version (2.2) in effect as of this writing. This section on LEED and controls will connect control methods discussed earlier in this chapter with LEED-NC. The credit areas that can be affected include:

- Water efficiency (WE)
- Energy and atmosphere (EA)
- Indoor environmental quality (EQ)

The remainder of this section discusses control interactions that can help meet LEED prerequisites or obtain credit points. If the control option was covered earlier in this chapter, the reader will be referred to that section.

### Water Efficient Landscaping—Reduce by 50%: WE Credit 1.1

Controls allow soil moisture levels to be sensed and watering to take place on a "demand" basis. This reduces the water consumption for landscaping to that needed for the specific site, weather, and vegetation requirements. The BAS can also monitor and trend the actual water usage, so as the building moves to LEED-EB, data will be available for submission. See "Control System Role in Delivering Water Efficiency" on page 317 of this chapter for more detail.

### Fundamental Commissioning—
### EA Prerequisite 1 and Advanced Commissioning: EA Credit 3

Both fundamental and advanced commissioning can be aided by the control system—especially if the equipment calibration has been verified. The BAS can also help in verifying proper calibration from the factory.

Along with verifying feature and system sequences of operation, calibration, and functionality, the CxA is responsible for applying appropriate sampling techniques to verify equipment start-up and operation. Although commissioning is a process and not a technology, it seems logical that the facility BAS system can assist in this sampling process. For example, a typical rooftop VAV system might be set up to display the system operating conditions in a manner similar to that shown in Figure 15-1. Such a control display might include:

- Supply air conditions including fan control pressures
- Space conditions
- Other load conditions

Using the control system, it is pretty easy to see that the mixed air temperature sensor is not working properly.

For a commissioning example, according to ASHRAE Standard 90.1, in a DDC VAV system, fan pressure optimization is required. Fulfilling this requirement requires having VAV damper positions available. Even on what most people consider a fairly simple system, there are quite a few control points that must be calibrated and in working order. With equipment manufacturer engineered and installed

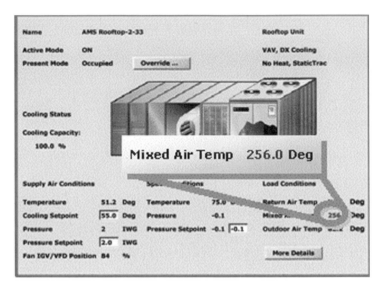

**Figure 15-1 Sample rooftop unit control display.**

controls, many of the points would be factory-commissioned. The BAS can easily be used to sample and determine if any points are not working properly.

For enhanced commissioning, the CxA is required to write an operations manual. The BAS can be used to enhance the building operator's use of this manual. See "Control System Commissioning Process" on page 319 of this chapter for additional detail.

### Energy Performance—
### EA Prerequisite 2 (Minimum) and Optimized: EA Credit 1

Minimum energy performance (MEP) is defined as meeting ASHRAE Standard 90.1-2004. For example, some of the Standard 90.1 requirements that require controls include:

- Economizer high-limit shutoffs
- Static pressure reset
- Optimum start controls
- Isolation zone control
- High-occupancy ventilation control
- Waterside heat recovery
- Service water heating temperature
- Automated lighting shutoff based on time of day or occupancy

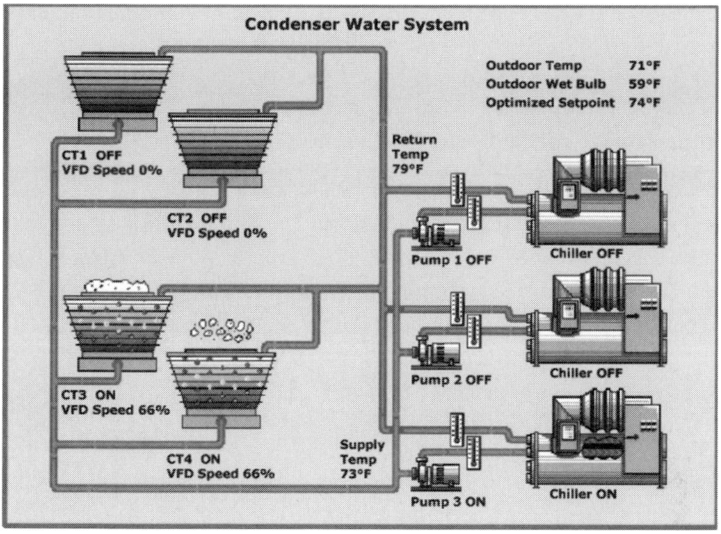

**Figure 15-2 Sample cooling tower fan control display.**

Credit points are accrued by reducing energy costs below ASHRAE Standard 90.1-2004. System strategies often require controls to optimize the energy performance. These strategies include:

- Supply temperature reset (air and hydronic)
- Demand shed controls
- Optimization of central plants
  - Chiller staging
  - Pump control and staging
  - Cooling tower fan control (see Figure 15-2)
  - Variable condenser water flow
- Dual minimums for VAV boxes
- Demand-controlled ventilation
- Economizer isolation zone control
- Occupancy-sensed HVAC and lighting

In addition, there are many mechanical systems that require well-operated controls for optimization. They include:

- Waterside heat recovery
- Variable primary-flow systems
- Low-flow systems
- Geothermal heat-pump systems

In addition to HVAC control, the BAS can automatically open and close blinds to reduce system load. There are a myriad of ways that automated controls can help reduce energy consumption and costs.

### Measurement and Verification: EA Credit 5

M&V of the building systems are developed according to the International Measurement and Verification Protocol (IPMVP). While a BAS cannot perform all the requirements to gain the M&V credit, it can greatly simplify tracking and trending the data, especially in the isolation of systems for verification (IPMVP Options A or B). Examples of data that can be gathered include:

- Economizer cycles
- Heat exchanger cycles
- Air distribution
  - Pressures
  - Volumes
- Process loads
- Water usage
- Lighting systems and control
- Motor loads
- VFD operation
- Chiller efficiency
- Boiler efficiency
- Cooling load

The data may also be presented using "trend logs" available from many control providers (see Figure 15-3).

### Indoor Environmental Quality—
### Outdoor Air Delivery Monitoring: EQ Credit 1

BAS controls can be used to monitor $CO_2$ as well as any direct airflow measurement devices required to achieve this credit (see Figure 15-4).

### Indoor Environmental Quality—Increased Ventilation: EQ Credit 2

This credit requires the ventilation airflow to be at least 30% above the ASHRAE Standard 62.1-2004 ventilation requirements. In many systems there are calculations that must be dynamically performed in response to system loads and ambient conditions. The BAS can monitor, perform the required calculations, and reset damper positions to maintain the additional ventilation required.

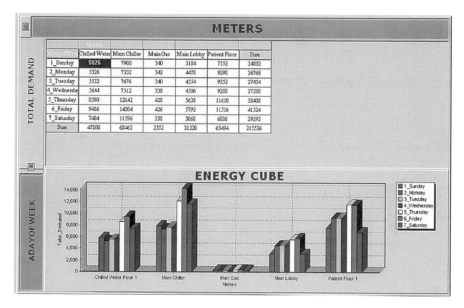

**Figure 15-3 Sample rooftop unit control display.**

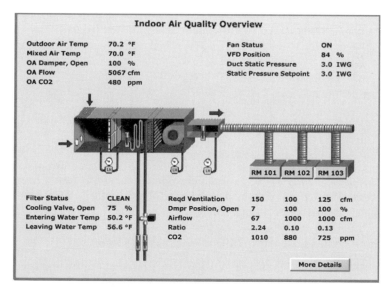

**Figure 15-4 Sample IAQ control display.**

### Indoor Environmental Quality— Construction IAQ Management Plan: EQ Credit 3.2

This credit requires flushing the building out either prior to occupancy or while the building is occupied. In either case the amount of ventilation air, as well as proper humidity control, are required if Option 1 of this credit is used. The BAS can monitor, trend, and sum the amount of outdoor air delivered to the space, as well as ensure the mechanical system is controlled to maintain the required humidity levels.

### Indoor Environmental Quality— Indoor Chemical and Pollutant Source Control: EQ Credit 5

The HVAC system must be operated to properly pressurize the space while exhausting sufficient air to provide pollutant removal. The BAS system can monitor exhaust air measurements, sense various space pressures, and perform control algorithms to maintain proper space pressure relationships.

### Controllability of Systems—Lighting: EQ Credit 6.1

The BAS can provide lighting controls that meet the mandatory provisions of Standard 90.1, such as automated lighting shutoff and independent control within the space.

### Controllability of Systems—Thermal Comfort: EQ Credit 6.2

If the HVAC system is properly designed, the BAS can provide individualized controls and maintain thermal comfort conditions, as described in ASHRAE Standard 55-2004 and illustrated in Figure 15-5. The primary factors for such a system will be:

- air temperature
- radiant temperature
- air speed
- humidity

According to the LEED-NC 2.2 Reference Guide, "Comfort system control, for the purposes of this credit, is defined as the provision of control over at least one of these primary factors in the occupant's local environment."

### Daylight and Views, Daylight 75% of Spaces: EQ Credit 8.1

The BAS system can change lighting levels in response to photocell-based sensors and reduce artificial lighting while still maintaining proper space illumination.

See "Control System Role in Delivering IEQ" on page 318 of this chapter for additional detail concerning the EQ credits discussed above.

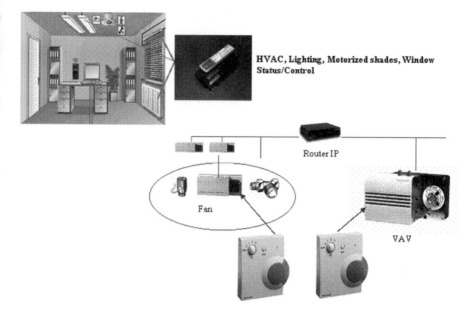

HVAC, Lighting, Motorized shades, Window
Status/Control

Router IP

Fan

VAV

**Figure 15-5 Sample IAQ control display.**

## Innovation in Design

BAS systems can aid in achieving exceptional performance in water reduction, energy efficiency, etc., and help attain innovation in design credit points.

## DESIGNING FOR SUSTAINED EFFICIENCY

This chapter contains a number of recommendations on how building and system controls can be used to obtain a good, green design. Getting a good design up front is important but is only the beginning. Sustainability, as the name implies, also includes continued efficient building operation over its entire lifetime. Two factors that are critical to sustaining the efficiency level of a new building are (1) a well-designed M&V process and (2) good operator training on the control system functions. The first factor allows building operators to monitor performance on a regular basis and intervene when problems are detected. As discussed previously in this chapter, the control system is essential in implementing a good M&V process. Good training can help ensure that all of the capabilities of the control system, including those related to M&V, are used to their full potential over the lifetime of the building.

## REFERENCES TO OTHER SOURCES OF INFORMATION

There are a number of references within ASHRAE and other organizations that are valuable sources of information related to building and equipment controls for the HVAC engineer. These include:

- *ANSI/ASHRAE Standard 55-2004, Thermal Environmental Conditions for Human Occupancy*
- *ANSI/ASHRAE Standard 62.1-2004, Ventilation for Acceptable Indoor Air Quality*
- *ANSI/ASHRAE Standard 90.1-2004, Energy Standard for Buildings Except Low-Rise Residential Buildings*
- *ASHRAE Guideline 0, The Commissioning Process*
- *ASHRAE Guideline 1, The HVAC Commissioning Process*
- *ASHRAE Guideline 13, Specifying Direct Digital Control Systems*
- *ASHRAE Guideline 14, Measurement of Energy and Demand Savings*
- DDC Online, www.ddc-online.org
- International Performance Measurement and Verification Protocol, www.ipmvp.org
- National Building Controls Information Program (NBCIP), www.buildingcontrols.org

## ASHRAE GreenTip #38: Mixed Air Temperature Reset

### GENERAL DESCRIPTION

Mixed air temperature (MAT), in this case, refers to the temperature of the mix of outdoor and return (recirculated) air that exists on an operating supply AHU prior to any "new" thermal energy being added to the airstream. In the days when constant air volume (CAV) systems were prevalent, it was customary to set the MAT controls to maintain a constant 55°F (13°C) nominally. (The controls would adjust the relative positions of outdoor and return air dampers to apportion the relative quantities of each airstream to satisfy the MAT setpoint, but never allowing less than the code-required minimum outdoor air.) In the "wintertime"—or heating season—when the outdoor air temperature was generally below 55°F (13°C), the MAT would be the "cooling" airstream or cold deck—the lowest-temperature air available for zones that needed cooling in this season. As heating was required, heat would be added at some point, either through a "hot deck" airstream within the AHU or through reheat by downstream coils.

The reset technique is based on the premise that the MAT from a supply air-handling system is colder than any one zone requires to maintain the set conditions of that zone. To the extent that this condition prevails, it means that the mixed (or cold) airstream must be mixed with some warm (hot deck) air to yield the proper supply air temperature to satisfy even the zone requiring the *lowest* temperature air supply. Since warmer air would need to be mixed in to do this, that would require "new" energy and is thus somewhat wasteful of heating energy (a form of simultaneous heating and cooling). In the heating season, cooling—being derived from outdoor air—is free.

The idea is to reset the MAT to a temperature that just satisfies the space with the lowest cold air demand. Reset controls involve raising the setpoint of the MAT controls based on input that indicates the demand of that zone needing the coldest air—limited still by the need to main the *minimum* quantity of outdoor air. This, in turn, requires sensors that can monitor that and other zone demands continuously; this input could come from hot deck/cold deck mixing dampers, mixing box damper positions, or thermostat output signals that indicate zone temperature demands. The goal would be to raise the MAT just enough so that the zone with the lowest supply air temperature demand was satisfied on a continuing basis. (As conditions change over time, that zone may change.)

### WHEN/WHERE IT'S APPLICABLE

As stated above, this technique, in most cases, *should only be used on CAV systems*. If it is used with VAV systems, it can often backfire since other energy variables (such as fan energy, in the case of air systems) may change in the opposite direction from heating energy saved, possibly resulting in a net increase in energy use or cost. Thus, if it *is* applied to VAV systems, it should come into play when any other affected variable is already at its minimum (e.g., fan already at its minimum turndown rate).

As the season becomes warmer and the outside temperature rises, this technique may become less and less effective, especially since the served zones may require more cooling and ever lower supply air temperatures.

Although there may not be a lot of CAV systems installed in new designs, there are still plenty operating in existing buildings (though it should not be applied to CAV systems converted to variable volume). This technique does lend itself well to retrofit, and since the controls are basically the same for large- or small-sized air-handling systems, the savings can be large for a relatively low capital cost.

## PROS AND CONS

### Pro

1. MAT reset saves heating energy and the associated operating cost.
2. It can yield a low payback, especially on larger air-handling systems.
3. It is relatively low in capital cost in the full spectrum of energy retrofits.

### Con

1. MAT reset may require greater attention to periodic controls calibration.
2. To be effective, there must be evidence that worst-zone demands will allow sufficient upward reset of temperature to realize appreciable savings.
3. Sampling of zone demands may be difficult to do in remote or scattered locations.
4. It is relatively easy to do as a retrofit on existing systems.

## KEY ELEMENTS OF COST

The following provides a possible breakdown of the various cost elements that might differentiate a mixed air temperature (MAT) reset system from a conventional one and an indication of whether the net cost for the alternative is likely to be lower (L), higher (H), or the same (S). This assessment is only a perception of what might be likely, but it obviously may not be correct in all situations. **There is no substitute for a detailed cost analysis as part of the design process.** The listings below may also provide some assistance in identifying the cost elements involved.

### First Cost

- Reset controls and installation          H
- Zone input sensors and connection to reset controls    H

**Recurring Cost**

- Heating energy (heating coil)          L
- Maintenance                            H
- Operator training                      H

## SOURCE OF FURTHER INFORMATION

ASHRAE. 2003. *2003 ASHRAE Handbook—HVAC Applications*. Atlanta: American Society of Heating, Refrigerating and Air-Conditioning Engineers, Inc.

## ASHRAE GreenTip #39:
## Cold Deck Temperature Reset with Humidity Override

### GENERAL DESCRIPTION

Cold deck temperature (CDT) reset is very similar to MAT reset, but it applies to the air temperature leaving a cooling coil—or CDT of a CAV AHU during "summertime," or the season when mechanical cooling is required. In this situation, the object is to save cooling energy supplied at the cooling coil by allowing the set CDT to "ride up" above a nominal design level (e.g., 55°F [13°C]) as long as all zone cooling needs are met.

The control needs and techniques are similar to MAT reset (reviewing GreenTip #38 is a prerequisite to considering this one) with one addition: a humidity override. When in the mechanical cooling mode in moderate-to-humid climates, mechanically cooled air serves the function of dehumidification (latent cooling) as well as sensible cooling. The degree of humidification achieved in the occupied zones served depends largely on the CDT being maintained. Thus, if CDT is allowed to ride up for cooling purposes, it should not be allowed to rise beyond the temperature needed to maintain comfortable humidity conditions. Thus, the occupied zone humidity parameter sets an upper limit for the reset function.

Zone humidity input can be sensed from a sampling of served zones themselves, or one could sense the return airstream at the AHU. The latter would yield an *average* humidity of all spaces served rather than the highest humidity of any one space. However, given that return air humidity would probably be a lot easier and less expensive to sense than that of several remote zones, it may be good enough to serve the purpose.

The upper limit of humidity chosen as the limiting factor for this reset technique would depend on what the building operator feels is within the comfort tolerance of the occupants. While a nominal relative humidity level of 50% is often the goal for cooling season comfort, higher levels can be tolerated, and sometimes an upper limit of 55% to 60% may be selected. Whatever is chosen, however, is easily adjustable. High-quality humidity-sensing equipment is recommended.

Reducing the CDT off the cooling coil can also result in savings at other "upstream" components of the building's cooling system, such as not-as-cold temperatures off a central chiller or reduced chilled-water flow in a variable-flow pumping system. (In fact, this is where the cost savings would actually be realized.) If considering this technique, the designer should ensure that the piping, valves, and control configurations are such that "up-the-line" energy and cost savings are indeed achievable.

## WHEN/WHERE IT'S APPLICABLE

The same constraints apply here as with MAT reset. Again, before doing this, the designer should be sure that there are likely to be opportunities for significant upward reset to take place. If it is found that there is just one space that is likely to always need the design CDT during the cooling season, regardless of weather or other changing conditions, then this technique is probably not a good bet.

## PROS AND CONS

### Pro

1. CDT reset saves cooling energy and associated operating cost.
2. It can yield a good payback when the situation is right, although not as low as MAT reset.
3. Capital cost is still relatively low (though more controls are required than with MAT reset).

### Con

1. There are the same drawbacks as with MAT reset, plus.
2. There could be added problems with excessive space humidity if the humidity sensing is not accurate.

## KEY ELEMENTS OF COST

The following provides a possible breakdown of the various cost elements that might differentiate a CDT reset system from a conventional one and an indication of whether the net incremental cost for the alternative option is likely to be lower (L), higher (H), or the same (S). This assessment is only a perception of what might be likely, but it obviously may not be correct in all situations. **There is no substitute for a detailed cost analysis as part of the design process.** The listings below may also provide some assistance in identifying the cost elements involved.

### First Cost

- Same elements as for MAT reset, plus      H
- Humidity sensor(s) and connection reset controls      H

### Recurring Cost

- Cooling energy      L
- Maintenance      H
- Operator Training      L

## SOURCE OF FURTHER INFORMATION

ASHRAE. 2003. *2003 ASHRAE Handbook—HVAC Applications.* Atlanta: American Society of Heating, Refrigerating and Air-Conditioning Engineers, Inc.

# 16

# COMPLETING DESIGN AND DOCUMENTATION FOR CONSTRUCTION

## DRAWINGS/DOCUMENTS STAGE

Once the project has reached the working drawing/construction document stage, the green design concepts and resulting configurations should be well set, and the task of incorporating them into the documents that contractors will use to build the project should be relatively routine. However, quality control at this stage is especially important in green design projects.

Many firms have a routine procedure to review the documents for quality before they are released to contractors, ensuring concepts are adequately depicted and described and catching errors and omissions. This process should also include a green design concept review, preferably by one or more design team members that were in on the early stages of the project. This is particularly true if those preparing the construction documents were not part of that process. This is not the time to allow an excellent green design to become diluted or slip away.

## SPECIFYING MATERIALS/EQUIPMENT

### Green Building Materials

Sources for guidance on selecting and specifying materials for a green project are:

*   Athena Institute, www.athenasmi.ca.
*   BuidingGreen, www.buildinggreen.com.

Further, as a guide that could prove helpful to some, refer to "One Design Firm's Materials Specification Checklist," on page 342.

### Controlling Construction Quality

It is far easier to control construction quality in the design and specification stage of a project than during its construction. During preparation of the final construction drawings and specifications, it is critical to be diligent in spelling out the quality expected in the field to carry through the green design concepts developed throughout the early design stages. Some further thoughts on this subject, many applicable to the design phase, are covered in Chapter 17, "Construction."

## COST ESTIMATING AND BUDGET RECONCILIATION

### Design Costs

While professional design fees are not normally a subject discussed in the ASHRAE world of standards and guidelines, it is one of the major potential impediments to green design. Now why would any consulting engineer, in business to do competent work and to stay afloat, agree to be paid to reduce the cost of the mechanical system based on mechanical system construction cost? If a good job were done, that engineer's fee would be reduced! It would be more in the engineer's interest to get high-efficiency chillers, additional controls, etc., installed at an additional cost; these components might have a short payback that could be justified, and some owners might accept them. So the engineer gets a bit more money, and project energy usage is reduced a bit. Life is good!

*But... the energy usage of the building did not really change that much!*

What if the fee structure were changed so that innovative design with significantly reduced energy usage was encouraged? Some possible alternative approaches follow:

**Fixed Fees.** If professional fees were fixed at the "normal" level, the fees would not be reduced if the cost of the mechanical work were reduced. While this is not a really positive incentive for green design work, it at least removes the disincentive. The designer actually has to work harder to do less (smaller mechanical system).

**Premium Fees.** If an engineering designer is actually doing more and contributing to a project beyond normal service, such as working in the "reduce the load" area, he or she is providing a higher and more valuable level of service. When a client goes into a "scary" area such as green design, engineers that can provide an added "value" service not only will be more likely to get the project but will get paid for it as well.

This is stepping beyond the world of "commodity service," where anyone can do the project, so the market goes for the lowest cost. The new model is the value-added world that will pay more for better engineering input. Just look at the range of cars on the road. Many of them are not the least expensive models available.

In the end, there has to be better value for the client if that client is going to go for a different structure. If capital costs are the same or similar and the operating costs are significantly reduced, there is a big return for the client for some amount of additional fee.

**Modeling Costs.** There are opportunities for additional fees in the area of energy performance computer modeling. This might be done at the initial "reduce the load" stage or later to evaluate design options, or in the code compliance submission stage (because you did not use the prescriptive route), or to demonstrate LEED compliance. The modeling could be done in house if capabilities are available or it could be contracted out.

Further, some form of modeling is usually required to demonstrate projected results when a design outside the standard design for the region is used.

**Performance Fees.** An opportunity for increased fees with some risk is performance-based fees. With this approach, a basic fee is paid for design and construction work. Should the project meet the reduced energy performance goals (adjusted for common hours of operation, weather, etc.), an agreed premium fee is paid. It further could be staged, depending upon actual performance.

**Programs.** Sometimes there are government or industry programs to support using innovative or energy-saving technologies or creating lower-energy-use building designs. The design team should become familiar with the programs available in their region and utilize them when appropriate. These programs often provide design cost incentives, capital cost incentives, or tax incentives, resulting in a better financial performance by the project and increased acceptance of the lower-energy-use approaches

## Capital Costs

**System/Equipment Reductions Due to Green Design.** The common perception is that green buildings will cost more. Most people's ingrained experience is that if you want something better, you have to pay more for it. When one goes to the store to buy something such as a car, stereo, or computer, the better performing models generally cost more.

For green buildings to become commonplace and eventually be the norm, they must be built without significant, if any, additional cost. To maintain traditional project cost levels (factoring out normal escalation over time), many practitioners have found that there is usually a cost transfer required from the mechanical system to other building elements (e.g., high-performance glass). A number of built examples of this phenomenon exist.

The key to proper cost analysis of sustainable buildings is broadening the scope of the cost analysis. As a further example, a radiant floor/displacement ventilation HVAC system may cost more to install by itself than a conventional system, but the elimination of dropped ceiling space and the resulting savings in slab-to-slab height required may translate into overall project savings.

As in this case, the economic advantage of green engineering is often found through the "no-build" offset. Since sustainably designed buildings are often life-cycle costed over longer time horizons, engineers (and owners) must similarly shift their focus.

### Recurring Costs/Benefits

**Operating/Maintenance.** Reduced on-site energy use and the attendant reduction in energy costs are obviously a key benefit of successful green design. This is accompanied by lowered use of the building's share of energy resources and other natural resources, which is one of the rationales for green design.

Maintenance costs may or may not be lowered on balance. While reduced maintenance would be expected to accompany smaller-sized equipment and systems, where such result from green design, the fact that some green systems or features must be kept in proper operating order to continue to be effective over the long term would tend to drive the cost of maintenance somewhat higher than it would be otherwise. Thus, it is difficult to generalize about such costs, and projections of maintenance cost benefits are best made on a project-by-project basis.

The impacts on operating and maintenance costs have been covered extensively throughout this document. A more elusive benefit of green design is its impact on human productivity.

**Human Productivity.** While difficult to measure, the benefits of improving the workplace environment (*all* aspects of IEQ) and workers' feeling of well-being can yield big gains in human productivity. Figure 16-1 shows the typical relationship among the various categories of cost of operating a business, with human costs outstripping all other costs several times over.

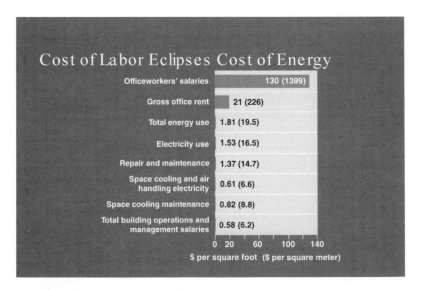

**Figure 16.1 People cost vs. other costs in business operation.**

Some case study results of improved productivity are summarized below.

- Lockheed
  - 15% rise in production
  - 15% drop in absenteeism
- West Bend Mutual Insurance
  - 16% increase in claims processes
- ING Bank
  - 15% drop in absenteeism
- Verifone
  - 5% increase in productivity
  - 40% drop in absenteeism

## Sources for Further Information

BetterBricks, www.betterbricks.com.
Greening the Building and the Bottom Line, Rocky Mountain Institute, Available from www.rmi.org.

## Intangibles

**Worth of Green Design.** Reviewing green design work done by others allows engineers to see how built projects have created new opportunities and can guide efforts to sell and provide green engineering services. Performing sustainable design can yield important benefits to design engineers individually and to their firms. Improved client service, more repeat work, improved market position, enhanced public relations, and better employee satisfaction and retention are among the many benefits that numerous architectural firms and several pioneering engineering firms have derived from informing their practices with sustainable design expertise.

On the other hand, failure to address client concerns about green design can harm reputation. Disproportionate start-up costs, risk, or other perceived obstacles (such as the educational investment and commitment required to change engineering thinking and culture) may be perceived as being associated with undertaking sustainable design. Managing each of these issues, communicating frankly, and crafting creative solutions as required can reduce exposure to these potentially negative issues.

## One Design Firm's Materials Specification Checklist

❑ Choose at least a portion of materials/products that are:

- Local and/or indigenous, reducing the environmental impacts resulting from transportation and supporting the local economy.
- Extracted, harvested, recovered, manufactured regionally within a radius of 500 miles (805 km).
- Have low embodied energy.
- Reused, recycled, and/or recyclable, reducing the impacts resulting from extraction of new resources.
- Salvaged building materials (lumber, millwork, plumbing fixtures, hardware, etc.)
- Post-consumer recycled content material and/or post-industrial recycled content (recovered) material.
- Nonhazardous to recycle, compost, or dispose of.
- Renewable and sustainably harvested (no old-growth timber), preferably with minimal associated environmental burdens, reducing the use and depletion of finite raw and long-cycle renewable materials.
- Rapidly renewable building materials, including any nonwood materials that are typically harvested within a ten-year or shorter cycle.
- Nontoxic/nonpolluting in manufacture, use, and disposal.

❑ Use finished materials, products, and furnishings that are free of known, probable, and suspected carcinogens, mutagens, teratogens, persistent toxic organic pollutants, and toxic heavy metals pursuant to EPA's *Toxicity Characteristic Leaching Procedure (TCLP) Test*, 40 CFR Part 260. (Based on EPA's *Universal Waste Rule* and Part 260, building owners and their contractors must use the TCLP to determine if they are generators of hazardous waste and thus subject to EPA and state hazardous waste regulations when disposing of mercury lamps and other waste products.)

❑ Specify lead-free solder for copper water supply tubing; do not use plastic for supply water.

❑ Avoid plastic foam insulation.

❑ Specify natural, nontoxic, low-VOC–emitting, nonsolvent-based finishes, paints, stains, and adhesives.

❑ When specifying painting:

- Consider surfaces that don't require painting.
- Choose paints that have been independently certified (such as Green Seal).
- Choose latex over oil-based paint.
- Increase direct-to-outdoors ventilation when painting.

- Dispose of oil-based paints like hazardous waste.
- Recycle latex paint or save for touch-ups.

❏ Avoid the use of materials containing or produced with ozone-depleting CFCs or HCFCs (often embedded in foam insulation and refrigeration/cooling systems).

❏ Avoid CFC-based refrigerants in new base-building HVAC&R systems. When reusing existing base-building HVAC equipment, complete a comprehensive CFC phaseout conversion.

❏ Commit to working with manufacturing teams who provide performance/service contracts of integrated systems for service delivery, product longevity, adaptability, and/or recycling.

❏ Favor suppliers who will minimize packaging and will take back excess packaging such as pallets, crates, cardboard, and excess building materials.

❏ Maximize the use of materials that retain a high value in future life cycles.

❏ Specify products or systems that extend manufacturer responsibility through a lease or take-back program that ensures future reuse or recycling into a product of similar or higher value.

❏ To guide material selection choices during the design process, educate designers and elicit life-cycle information from manufacturers to encourage selection of building materials and furnishings with favorable life-cycle performance.

❏ Select materials from manufacturers that are committed to improving their overall environmental performance at their manufacturing facilities and their suppliers' facilities.

# Section 3: Post-Design—Construction to Demolition

# 17

# CONSTRUCTION

"There is many a slip 'twixt the cup and the lip" is an old proverb that may be applicable to the design/construction process. To minimize gaps between design intent and what is actually built, the green-conscious engineer should provide a level of construction administration above the norm. That engineer (or an independent agent) may also provide mechanical/electrical system commissioning. (See Chapter 3, "Commissioning," and Chapter 18, "Operation/Maintenance/Performance Evaluation," as well as a related section in Chapter 5 on project delivery methods and contractor selection.)

## CONSTRUCTION PRACTICES AND METHODS

In design, the specifications should prescribe that certain construction methods and procedures must be followed to ensure a fully realized green project. This would include topics such as reduced site disturbance, handling of construction waste, control of rainwater runoff, and IAQ management during construction. Some items for consideration, which may or may not be applicable to the mechanical/electrical contractors on site, are listed on page 350 at the end of this chapter.

It is recommended that the engineer and architect work with the construction manager or general contractor and associated subcontractors to develop a Construction IAQ Management Plan that the contractor would be required to follow. SMACNA's *IAQ Guideline for Occupied Buildings under Construction* is cited by the USGBC as a source document. The contractor should particularly protect installed or stored absorptive materials, such as insulation or sheetrock, from water damage or other contamination. Water damage is especially insidious if materials get wet, are installed wet, and are then covered up. If AHUs run during construction, the construction manager and associated subcontractor should be required to protect the unit components and the duct distribution system from dirt and debris, clean any components and ductwork that are damaged, and replace any filters before occupancy. (ASHRAE Standard 52.2 deals with this subject.)

## THE ENGINEER'S ROLE IN CONSTRUCTION QUALITY

In construction, the engineer's key roles encompass shop drawing review, review of equipment substitutions, handling of change order and value engineering requests, and site visits and inspections. LEED or other green design process documentation coordination and processing are sometimes required when the project team has decided to achieve LEED certification (or some other green design process recognition) for the building.

### Shop Drawing Review

There should be a thorough review of shop drawing submittals. The specifications should require their timely submission to allow time for proper review and in order that the review process does not delay the project. The purpose of the review is to ensure that the contractor is correctly interpreting the specifications and drawings and is not missing any important details. Areas for emphasis include checking motor horsepower and efficiency ratings, checking air filter details, checking for proper clearances around equipment for servicing, and checking anything else relevant on the project concerning green design elements. Thorough review of all control sequences submitted by the BAS subcontractor is critical.

### Alternative Equipment Substitutions and "Or Equivalent"

Specifications may be written with several named manufacturers (often three) of a given product, or it may name one and say "or equivalent." In both cases, the system designed is usually based on a single manufacturer's product. If others are named, it usually means they are acceptable if they meet the specified conditions and if they fit in the space provided on the project, including all required service and maintenance access clearances. The shop drawing reviewer must determine whether any alternative equipment is truly equal to the base design, including meeting the green design aspects of the equipment and systems.

If "or equivalent" language is used to specify equipment, determining equivalence can be more difficult. It is advisable to be certain that what the contractor proposes is truly equivalent, if not superior, to the specified components or systems. Contractors may have legitimate reasons for proposing alternative products, such as better delivery times or negative experiences with the products or manufacturers specified. Such "equivalent," or outright substitution, proposals from the contractor should be treated seriously and examined carefully to ascertain that they will not adversely impact the project goals. If they do not meet the requirements, the objections should first be discussed with the construction manager, owner's representative, and subcontractor. Only if no satisfactory alternative can be agreed on should they then be rejected. It is very important to clearly identify the substitution process requirements when specifying the equipment during the design phase of the project.

## Change Order Requests

Many change orders are legitimate, such as an owner changing the scope, expansion or reduction of project scope, or the construction manager and/or subcontractors encountering unforeseen conditions. However, any change order that only cheapens the project or lowers the project's green design standards without counteracting benefits, should be regarded with skepticism. Likewise any "value engineering" (VE) offers should be carefully studied for their impact on the project's green design goals. (VE is often offered under the assumption that first-cost savings are paramount to the owner and project team.) The need for careful study remains true even in the case of genuine VE done by trained professionals who perform real trade-off analyses to arrive at the "best value" for a project. Before beginning such a VE exercise, the VE facilitator should clearly understand the green design objectives in order to ensure that his/her suggestions and recommendations are consistent with the project goals, as well as the priority those goals have with respect to first cost or life-cycle cost. Each VE item must include the first-cost impact in order to properly evaluate the suggestion.

## Site Visits/Observations

Site visits should be planned for key times and involve the engineer's best personnel. HVAC work should be viewed before it is covered up by ceiling, floor, and wall installations. Check that equipment nameplates are correct. For example, consider fan motors: if high-efficiency motors were specified, check in the field to be certain that is what is being installed. Look up the manufacturer's data on motor efficiency and see that it matches the specification. Check that absorptive construction materials (e.g., insulation) are being stored in accord with the IAQ Construction Management Plan and that no contamination has occurred. Check for air filters in any operating AHUs. Check that air-handling equipment has been stored properly, cleaned (internals), and that filters were installed prior to start-up. If AHUs are used for construction activities, check that return air paths have proper temporary filtration at the inlets to the ductwork.

Work with the CxA to review their site observations and address any design-based questions that arise during the final commissioning process. More information on this is provided in the following chapter.

Final punch list preparation, follow-up, and the final "sign-off" observation are particularly important. If the earlier site visits and observations were done thoroughly and at appropriate intervals, and if the construction manager and subcontractors have been "part of the team" and in accord with the green design goals, the final punch list should be minimal and the final observation should go smoothly.

## Construction Factors to Consider in a Green Design

- Determine locations for construction vehicle parking, temporary piling of topsoil, and building material storage to minimize soil compaction and other site impacts.
- Control erosion to reduce negative impacts on water and air quality.
- Design to a site sediment and erosion control plan that conforms to best management practices in the EPA's *Storm Water Management for Construction Activities*, EPA Document No. EPA-832-R-92-005, or local erosion and sedimentation control standards and codes, whichever is more stringent.
- Conserve existing natural areas and restore damaged areas to provide habitat and promote biodiversity.
- Schedule construction carefully to minimize impacts.
- Avoid leaving disturbed soil exposed for extended periods.
- Fill trenches quickly to minimize damage to severed tree roots.
- Avoid building when the ground is saturated and easily damaged.
- Carefully estimate the amount of material needed to avoid excess.
- Design to accommodate standard lumber and drywall sizes.
- Assess construction site waste stream to determine which materials can be reduced, reused, and recycled.
- Conduct a waste audit, quantifying material diversion by weight.
- Recycle and/or salvage construction and demolition debris.
- Specify materials that minimize waste and reduce shipping impacts through bulk packaging, dry-mix shipping, reused bulk packaging, recycled-content packaging, or elimination of packaging.
- Develop and implement a waste management plan, quantifying material diversion by weight.
- Research markets in area for salvaged materials.
- Establish on-site construction material recycling areas and recycle and/or salvage construction, demolition, and land clearing waste.
- Contract with licensed haulers and processors of recyclables.
- Require subcontractors to be responsible for their waste (including lunch wastes); create incentives for minimizing waste.
- Educate employees and subcontractors.
- Monitor and evaluate waste/recycling program.
- Review the IAQ Construction Management Plan to ensure that all subcontractors understand the process and goals desired.

# 18

# Operation/Maintenance/ Performance Evaluation

## COMMISSIONING FOR BUILDING OCCUPANCY

Though the HVAC&R designer will not be the building operator, he or she can greatly facilitate the building's proper operation and maintenance by doing a good job of turning over the building to those who will operate it by assisting with the education of the building operators. This is an essential part of ensuring that a building designed to be sustainable/green actually operates correctly, thus causing minimum impact to the environment and maximum benefit to society and the owner. A proper commissioning process (covered in Chapter 3 and ASHRAE Guideline 0) is critical to meeting sustainable/green goals and is an essential component for success. The commissioning process includes the education of the operators and putting into place tools such as M&V to monitor building performance.

Educating the operators to understand building operation reduces the chances that they will inadvertently negatively modify or override building systems essential to maintaining performance. The CxA working with the designers provides great value to the owner and society by identifying the specific training requirements that meet the operational staff's educational needs and following up during warranty to verify the operators truly understand how to operate the building. The commissioning process requires the CxA to monitor building performance during the warranty period, conduct seasonal testing if necessary, and identify systems that are not performing to specification to help ensure that peak performance is maintained. Modifications that vary from the basis of design that are needed during the warranty period to improve performance are reviewed with the designers and owner for approval and implemented with the operations staff.

Use of M&V tools provides operators and owners a benchmark for continually assessing building performance under dynamic conditions. It also notifies operators and owners when building performance has drifted from peak performance, signaling that action is need. Construction phase commissioning focuses on identifying which system or systems are causing a reduction in performance and development of a plan to correct performance deficiencies through repair/ replacement and recommissioning.

Competent and well-trained operations/maintenance staff is essential to sustainable/green buildings. The designer can help owners make the important connection between the building performance and the competence/skills of the building operations staff. Building personnel can do much good—or inadvertent great harm—depending on whether they understand the impact of their actions on the building's operating efficiency. Obtaining the maximum benefit and performance from the design team's vision and the owner's goals requires not only that the systems operate as intended at the end of commissioning but also over the entire building's life. This requires trained operators with tools such as M&V. Systems whose efficiencies wane are identified, repaired, and recommissioned, providing greater assurance that a building designed to be sustainable/green will realize its full sustainable/green potential over the long term.

Monitoring performance is essential and required to maintain performance and is typically beyond the owner's capacity to perform. Operational staff can input utility information, but engineering knowledge is required to calculate performance. Monitoring and calculating performance is beyond the scope of normal professional fees and is usually not done. However, the building industry needs this kind of follow-up data to verify that new design techniques are working and that change is occurring in how buildings are designed; in short, they need to know that green design is working. It is certainly worth the effort by the designer or CxA to convince the owner that this could be a worthwhile extra service.

## RETROCOMMISSIONING (SM)

Retrocommissioning focuses on optimizing HVAC system operation and control in existing buildings that were not previously commissioned. Retrocommissioning has the greatest promise for substantially reducing energy consumption in the country and reducing dependency on foreign oil. Most existing buildings could immediately reduce energy consumption by 10% to 50% and realize a return on investment in as little as six months to two years. The retrocommissioning process also identifies improvements that can further reduce energy usage and carbon emissions that are directly related to the greenhouse effect so that owners can place these improvements into their capital improvement plans.

After being commissioned, building performance generally declines after two to five years. As a result, recommissioning (sometimes referred to as *continuous commissioning*) is required to return the building to its peak performance. It is important that designers have a deeper understanding of why performance wanes and how design decisions affect building operation efficiencies. Information on recommissioning is available through various organizations, including the federal government.

## Reference

*The Federal Energy Management Program Continuous Commissioning (SM) Guidebook.* www.eere.energy.gov/femp/operations_maintenance/commissioning _guidebook.cfm

## OCCUPANT SURVEYS

Occupant surveys can alert owners to specific problems with the IEQ, which could potentially have implications for occupant health and productivity

Occupants are an important, and often underutilized, source of information about IEQ and its effect on comfort, satisfaction, and productivity. Poor thermal comfort is often the more common occupant complaint in buildings. But relying on complaint logs only provides an indication of local, personal, or sporadic dissatisfaction. Surveys are much more effective by providing a systematic mechanism for occupants to provide feedback about a broader range of aspects of the indoor environment. They are an important tool for identifying the specific nature and location of any problems, guiding post-occupancy retrocommissioning and corrective actions, and helping owners decide how to prioritize their investments in building improvements. Everyone in the building process benefits from learning how a building actually performs in practice. Occupant surveys are an informative and critical link in closing that feedback loop, allowing building owners, facility managers, and the design team to understand more completely how the building design and operation is affecting occupant productivity and well-being.

A survey should be designed so that participation is voluntary, occupants' responses remain anonymous, and results are reported only in aggregate. The survey should ask occupants for general location information only, so one can identify if problems are occurring in particular zones of the building. After asking about basic demographics and workstation characteristics, a survey should then ideally address a wide variety of IEQ features, including thermal comfort, air quality, lighting, acoustics, office layout and furnishings, and building cleanliness and maintenance. A common form of satisfaction question asks occupants to respond on a seven-point satisfaction scale ranging from very dissatisfied, to neutral, to very satisfied. Ideally, occupants who are dissatisfied with a particular aspect of their environment are presented with follow-up questions that allow them to identify more specifically the nature and potential source of their dissatisfaction. This is important for providing diagnostic information, and helping the building operators become more informed about how to respond.

Surveys can be administered in a variety of ways, but once the method is selected it should be consistently applied and available for all normal occupants of the building. Surveys can be administered directly, either by phone or in person, although this is very time-intensive and raises potential issues about privacy and accuracy of results. Web-based surveys are becoming a more common alternative to the traditional paper-based surveys, and offer many advantages. They can be far less expensive to administer to a large number of people or to multiple buildings. They avoid the cost and potential errors of manually entering data from a paper survey. They allow for more interactive branching features that provide diagnostic information while keeping the survey to a reasonable length, and they offer the potential for automated reporting so that building owners and professionals can get quick access to the survey results.

Sources of sample surveys include, but are not limited to:

- Center for the Built Environment (CBE) Occupant Indoor Environmental Quality (IEQ) Survey, www.cbesurvey.org.
- Usable Buildings Trust, www.usablebuildings.co.uk.

### References

Leaman, A., B. and Bordass. 1993. Building design, complexity and manageability. *Facilities*, Vol. 11.

Zagreus, L., C. Huizenga, E. Arens, and D. Lehrer. 2004. Listening to the occupants: A Web-based indoor environmental quality survey. *Indoor Air 2004* 14(suppl 8):65–74.

### DISPOSAL/RECYCLING

This is an area in which the design engineer normally has little impact. During design, however, he or she could anticipate eventual maintenance, demolition, and removal requirements of the materials, equipment, and systems for which he or she responsible.

The USGBC's LEED Green Building Rating System® provides a good general guide on this topic.

## One Design Firm's Operations, Maintenance, and Performance Evaluation Checklist

☐ Provide for the ongoing accountability and optimization of building energy and water consumption performance over time.

☐ Design and specify equipment to be installed in base building systems to allow for comparison, management, and optimization of actual vs. estimated energy and water performance.

☐ Provide for the ongoing accountability and optimization of building energy and water consumption performance over time.

☐ Comply with the installed equipment requirements for continuous metering as stated in *Option B: Methods by Technology* of the US DOE's IPMVP for the following:
 - Lighting systems and controls
 - Constant and variable motor loads
 - VFD operation
 - Chiller efficiency at variable loads (kW/ton)
 - Cooling load
 - Air and water economizer and heat recovery cycles
 - Air distribution static pressures and ventilation air volumes
 - Boiler efficiencies
 - Building specific process energy efficiency systems and equipment
 - Indoor water risers and outdoor irrigation systems.

☐ Allocate an appropriate percentage of building funds for ongoing monitoring of environmental performance, product purchasing, maintenance, and improvements.

☐ Provide for the ongoing accountability of waste streams, including hazardous pollutants.

☐ Use environmentally safe cleaning materials.

☐ Educate operation and maintenance workers.

☐ Facilitate the reduction of waste generated by building occupants.
 - Provide an easily accessible dedicated area, serving the project, for the collection and storage of materials for recycling, including paper, glass, plastics, metals, and hazardous substances.

☐ After six months, evaluate existing ecosystems to determine if they have remained undisturbed.

☐ Assess building energy use to ensure it is at predicted levels.

- ❑ Determine if IAQ levels are at predicted levels (particularly $CO_2$ and airborne particulates).
- ❑ Measure water consumption and evaluate against target usage in original plan.
- ❑ Monitor water levels and determine if recycling and reuse of materials meet expectations.
- ❑ Monitor and evaluate additional sustainability goals for project.
  - Building specific process, energy-efficient systems and equipment.
  - Indoor water risers and outdoor irrigation systems.

# Afterword: Background to the Development of This Guide

## Preface to the First Edition

by David L. Grumman, PE

### BACKGROUND ON THE *ASHRAE GREENGUIDE*'S DEVELOPMENT

Comments heard by 1999/2000 ASHRAE President Jim Wolf, as he traveled around to chapters talking to grassroots members during his term were that ASHRAE was "missing the boat" in the "green" building design area. Many sister organizations had already published documents addressing this issue for their members. President-elect Bill Coad was assigned to follow up, and he requested TC 1.10 (Energy Resources) to form a subcommittee to develop a handbook or guide on green or sustainable building design, specifically for ASHRAE members.

The effort was kicked off with a conference call of interested TC 1.10 members on November 20, 2000, organized by then-TC Chair Sheila Hayter. During that call, numerous ideas were expressed about what the publication should and should not be and what it should contain. A green guide subcommittee was set up under the chairmanship of TC 1.10 member David L. Grumman, with Sheila Hayter and Jordan Heiman being the other members. The subcommittee then set up several meetings at the January 2001 Atlanta Winter Meeting to discuss the direction such a green guide should take. As many interested ASHRAE members coming to Atlanta as could be identified in a short time frame were invited to attend the first meeting, held just prior to the regular TC meeting. As a result of that meeting and subsequent actions, a direction was established, a content outline prepared, and interested and qualified authors sought.

As the authors' contributions came in, they were placed into the outline format and the result edited by the subcommittee members. Difficulties were experienced in achieving consistency in the pieces submitted and in obtaining authors for areas initially not covered, which slowed the process. In addition, work by the authors and editors was all voluntary.

### WHAT THE *ASHRAE GREENGUIDE* IS AND IS NOT

When the subcommittee started its work, it set forth some characteristics of what the guide was to be. One was that it have a well-defined purpose. That purpose, unlike many treatises written on this subject, was to provide guidance on how to apply green design techniques, not necessarily to motivate the use of them. Much has been written on the need for green building design, and this aspect is covered briefly herein. The reader should assume that when the HVAC&R designer finds himself or herself in a situation where a green design is to be done, this guide will help answer the question: "What do I do now?"

Other characteristics sought were that it be relevant to the target audience, useful and practical, concise and succinct, well organized and logical. Further, we wanted it to encourage team effort and to stimulate innovative ideas and independent thought. Finally, we wanted the reader to be able to find information easily (a good reference) and find the guide graphically pleasing to the eye.

This guide should be regarded as the collected views and opinions of experts in the respective technical fields they are addressing. It is not a consensus document. It has not been widely reviewed and commented on, except by a limited group (primarily the authors and a review panel) and all members of the developing subcommittee and its technical committee and/or meeting attendees. One does not have to agree with everything written here for this guide to be helpful.

## WHO SHOULD USE THE *ASHRAE GREENGUIDE*

The *ASHRAE GreenGuide* is primarily for HVAC&R designers. Architects, owners, building managers, operators, contractors, and others in the building industry, while perhaps finding certain areas of interest, are likely to find the guide less useful. Writing it for a specific audience keeps the document concise and focused. Nevertheless, from the design standpoint, the reader will see that considerable emphasis is placed on close coordination between parties and on teamwork.

The guide is not intended to be "the last word" or "the complete reference" on green design for engineers—nor, for that matter, a design guide proper. Liberal use of references to other sources is included, especially for details. For instance, throughout the guide, numerous techniques, processes, measures, or special systems are described succinctly in a modified outline or bullet form—but always in the same format. These are called **ASHRAE GreenTips**. The purpose of these is to give the designer enough familiarity with the subject to enable him or her to determine whether it might be suitable for the project being worked on. Each GreenTip concludes with a listing of other sources that may be referenced for greater detail. That source could be another book, a magazine article, a research paper, an organization, a Web site, or some other resource to which the designer could turn to seek more information on a subject that has promise of being applicable to the particular project at hand. Thus, there is a substantial bibliography included in this guide.

Even though this guide is intended for HVAC&R designers, related engineering disciplines are not omitted. The areas of plumbing and water management are definitely included (many HVAC&R types do plumbing design too). Also, because lighting systems have such a big impact on energy use, this subject is covered to the extent of lighting's impact on energy use and the building's HVAC&R system. (The aesthetic aspects of lighting design would not be appropriate here, however, and are thus excluded.) Likewise, there are many architectural features that have a large impact on energy use, and the engineer—even though not usually the designer of those features—should be generally familiar with these impacts. A brief chapter on architectural impacts is thus included.

This was intended to be a document that a design engineer, about to embark on a green building design project as part of a team, could pick up and *immediately get ideas and guidance* of what to do, where to turn, what to advise, and how to interact with other team members in a productive way. We want that engineer to be able to find information with ease on a subject of green-design relevance that may arise. (There is included a comprehensive index to seek out info on a given subject rapidly.)

Finally, although this guide has many specific and practical suggestions and tips on achieving green design, a designer who just tries to incorporate as many of these as possible and does little else is *not* assured of success. Considerable stress is placed on *coordinating efforts* with other members of the project team and seeking mutually acceptable solutions to the green design challenge. It will take innovative and synergistic designs that cut across design disciplines to achieve true success.

## HOW TO USE THIS *ASHRAE GREENGUIDE*

This document is intended to be used more as a *reference* than as something one would read in sequence from beginning to end. Depending on the needs and experience level of the reader, there are portions that may warrant more attention than others; indeed, there may well be portions that certain readers might skip altogether. The following is intended to provide some helpful guidelines for using this document effectively.

First, the table of contents is the best place for any reader to get an overall view of what is covered in this document. Next, all readers should take the time to read this preface and chapter 1, "Green Design, Sustainability and 'Good' Design," because they cover the background and philosophy behind this effort and provide some essential definitions and meanings of key terms.

Chapters 2 and 3, "Motivation" and "Background and Fundamentals," respectively, might well be skipped by the more experienced designer/readers. The former covers the "why" of green design—those factors that drive owners to undertake such a project. The latter covers the background of the green design movement and what other organizations have done. The second part of chapter 3 then reviews some engineering fundamentals that govern the technical engineering aspects of green design. The reader may find this part intimidating—perhaps offering more than he or she wants to know. Thus, only the engineering designer who feels a need for a review of such technical fundamentals should undertake to read through it. (Of course, it is always there for later reference, as the need may dictate.)

The reader who is interested in how the green design process works, however, should read chapter 4, "The Design Process—Early Stages." It covers such topics as creating a green design team, identifying the "players," defining each of their roles, and outlining some green design processes that others have found to work. It is not until after this chapter that the reader gets into the components of what makes up green building design proper.

The reader who wants to go directly to the practical things that can be done to implement a green design should start at chapter 5, "Architectural Design Impacts," which covers how architectural features impact green design. The nitty-gritty engineering aspects, however, start at chapter 6 and run through chapter 13. *This is where the reader will find virtually all the practical suggestions for possible incorporation in a green design*, The ASHRAE GreenTips. (A listing of all GreenTips and their page numbers, for quick and easy reference, can be found in the table of contents.)

Chapters 14 and 15 cover more aspects of the design process—including some of the analysis tools available. Again, these sections are probably better suited for occasional reference rather than "must" reading first time through.

The remainder of the *GreenGuide*, Chapters 16–19, covers what happens after the project's design is done—that is, during construction and after. There are some sound advice and helpful tips in that section, and the designer would be well served to be aware of the information contained therein early on in design. Thus, even though it covers a post-design time frame, *reading that section should not be put off until construction begins*.

For quick and ready reference, the *GreenGuide* includes up front not only a listing (with page numbers) of GreenTips but also all figures, charts and sidebars—and then at the back end a comprehensive bibliography, which compiles all the sources mentioned throughout the guide (and then some)—and an index for rapid location of a particular subject of interest.

# BIBLIOGRAPHY

*These listings do not necessarily imply endorsement or agreement by ASHRAE or the authors with the information contained in the documents.*

## BOOKS AND SOFTWARE PROGRAMS

ACGIH. 1999. *Bioaerosols: Assessment and Control.* J. Nacher, ed. American Conference of Governmental Industrial Hygienists. www.acgih.org.

AIA. 1996. *Environmental Resource Guide.* J. Demkin, ed. New York: John Wiley & Sons.

AIA. 2006. *Guidelines for Design and Construction of Health Care Facilities.* Washington, DC: American Institute of Architects.

Alvarez, S., E. Dascalaki, G. Guarracino, E. Maldonado, S. Sciuto, and L. Vandaele. 1998. *Natural Ventilation of Buildings—A Design Handbook.* F. Allard, ed. London: James & James Science Publishers Ltd.

AGCC. 1994. *Applications Engineering Manual for Direct-Fired Absorption.* American Gas Cooling Center.

Argiriou, A., A. Dimoudi, C.A. Balaras, D. Mantas, E. Dascalaki, and I. Tselepidaki. 1996. *Passive Cooling of Buildings.* M. Santamouris and A. Asimakopoulos, eds. London: James & James Science Publishers Ltd.

ARI. 1997. *ARI STANDARD 275-97, Standard for Application of Sound Rating Levels of Outdoor Unitary Equipment.* Arlington, VA: Air-Conditioning and Refrigeration Institute.

Arthus-Bertrand, Y. 2001. *Earth from Above: 365 Days.* New York: Harry N. Abrams.

Arthus-Bertrand, Y. 2002. *Earth From Above.* New York: Harry N. Abrams.

ASHE. *Green Guide for Health Care.* American Society of Healthcare Engineering. Available at www.gghc.org.

ASHRAE. 1988. *Active Solar Heating Systems Design Manual.* Atlanta: American Society of Heating, Refrigerating and Air-Conditioning Engineers, Inc.

ASHRAE. 1995. *Commercial/Institutional Ground-Source Heat Pump Engineering Manual.* Atlanta: American Society of Heating, Refrigerating and Air-Conditioning Engineers, Inc.

ASHRAE. 1996. *ASHRAE Guideline 1-1996, The HVAC Commissioning Process.* Atlanta: American Society of Heating, Refrigerating and Air-Conditioning Engineers, Inc.

ASHRAE. 2000. *2000 ASHRAE Handbook—HVAC Systems and Equipment*, p. 4.1. Atlanta: American Society of Heating, Refrigerating and Air-Conditioning Engineers, Inc.

ASHRAE. 2000. *ASHRAE Guideline 13, Specifying Direct Digital Control Systems.* Atlanta: American Society of Heating, Refrigerating and Air-Conditioning Engineers, Inc.

ASHRAE. 2000. *ANSI/ASHRAE Standard 52.2-1999, Method of Testing General Ventilation Air-Cleaning Devices for Removal Efficiency by Particle Size.* Atlanta: American Society of Heating, Refrigerating and Air-Conditioning Engineers, Inc.

ASHRAE. 2000. Interpretation IC-62-1999-30 (August) of *ANSI/ASHRAE Standard 62-1999, Ventilation for Acceptable Indoor Air Quality.* Atlanta: American Society of Heating, Refrigerating and Air-Conditioning Engineers, Inc.

ASHRAE. 2002. *2002 ASHRAE Handbook—Refrigeration.* Atlanta: American Society of Heating, Refrigerating and Air-Conditioning Engineers, Inc.

ASHRAE. 2002. *ASHRAE Guideline 14, Measurement of Energy and Demand Savings.* Atlanta: American Society of Heating, Refrigerating and Air-Conditioning Engineers, Inc.

ASHRAE. 2003. *2003 ASHRAE Handbook—HVAC Applications.* Atlanta: American Society of Heating, Refrigerating and Air-Conditioning Engineers, Inc.

ASHRAE. 2003. *2003 ASHRAE Handbook—HVAC Systems and Equipment.* Atlanta: American Society of Heating, Refrigerating and Air-Conditioning Engineers, Inc.

ASHRAE. 2003. *HVAC Design Manual for Hospitals and Clinics.* Atlanta: American Society of Heating, Refrigerating and Air-Conditioning Engineers, Inc.

ASHRAE. 2004. *2004 ASHRAE Handbook—Systems and Equipment.* Atlanta: American Society of Heating, Refrigerating and Air-Conditioning Engineers, Inc.

ASHRAE. 2004. *Advanced Energy Design Guide for Small Office Buildings.* Atlanta: American Society of Heating, Refrigerating and Air-Conditioning Engineers, Inc.

ASHRAE. 2004. *ANSI/ASHRAE Standard 55-2004, Thermal Environmental Conditions for Human Occupancy.* Atlanta: American Society of Heating, Refrigerating and Air-Conditioning Engineers, Inc.

ASHRAE. 2004. *ANSI/ASHRAE Standard 62.1-2004, Ventilation for Acceptable Indoor Air Quality* (plus interpretations). Atlanta: American Society of Heating, Refrigerating and Air-Conditioning Engineers, Inc.

ASHRAE. 2004. *ANSI/ASHRAE/IESNA Standard 90.1-2004, Energy Standard for Buildings Except Low-Rise Residential Buildings*. Atlanta: American Society of Heating, Refrigerating and Air-Conditioning Engineers, Inc.

ASHRAE. 2005. *2005 ASHRAE Handbook—Fundamentals*. Atlanta: American Society of Heating, Refrigerating and Air-Conditioning Engineers, Inc.

ASHRAE. 2005. *ASHRAE Guideline 0, The Commissioning Process*. Atlanta: American Society of Heating, Refrigerating and Air-Conditioning Engineers, Inc.

ASHRAE. 2006. *Advanced Energy Design Guide for Small Retail Buildings*. Atlanta: American Society of Heating, Refrigerating and Air-Conditioning Engineers, Inc.

ASHRAE. 1998. *Fundamentals of Water System Design*. Atlanta: American Society of Heating, Refrigerating and Air-Conditioning Engineers, Inc.

ASPE. 1998. *Domestic Water Heating Design Manual*. Chicago: American Society of Plumbing Engineers.

ASTM. 1998. *ASTM D 6245-1998: Standard Guide for Using Indoor Carbon Dioxide Concentrations to Evaluate Indoor Air Quality and Ventilation*. American Society for Testing and Materials.

Baker, N. 1995. *Light and Shade: Optimising Daylight Design*. European Directory of sustainable and energy efficient building. London: James & James Science Publishers Ltd.

Balcomb, J.D. 1984. *Passive Solar Heating Analysis: A Design Manual*. Atlanta: American Society of Heating, Refrigerating and Air-Conditioning Engineers, Inc.

Bauman, F.S., and A. Daly. 2003. *Underfloor Air Distribution Design Guide*. Atlanta: American Society of Heating, Refrigerating and Air-Conditioning Engineers, Inc.

Beckman, W., S.A. Klein, and J.A. Duffie. 1977. *Solar Heating Design*. New York: John Wiley and Sons.

Bisbee, D. 2003. *Pulse-Power Water Treatment Systems for Cooling Towers*. Energy Efficiency & Customer Research & Development, Sacramento Municipal Utility District, November 10, 2003.

Brand, S. 1995. *How Buildings Learn: What Happens After They're Built*. New York: Viking Penguin USA.

BRESCU, BRE. 1999. *Natural Ventilation for Offices Guide and CD-ROM*. ÓBRE on behalf of the NatVent Consortium, Garston, Watford, UK, March.

California Energy Commission. 1996. *Source Energy and Environmental Impacts of Thermal Energy Storage*. Tabors, Caramanis & Assoc.

Caneta. GS-2000TM (a computer program for designing and sizing ground heat exchangers for these systems). Caneta Research Inc., Mississauga, ON, CAN L5N 6J7.

CEC. 1996. *Source Energy and Environmental Impacts of Thermal Energy Storage.* Tabors, Caramanis & Assoc. California Energy Commission.

CEC. 2001. *Nonresidential Alternative Calculations Methods Manual.* Sacramento: California Energy Commission.

CEC. 2002. *Part II: Measure Analysis and Life-Cycle Cost 2005, California Building Energy Standards*, P400-02-012. Sacramento: California Energy Commission.

CEC. 2004. Monitoring the energy-use effects of cool roofs on California commercial buildings, CEC PIER study report. California Energy Commission. www.energy.ca.gov/pier/final_project_reports/500-04-046.html.

Crosbie, M.J., ed. 1998. *The Passive Solar Design and Construction Handbook*, Steven Winter Associates. New York: John Wiley & Sons.

Del Porto, D., and C. Steinfeld. 1999. *The Composting Toilet System Book.* The Center for Ecological Pollution Prevention.

DOE. 2002. *Energy Tip Sheet #1*, May. US Department of Energy, Office of Industrial Technologies, Energy Efficiency, and Renewable Energy.

DOE. *The International Performance Measurement and Performance Protocol* (IPMPP). US Department of Energy, Office of Energy Efficiency and Renewable Energy.

Dorgan, C., and J.S. Elleson. 1993. *Design Guide for Cool Thermal Storage.* Atlanta: American Society of Heating, Refrigerating and Air-Conditioning Engineers, Inc.

Elleson, J.S. 1996. *Successful Cool Storage Projects: From Planning to Operation.* Atlanta: American Society of Heating, Refrigerating and Air-Conditioning Engineers, Inc.

EPA. *Guide to Purchasing Green Power.* www.epa.gov/grnpower/buygreenpower/guide.htm.

EPA. *How to Conserve Water and Use It Effectively.* www.epa.gov/OW/you/chap3.html. US Environmental Protection Agency.

EPA. *Toxicity Characteristic Leaching Procedure (TCLP) Test*, 40 CFR Part 260. US Environmental Protection Agency.

EPA. *Universal Waste Rule*, 40 CFR Part 260. US Environmental Protection Agency.

Evans, B. 1997. Daylighting design. *In Time-Saver Standards for Architectural Design Data.* New York: McGraw-Hill.

*The Federal Energy Management Program Continuous Commissioning (SM) Guidebook.* www.eere.energy.gov/femp/operations_maintenance/commissioning_guidebook.cfm.

Gerston, J. *Rainwater Harvesting: A New Water Source.* twri.tamu.edu.

Givoni, B. 1994. *Passive and Low Energy Cooling of Buildings.* New York: Van Nostrand Reinhold.

Gottfried, D. (Continuous updating). *Sustainable Building Technical Manual, Green Building Design, Construction, and Operations*. US Green Building Council. www.usgbc.org.

Hawkin, P., A. Lovins, and L.H. Lovins. 1999. *Natural Capitalism*. Little Brown.

Henning. H-M., ed. 2004. *Solar Assisted Air-Conditioning in Buildings—A Handbook For Planners*. Wien: Springer-Verlag.

Houghton, J.T., G.J. Jenkins, and J.J. Ephraums, eds. 1990. *Climate Change: The IPCC Scientific Assessment*. Intergovernmental Panel on Climate Change. New York: Cambridge UP.

Howell, J., R. Bannerot, and G. Vliet. 1982. *Solar-Thermal Energy Systems: Analysis and Design*. New York: McGraw-Hill.

IECC. 2006. *International Energy Conservation Code*. Available from www.iccsafe.org.

IESNA. 1998. *RP-33-99, Lighting for Exterior Environments; RP-20-98, Lighting for Parking Facilities*. New York: Illuminating Engineering Society of North America. www.iesna.org.

IESNA. 2002. Proposed revisions to ASHRAE/IESNA Standard 90.1-2001. Illuminating Engineering Society of North America, Energy Management Committee.

IESNA. *Lighting Handbook*. Illuminating Engineering Society of North America. www.iesna.org.

Kavanaugh, S.P., and K. Rafferty. 1997. *Ground-Source Heat Pumps: Design of Geothermal Systems for Commercial and Institutional Buildings*. Atlanta: American Society of Heating, Refrigerating and Air-Conditioning Engineers, Inc.

Kinsley, M. 1997. *Economic Renewal Guide*. Rocky Mountain Institute.

Klein, S., and W. Beckman. 2001. *F-Chart Software*. Madison, WI: The F-Chart Software Company. Available from www.fchart.com/index.shtml.

Kowalski, W.J. 2006. *Aerobiological Engineering Handbook: Airborne Disease and Control Technologies*. New York: McGraw-Hill.

Kreith, F., and J. Kreider. 1989. *Principles of Solar Engineering*, 2d ed. Hemisphere Pub.

*LEED Reference Guide*. www.usgbc.org.

Ludwig, A. 1997. *Builder's Greywater Guide and Create an Oasis with Greywater*. Oasis Design.

Massachusetts State Building Code. www.mass.gov/bbrs/code.htm.

McIntosh, I.B.D., C.B. Dorgan, and C.E. Dorgan. 2002. *ASHRAE Laboratory Design Guide*. Atlanta: American Society of Heating, Refrigerating and Air-Conditioning Engineers, Inc.

Mendler, S.F., and W. Odell. 2000. *HOK Guidebook to Sustainable Design*. New York: John Wiley & Sons, Inc.

NADCA. *General Specifications for the Cleaning of Commercial Heating, Ventilating and Air Conditioning Systems*. National Air Duct Cleaners Association. www.nadca.com.

Nattrass, B., and M. Altomare. *The Natural Step for Business; Wealth, Ecology and the Evolutionary Corporation.* Gabriola Island, British Columbia: New Society Publishers.

NBI. 1998. *Gas Engine Driven Chillers Guideline.* Fair Oaks, CA: New Buildings Institute and Southern California Gas Company. www.newbuildings.org/ downloads/guidelines/GasEngine.pdf.

NBI. 2001. *Advanced Lighting Guidelines.* New Buildings Institute. www.newbuildings.org/lighting.htm.

NFPA. 2004. *NFPA 45, Standard on Fire Protection for Laboratories using Chemicals.* Quincy, MA: National Fire Protection Association.

NFPA. 2005. *NFPA 99, Standard for Health Care Facilities.* Quincy, MA: National Fire Protection Agency.

NRC. *Photovoltaic Systems Design Manual.* Natural Resources Canada, Office of Coordination and Technical Information, Ottawa, Ontario, Canada.

NRC. RETScreen (Renewable Energy Analysis Software). Natural Resources Canada, Energy Diversification Research Laboratory, Varennes PQ CAN J3X 1S6. www.retscreen.net.

NSF. 2000. *NSF/ANSI 5-2000e: Water Heaters, Hot Water Supply Boilers, and Heat Recovery Equipment.* Ann Arbor, MI: National Sanitation Foundation.

Outdoor Lighting Research, California Outdoor Lighting Standards, June 6, 2002. www.cec.org.

PG&E. 2002. *Codes and Standards Enhancement Report, Code Change Proposal for Cooling Towers.* Pacific Gas and Electric.

Schaffer, M. 2004. *Practical Guide to Noise and Vibration Control for HVAC Systems, Second Edition.* Atlanta: American Society of Heating, Refrigerating and Air-Conditioning Engineers, Inc.

SMACNA. *Indoor Air Quality Guideline for Occupied Buildings Under Construction.* Chantilly, VA: Sheet Metal and Air Conditioning National Association, Inc.

Stack, A., J. Goulding, and J.O. Lewis. *Solar Shading.* European Commission, DG TREN, Brussels. http://erg.ucd.ie/erg_downloads.html.

*Sustainable Building Technical Manual—Green Building Design, Construction and Operations.* 1996. Public Technology Inc., US Department of Energy and the US Green Building Council. www.epa.gov/OW/you/chap3.htm and www.usgbc.org.

Taylor, S., P. Dupont, M. Hydeman, B. Jones, and T. Hartman. 1999. *The CoolTools Chilled Water Plant Design and Performance Specification Guide.* San Francisco, CA: PG&E Pacific Energy Center.

Trane. 1999. *Trane Applications Engineering Manual, Absorption Chiller System Design*, SYS-AM-13. Lacrosse, WI: Trane Company.

*Trane Installation, Operation, and Maintenance Manual*, RTAA-IOM-3. Lacrosse, WI: Trane Company.

Trane. 1994. *Water-Source Heat Pump System Design*. Trane Applications Engineering Manual, SYS-AM-7. Lacrosse, WI: Trane Company.

Trane. 2000. *Multiple-Chiller-System Design and Control Manual*. Lacrosse, WI: Trane Company.

UG. *Combo Heating Systems: A Design Guide*. Chatham, ON: Union Gas.

USGBC. LEED-NC v 2.2 Rating System. US Green Building Council.

USGBC. 2005. *LEED Reference Guide, Version 2.2*. US Green Building Council. www.usgbc.org.

Waterfall, P.H. 1998. *Harvesting Rainwater for Landscape Use*. http://ag.arizona.edu/pubs/water/az1052/.

Watson, D., and G. Buchanan. 1993. *Designing Healthy Buildings*. Washington DC: American Institute of Architects.

Watston, R.D., and K.S. Chapman. 2002. *Radiant Heating and Cooling Handbook*. McGraw-Hill.

WATSUN-PV (simulation software). University of Waterloo, Waterloo, Ontario, Canada.

Waye, K.P., J. Bengtsson, R. Rylander, F. Hucklebridge, P. Evans, A. Clow. 2002. Low frequency noise enhances cortisol among noise sensitive subjects during work performance. *Life Science* 70(7):745–58.

Weiss, W., I. Bergmann, and G. Faninger. 2006. *Solar Heat Worldwide*. Solar Heat & Cooling Programme, International Energy Agency, www.iea-shc.org/welcome/IEASHCSolarHeatingWorldwide2006.pdf.

Wolverton, B.C. 1996. *How to Grow Fresh Air: 50 Plants That Purify Your Home of Office*. Baltimore, MD: Penguin Books.

## PERIODICALS AND REPORTS

Adamson, R. 2002. Mariah Heat Plus Power™ Packaged CHP, Applications and Economics. Presented at the 2nd Annual DOE/CETC/CANDRA Workshop on Microturbine Applications at the University of Maryland, College Park.

Andersson, L.O., K.G. Bernander, E. Isfält, and A.H. Rosenfeld. 1979. Storage of heat and coolth in hollow-core concrete slabs. Swedish experience and application to large, American style buildings. Second International Conference on Energy Use and Management, Lawrence Berkeley National Laboratory, LBL-8913.

ASHE. 2002. Green Healthcare Design Guidance Statement. American Society of Healthcare Engineering.

Bahnfleth, W.P., and W.S. Joyce. 1994. Energy use in a district cooling system with stratified chilled water storage. *ASHRAE Transactions* 100(1):1767–78.

Balaras, C.A. 1995. The role of thermal mass on the cooling load of buildings. An overview of computational methods. *Energy and Buildings* 24(1):1–10.

Balaras, C.A., H.-M. Henning, E. Wiemken, G. Grossman, E. Podesser, and C.A. Infante Ferreira. 2006. Solar cooling—An overview of european applications and design guidelines. *ASHRAE Journal* 48(6):14–22.

Bauman, F., and T. Webster. 2001. Outlook for underfloor air distribution. *ASHRAE Journal* 43(6).

Beebe, J.M. 1959. Stability of disseminated aerosols of *Pastuerella tularensis* subjected to simulated solar radiations at various humidities. *Journal of Bacteriology* 78:18–24.

Braun, J.E. 1990. Reducing energy costs and peak electrical demand through optimal control of building thermal storage. *ASHRAE Transactions* 96(2):876–88.

CEC. 2005. California Energy Commission, Title 24 2005 Reports and Proceedings, www.energy.ca.gov.

Chapman, K.S. 2002. Development of radiation transfer equation that encompasses shading and obstacles. *ASHRAE Transactions* 108(2):997–1004.

Chapman, K.S. ASHRAE RP-907, Final report, Development of design factors for the combination of radiant and convective in-space heating and cooling systems. Atlanta: American Society of Heating, Refrigerating and Air-Conditioning Engineers, Inc.

Chapman, K.S., J. Rutler, and R.D. Watson. 2000. Impact of heating systems and wall surface temperatures on room operative temperature fields. *ASHRAE Transactions* 106(1):506–14.

Chapman, K.S., J.E. Howell, and R.D. Watson. 2001. Radiant panel surface temperature over a range of ambient temperatures. *ASHRAE Transactions* 107(1):383–89.

Chapman, K.S., J.M. DeGreef, and R.D. Watson. 1997. Thermal comfort analysis using BCAP for retrofitting a radiantly heated residence. *ASHRAE Transactions* 103(1):959–65.

Cho, Y.I., S. Lee, and W. Kim. 2003. Physical water treatment for the mitigation of mineral fouling in cooling-tower water applications. *ASHRAE Transactions* 109(1):346–57.

Coad, W.J. 1999. Conditioning ventilation air for improved performance and air quality. *Heating/Piping/Air Conditioning*, September.

Darlington, A., M. Chan, D. Malloch, C. Pilger, and M.A. Dixon. 2000. The biofiltration of indoor air: Implications for air quality. *Indoor Air 2000* 10(1):39–46.

Darlington, A., M.A. Dixon, and C. Pilger. 1998. The use of biofilters to improve indoor air quality: The removal of toluene, TCE, and formaldehyde. *Life Support Biosph Sci* 5(1):63–69.

Darlington, A.B., J.F. Dat, and M.A. Dixon. 2001. The biofiltration of indoor air: Air flux and temperature influences the removal of toluene, ethylbenzene, and xylene. *Environ Sci Technol* 35(1):240–46.

DeGreef, J.M., and K.S. Chapman. 1998. Simplified thermal comfort evaluation of MRT gradients and power consumption predicted with the BCAP methodology. *ASHRAE Transactions* 104(1):1090–97.

Del Porto, D., and C. Steinfeld. 1999. *The Composting Toilet System Book*. Center for Ecological Pollution Prevention.

DOE. *Transpired Air Collectors*. Energy Efficiency and Renewable Energy, US Department of Energy. www.eere.energy.gov/de/transpired_air.html.

DOE. 2002. Energy efficiency and renewable energy. *Energy Tip Sheet #1*, May. US Department of Energy, Office of Industrial Technologies.

Duffy, G. 1992. Thermal storage shifts to saving energy. *Engineering Systems*.

Ehrt, D., M. Carl, T. Kittel, M. Muller, and W. Seeber. 1994. High performance glass for the deep ultraviolet range. *Journal of Non-Crystalline Solids* 177:405–19.

El-Adhami, W., S. Daly, and P.R. Stewart. 1994. Biochemical studies on the lethal effects of solar and artificial ultraviolet radiation on *Staphylococcus aureus*. *Arch Microbiol* 161:82–87.

Ellis, M.W., and M.B. Gunes. 2002. Status of fuel cell systems for combined heat and power applications in buildings. *ASHRAE Transactions* 108(1):1032–44.

EPA. Storm water management for construction activities. EPA document No. EPA-8320R-92-005. US Environmental Protection Agency.

Erickson, D.C., and M. Rane. 1994. Advanced absorption cycle: Vapor exchange GAX. *Proceedings of the ASME International Absorption Heat Pump Conference, New Orleans, LA, Jan. 19–21*.

Fagan, D. 2001. A comparison of storage-type and instantaneous heaters for commercial use. *Heating/Piping/Air Conditioning Engineering*, April.

Fernandez, R.O. 1996. Lethal effect induced in *Pseudomonas aeruginosa* exposed to ultraviolet-A radiation. *Photochem & Photobiol* 64(2):334–39.

Fiorino, D.P. 1994. Energy conservation with stratified chilled water storage. *ASHRAE Transactions* 100(1):1754–66.

Fiorino, D.P. 2000. Six conservation and efficiency measures reducing steam costs. *ASHRAE Journal* 42(2):31–39.

Galuska, E.J. 1994. Thermal storage system reduces costs of manufacturing facility (Technology Award case study). *ASHRAE Journal*, March.

Gansler, R.A., D.T. Teindil, and T.B. Jekel. 2001. Simulation of source energy utilization and emissions for HVAC systems. *ASHRAE Transactions* 107(1):39–51.

Gerston, J. Rainwater harvesting: A new water source. http://twri.tamu.edu.

Goss, J.O., L. Hyman, and J. Corbett. 1996. Integrated heating, cooling and thermal energy storage with heat pump provides economic and environmental solutions at California State University, Fullerton. *EPRI International Conference on Sustainable Thermal Energy Storage*, pp. 163–67.

Grondzik, W.T. 2001. The (mechanical) engineer's role in sustainable design: Indoor environmental quality issues in sustainable design. HTML presentation available at www.polaris.net/~gzik/ieq/ieq.htm.

Gupta, A. 2002. Director of Energy, NRDC. *New York Times*, March 17.

Hayter, S., P. Torcellini, and R. Judkoff. 1999. Optimizing building and HVAC systems. *ASHRAE Journal* 41(12):46–49.

HPAC. 2004. Innovative grocery store seeks LEED certification. *HPAC Engineering* 27:31.

I.C. Young, S.H. Lee, and W. Kim. 2003. Physical water treatment for the mitigation of mineral fouling in cooling tower water applications. *ASHRAE Transactions* 109(1):346–57.

IUVA. 2005. General guideline for UVGI air and surface disinfection systems. IUVA-G01A-2005 International Ultraviolet Association. Ayr, Ontario, Canada, www.iuva.org.

Jones, B., and K.S. Chapman. ASHRAE RP-657, Final report, Simplified method to factor mean radiant temperature (MRT) into building and HVAC system design. Atlanta: American Society of Heating, Refrigerating and Air-Conditioning Engineers, Inc.

Kainlauri, E.O., and M.P. Vilmain. 1993. Atrium design criteria resulting from comparative studies of atriums with different orientation and complex interfacing of environmental systems. *ASHRAE Transactions* 99(1):1061–69.

Keeney, K.R., and J.E. Braun. 1997. Application of building precooling to reduce peak cooling requirements. *ASHRAE Transactions* 103(1):463–69.

Kintner-Meyer, M., and A.F. Emery. 1995. Optimal control of an HVAC system using cold storage and building thermal capacitance. *Energy and Buildings* 23:19–31.

Kosik, W.J. 2001. Design strategies for hybrid ventilation. *ASHRAE Journal* 43(10):18–19, 22–24.

Lawrence, T.M. 2004. Demand-controlled ventilation and sustainability. *ASHRAE Journal* 46(12):117–21.

Lawson, 1988. Computer facility keeps cool with ice storage. *HPAC*, August.

Leaman, A., and B. Bordass. 1993. Building design, complexity and manageability. *Facilities*, Vol. 11.

LeMar, P. 2002. Integrated energy systems (IES) for buildings: A market assessment. Final report by Resource Dynamics Corporation for Oak Ridge National Laboratory, Contract No. DE-AC05-00OR22725, September 2002.

Mathaudhu, S.S. 1999. Energy conservation showcase. *ASHRAE Journal* 41(4):44–46.

McDonough, W. 1992. The Hannover principles: Design for sustainability. Presentation, Earth Summit, Brazil.

McKurdy, G., S.J. Harrison, and R. Cooke. Preliminary evaluation of cylindrical skylights. 23rd Annual Conference of the Solar Energy Society of Canada Inc., Vancouver, British Columbia. www.solarenergysociety.ca.

Morris, W. 2003. The ABCs of DOAS. *ASHRAE Journal* 45(5).

Mumma, S.A. 2001. Designing dedicated outdoor air systems. *ASHRAE Journal* 43(5).

O'Neal, E.J. 1996. Thermal storage system achieves operating and first-cost savings (Technology Award case study). *ASHRAE Journal* 38(4).

Palmer, J.M., and K.S. Chapman. 2000. Direct calculation of mean radiant temperature using radiant intensities. *ASHRAE Transactions* 106(1):477–86.

Patnaik, V. 2004. Experimental verification of an absorption chiller for BCHP applications. *ASHRAE Transactions* 110(1):503–507.

Pearson, F. 2003. ICR 06 Plenary Washington. *ASHRAE Journal* 45(10).

Rautiala, S., S. Haatainen, H. Kallunki, L. Kujanpaa, S. Laitinen, A. Miihkinen, M. Reiman, and M. Seuri. 1999. Do plants in office have any effect on indoor air microorganisms? *Indoor Air 99: Proceedings of the 8th International Conference on Indoor Air Quality and Climate, Edinburgh, Scotland*, pp. 704–709.

Reindl, D.T., D.E. Knebel, and R.A. Gansler. 1995. Characterising the marginal basis source energy and emissions associated with comfort cooling systems. *ASHRAE Transactions* 101(1):1353–63.

Reindl, D.T., R.A. Gansler, and T.B. Jekel. 2001. Simulation of source energy utilization and emissions for HVAC systems. *ASHRAE Transactions* 107(1):39–51.

Rosenbaum, M. 2002. A green building on campus. *ASHRAE Journal* 44(1):41–44.

Rosfjord, T., T. Wagner, and B. Knight. 2004. UTC microturbine CHP product development and launch. Presented at the 2004 DOE/CETC Annual Workshop on Microturbine Applications.

Ruud, M.D., J.W. Mitchell, and S.A. Klein. 1990. Use of building thermal mass to offset cooling loads. *ASHRAE Transactions* 96(2):820–29.

Ryan, W.A. CHP: The concept. Midwest CHP Application Center, Energy Resources Center, University of Illinois, Chicago. www.chpcentermw.org/ presentations/WI-Focus-on-Energy-Presentation-05212003.pdf.

Scheatzle, David. ASHRAE RP-1140 Final report, Establishing a base-line data set for the evaluation of hybrid HVAC systems. Atlanta: American Society of Heating, Refrigerating and Air-Conditioning Engineers, Inc.

Schell, M., and D. Int-Hout. 2001. Demand control ventilation using $CO_2$. *ASHRAE Journal* 43(2).

Svensson C., and S.A. Aggerholm. 1998. A design tool for natural ventilation. Presented at *IAQ '98 Conference, New Orleans, LA, October 24–27*.

Tabors, Caramanis, and Associates. 1996. *Source energy and environmental impacts of thermal energy storage.* California Energy Commission, February.

Taylor, S. 2002. Primary-only vs. primary-secondary variable flow chilled water systems. *ASHRAE Journal* 44(2):25–29.

Taylor, S., and J. Stein. 2002. Balancing variable flow hydronic systems. *ASHRAE Journal* 44(10):17–24.

Taylor, S., P. Dupont, M. Hydeman, B. Jones, and T. Hartman. 1999. *The CoolTools Chilled Water Plant Design and Performance Specification Guide.* San Francisco, CA: PG&E Pacific Energy Center.

Taylor, S.T. 2005. LEED and Standard 62.1. *ASHRAE Journal Sustainability Supplement*, September.

Torcellini, P.A., N. Long, and R. Judkoff. 2004. Consumptive water use for U.S. power production. *ASHRAE Transactions* 110(1):96–100.

Torcellini, P., N. Long, S. Pless, and R. Judkoff. 2005. Evaluation of the low-energy design and energy performance of the Zion National Park Visitor Center, NREL Report No. TP-550-34607. National Renewable Energy Laboratory. www.eere.energy.gov/buildings/highperformance/research_reports.html.

Trane. 2000. Energy conscious design ideas—Air-to-air energy recovery. *Engineers Newsletter* 29(5). Publication ENEWS-29/5. Lacrosse, WI: Trane Company.

Trane. *A Guide to Understanding ASHRAE Standard 62-2001*. Publication IAQ-TS-1, July. Lacrosse, WI: Trane Company. http://trane.com/commercial/issues/iaq/ashrae2001.asp.

Waterfall, P.H. 1998. Harvesting rainwater for landscape use. http://ag.arizona.edu/pubs/water/az1052/.

Watson, R.D., K.S. Chapman, and J.M. DeGreef. 1998. Case study: Seven-system analysis of thermal comfort and energy use for a fast-acting radiant heating system. *ASHRAE Transactions* 104(1).

Watson, R.D., K.S. Chapman, and L. Wiggington. 2001. Impact of dual utility selection on 305 m$^2$ (1000 ft$^2$) residences. *ASHRAE Transactions* 107(1):365–70.

Zagreus, L., C. Huizenga, E. Arens, and D. Lehrer. 2004. Listening to the occupants: A Web-based indoor environmental quality survey. *Indoor Air 2004* 14(suppl 8):65–74.

## WEB SITES

Advanced Buildings Benchmark™ (E-Benchmark, version 1.1)
www.poweryourdesign.com/benchmark.htm

Advanced Buildings Technologies and Practices
www.advancedbuildings.org

Advanced Lighting Guidelines
www.newbuildings.org/lighting.htm

AGORES—A Global Overview of Renewable Energy Sources (the official European Commission renewable energy information centre and knowledge gateway, with a global overview of RES)
www.agores.org

Air-Conditioning and Refrigeration Institute
www.ari.org

Alliance to Save Energy
www.ase.org

American Council for an Energy-Efficient Economy
www.aceee.org

American Gas Association
www.aga.org
American Institute of Architects
www.aia.org
The American Society of Plumbing Engineers
www.aspe.org
American Water Works Association, WaterWiser, The Water Efficiency
Clearinghouse
www.waterwiser.org
Architectural Energy Corporation
www.archenergy.com
Armstrong Intelligent Systems Solutions
www.armstrong-intl.com
ASES—American Solar Energy Society
www.ases.org
ASHE—American Society of Healthcare Engineering
www.ashe.org
ASHE's *Green Guide for Health Care*
www.gghc.org
ASHRAE
www.ashrae.org
ASHRAE's Sustainability Roadmap
http://images.ashrae.biz/renovation/documents/sust_roadmap.pdf
ASHRAE TC 8.3, recently sponsored programs and presentations
http://tc83.ashraetcs.org/programs.html
Athena Institute
www.athenasmi.ca
BEER—Building Energy Efficiency Research Project at the Department of Architecture, The University of Hong Kong
www.arch.hku.hk/research/beer/
BetterBricks
www.betterbricks.com
BREEAM® (Building Research Establishment Environmental
Assessment Method)
www.breeam.org
Building Energy Analysis Tool, Environmental Protection Agency (see eQUEST)
www.energydesignresources.com/resource/130/
BuildingGreen
www.greenbuildingadvisor.com
CADDET—Centre for Analysis and Dissemination of Demonstrated Energy Technologies, IEA
www.caddet.org

California Energy Commission
www.energy.ca.gov

California Energy Commission, Part II: Measure Analysis and Life-Cycle Cost 2005, *California Building Energy Efficiency Standards, Part IV,*
www.energy.ca.gov/2005_standards/documents/2002-08-27_workshop/2002-08-14_4th_GROUP_ELEY.pdf

California Energy Commission, Renewable Energy Program
www.energy.ca.gov/renewables/

California Energy Commission, Title 24 2005 Reports and Proceedings
www.energy.ca.gov

Canadian Earth Energy Association
www.earthenergy.org

Canadian Renewable Energy Network
www.canren.gc.ca/

The Center for Health Design
www.healthdesign.org

Center for the Built Environment (CBE) Occupant Indoor Environmental Quality (IEQ) Survey
www.cbesurvey.org

Center of Excellence for Sustainable Development (CESD)
www.sustainable.doe.gov

Center of Excellence for Sustainable Development (CESD),
Smart Communities Network
www.smartcommunities.ncat.org

"Centex—Most efficient building in U.S. in 1999"
www.energystar.gov/index.cfm?fuseaction=labeled_buildings.showProfile&profile_id=1306.

"CHP: The Concept," by W.A. Ryan
www.chpcentermw.org/presentations/WI-Focus-on-Energy-Presentation-05212003.pdf

City of Chicago Green Roof Policy (summary)
http://egov.cityofchicago.org/webportal/COCWebPortal/COC_EDITORIAL/Green_Roof_Policy_Matrix_revised.pdf

Closed Water-Loop Heat Pump, Tri-State Generation and Transmission Association Inc.
http://tristate.apogee.net/cool/cchc.asp

The Collaborative for High Performance Schools
www.chps.net

Commercial Building Incentives Program (CBIP) for New Buildings
oee.nrcan.gc.ca/commercial/financial-assistance/new-buildings/index.cfm?attr=20

Cool Roof Rating Council
www.coolroofs.org

*CoolTools Chilled Water Plant Design Guide*
www.hvacexchange.com/cooltools/

DDC Online (direct digital controls)
www.ddc-online.org

DDC (New York City Department of Design and Construction),
Sustainable Design
home.nyc.gov/html/ddc/html/ddcgreen/

DOE Building Technologies Program: Building Commissioning
www.eere.energy.gov/buildings/operate/buildingcommissioning.cfm

*EN4M Energy in Commercial Buildings* (software tool)
www.eere.energy.gov/buildings/tools_directory/software.cfm/ID=299/

Energy Design Resources
www.energydesignresources.com/publication/gd/

Energy Efficiency and Renewable Energy Network (EREN)
www.eere.energy.gov

Energy Efficiency and Renewable Energy Network (EREN), High Performance
Buildings Research Initiative
www.eere.energy.gov/buildings/highperformance

ENERGY STAR®
www.energystar.gov

Energy Trust of Oregon
www.energytrust.org

Engineering for Sustainability, ASHRAE
www.engineeringforsustainability.org/

EPIA—European Photovoltaic Industry Association
www.epia.org

eQUEST (Building Energy Analysis Tool, Environmental Protection Agency)
www.energydesignresources.com/resource/130/

ESTIF—European Solar Thermal Industry
www.estif.org

Eugene, Green Energy Standard in Europe
www.eugenestandard.org

EUREC—European Renewable Energy Centres Agency
www.eurec.be

European Commission (DG TREN), information on eco-building demonstration
projects throughout Europe
www.sara-project.net

The European Green Electricity Network (Eugene) (provides contact information
for green energy suppliers throughout Europe)
www.eugenestandard.org

EUROSOLAR—The European Association for Renewable Energies e.V.
www.eurosolar.org

EWEA—The European Wind Energy Association
www.ewea.org

"Evaluation of the Low-Energy Design and Energy Performance of the Zion National Park Visitor Center" (NREL Report No. TP-550-34607)
www.eere.energy.gov/buildings/highperformance/research_reports.html

EZ-Conserve
www.ezconserve.com

F-Chart Software
www.fchart.com/index.shtml

Geoexchange, Geothermal Heat Pump Consortium
www.geoexchange.org

Green Builder Magazine
www.greenbuilder.com/sourcebook/GasWaterHeat.html

The Green Building Advisor
www.greenbuildingadvisor.com

Green Building Challenge
www.greenbuilding.ca

Green Energy In Europe
www.greenprices.com

Green Globes
www.greenglobes.com

Green Roof Industry Information Clearinghouse and Database
www.greenroofs.com

Green Venture
www.greenventure.on.ca

Greening the Building and the Bottom Line, Rocky Mountain Institute
www.rmi.org

GreenSpec®
www.buildinggreen.com/menus/index.cfm

Guideline for Planning, Execution and Upkeep of Green-Roof Sites (German standard similar in nature to ASHRAE standard specifically focused on green roof design issues)
www.f-l-l.de/english.html

The Hannover Principles
www.mindfully.org/Sustainability/Hannover-Principles.htm

Heschong-Mahone Group
www.h-m-g.com

Heschong Mahone Group for Pacific Gas and Electric Company
www.h-m-g.com

Heshong-Mahone Group, Sky lighting
    www.h-m-g.com/skylighting/skycalcreg.htm
Hewlett Foundation, Link to LEED Gold Building
    www.usgbc.org/Docs/Certified_Projects/Cert_Reg67.pdf
High Performance Buildings Research Initiative, EERE
    www.eere.energy.gov/buildings/highperformance
ICC—International Code Council
    www.iccsafe.org
IEA Heat Pump Centre
    www.heatpumpcentre.org
IEA Photovoltaic Power Systems Programme
    www.iea-pvps.org
IEA Solar Heating And Cooling Programme
    www.iea-shc.org
Illuminating Engineering Society of North America (IESNA)
    www.iesna.org
International Performance Measurement and Verification Protocol
    www.ipmvp.org
Irrigation Association
    www.irrigation.org
ISES—International Solar Energy Society
    www.ises.org
Labs 21 Environmental Performance Criteria
    www.labs21century.gov
Lawrence Berkeley National Laboratories
    www.arch.ced.berkeley.edu/vitalsigns
Lawrence Berkeley National Laboratories, Environmental Energy
    Technologies Division
    http://eetd.lbl.gov/
Lawrence Berkeley National Laboratory, Radiance Home
    http://radsite.lbl.gov/radiance/HOME.html
Minnesota Sustainable Design Guide
    www.sustainabledesignguide.umn.edu
"Monitoring the Energy-Use Effects of Cool Roofs on California Commercial
    Buildings," CEC PIER study report
    www.energy.ca.gov/pier/final_project_reports/500-04-046.html
National Building Controls Information Program (NBCIP)
    www.buildingcontrols.org
National Oceanic and Atmospheric Administration (NOAA)
    www.noaa.gov
National Renewable Energy Laboratory, (NREL)
    www.nrel.gov

National Renewable Energy Laboratory, Center for Buildings and
    Thermal Systems
    www.nrel.gov/buildings_thermal/

National Renewable Energy Laboratory, The Center for Buildings and Thermal Systems, Solar Energy Research
    www.nrel.gov/buildings_thermal/solar.html

National Renewable Energy Laboratory, PV for Buildings
    www.nrel.gov/buildings/pv/index.html

NAVFAC, Design of Sustainable Facilities and Infrastructure
    www.navfac.navy.mil/safety/

New Buildings Institute
    www.newbuildings.org

New York City High Performance Building Design Guidelines (1999)
    www.nyc.gov/html/ddc/html/ddcgreen/documents/guidelines.pdf

New York City Department of Design and Construction (DDC),
    Sustainable Design
    home.nyc.gov/html/ddc/html/ddcgreen/

New York's Battery Park City Authority
    www.batteryparkcity.org/guidelines.htm

NOAA—National Oceanic and Atmospheric Administration
    www.noaa.gov

North Carolina Cooperative Extension Service: Water Quality and
    Waste Management
    www.bae.ncsu.edu/programs/extension/publicat/wqwm

Oikos: Green Building Source
    www.oikos.com

Photovoltaic Resource Site
    www.pvpower.com

Photovoltaics, EERE, US DOE
    www1.eere.energy.gov/solar/photovoltaics.html

"PNNL Scientist Studies Worker Productivity in Energy-Efficient Buildings"
    www.eere.energy.gov/femp/newsevents/femp_focus/
    sept01_pnnlscientist_studies.html

PV Power Resource Site
    www.pvpower.com

QuikWater, High Efficiency Direct Contact Water Heaters
    www.quikwater.com

*Rainwater Harvesting: A New Water Source* by J. Gerston
    http://twri.tamu.edu

RealWinWin, Inc.
    www.realwinwin.com

Regional Climate Data
www.wrcc.dri.edu/rcc.html

Renewable Energy Deployment Initiative (REDI) (a Canadian federal program that supports the deployment of renewable technologies; some technologies qualify for incentives)
www.nrcan.gc.ca/redi

RETScreen (software for renewable energy analysis)
www.retscreen.net

Rising Sun Enterprises
www.rselight.com

Rocky Mountain Institute
www.rmi.org

Safe Drinking Water Act (SDWA), US Environmental Protection Agency, Regulations and Guidance
www.epa.gov/safewater/regs.html

Savings by Design
www.savingsbydesign.com

School of Photovoltaic and Renewable Energy Engineering, University of New South Wales
www.pv.unsw.edu.au

SEIA—Solar Energy Industries Association
www.seia.org

Sheet Metal and Air Conditioning Contractors' National Association
www.smacna.org

SkyCalc
www.energydesignresources.com/resource/129/

Solar Energy Industries Association (SEIA)
www.seia.org

Solar Energy Society of Canada, Inc.
www.solarenergysociety.ca

Spirx Sarco Design of Fluid Systems—Steam Learning Module
www.spiraxsarco.com/learn/modules.asp

Sustainable Building Guidelines
www.ciwmb.ca.gov/GreenBuilding/Design/Guidelines.htm

Sustainable Buildings Industry Council
www.sbicouncil.org

*Sustainable Building Technical Manual—Green Building Design, Construction and Operations.* Public Technology, Inc., US Green Building Council, US Department of Energy
www.epa.gov/OW/you/chap3.htm and www.usgbc.org

Sustainable Communities Network (SCN)
www.sustainable.org

Sustainable Communities Network (SCN), Sustainable Building Resource Directory (focused on Mid-Atlantic region of the US)
www.sbrd.org

Sustainable Sources
www.greenbuilder.com

Sustainable Sources Bookstore
bookstore.greenbuilder.com/index.books

*Tips for Daylighting with Windows: The Integrated Approach*
http://windows.lbl.gov/pub/designguide/dlg.pdf

Trane Company, *Multiple-Chiller-System Design and Control Manual*
www.trane.com

US Department of Energy, Consumer Information Program
www.eere.energy.gov/

US Department of Energy, Federal Energy Management Program
www.eere.energy.gov/femp/

US Department of Energy, High Performance Buildings Institute
www.highperformancebuildings.gov

US Environmental Protection Agency, Green Chemistry
www.epa.gov/greenchemistry/

US Green Building Council LEED Green Building Rating System®
www.usgbc.org/DisplayPage.aspx?CategoryID=19

US Green Building Council
www.usgbc.org

Usable Buildings Trust
www.usablebuildings.co.uk

WaterWiser®: The Water Efficiency Clearinghouse, American Water Works Association
www.waterwiser.org

The Whole Building Design Guide
www.wbdg.org

The World's Water
www.worldwater.org

WRDC—World Radiation Data Centre, maintained for the World Meteorological Organization (WMO); provides solar radiation and radiation balance data (the world network)
http://wrdc-mgo.nrel.gov/

# Terms, Definitions, and Acronyms

| | | |
|---|---|---|
| Δ | = | change or change in |
| A/C | = | air-conditioning |
| AFF | = | above finished floor |
| AHU | = | air-handling unit |
| ANSI | = | American National Standards Institute |
| ASHRAE | = | American Society of Heating, Refrigerating and Air-Conditioning Engineers |
| ASPE | = | American Society for Plumbing Engineers |
| BAS | = | building automation system |
| BF | = | ballast factor |
| BIM | = | building information modeling |
| BIPV | = | building-integrated photovoltaics |
| BREEAM® | = | Building Research Establishment Environmental Assessment Method |
| brownfield | = | real estate property that is, or potentially is, contaminated |
| Btu | = | British thermal unit |
| C | = | centigrade (temperature scale) |
| C-2000 | = | Canadian Integrated Design Process program |
| CAN | = | Canada |
| CAV | = | constant air volume |
| CBIP | = | Commercial Buildings Incentive Program |
| CCHP | = | combined cooling, heating, and power |
| CDT | = | cold deck temperature |
| CETC | = | Canmet Energy Technology Centre, a division of Natural Resources Canada, Canadian Energy Ministry |
| CFC | = | chlorofluorocarbon |
| cfm | = | cubic feet per minute |
| charette | = | intense effort to solve a design problem within a limited time |

| | | |
|---|---|---|
| CHP | = | combined heating and power |
| CIR | = | credit interpretation ruling |
| CO | = | carbon monoxide |
| $CO_2$ | = | carbon dioxide |
| condenser | = | device to dissipate (get rid of) excess energy in A/C systems |
| COP | = | coefficient of performance |
| CRAC | = | computer room air conditioner |
| CxA | = | commissioning authority |
| daylighting | = | lighting (of a building) using daylight directly or indirectly from the sun |
| D/B | = | design/build |
| dB(A) | = | A-weighting |
| D/B-B | = | design/bid-build |
| DC | = | district cooling |
| DDC | = | direct digital control |
| DE | = | district energy |
| delta-T or $\Delta t$ | = | change or change in temperature |
| DG | = | distributed generation |
| DH | = | district heating |
| DOAS | = | dedicated outdoor air system |
| DOE | = | Department of Energy |
| E | = | ventilation effectiveness |
| EA | = | energy and atmosphere |
| ECB | = | energy cost budget |
| ECM | = | electronically commutated motor |
| EDC | = | environmental design consultant |
| EDG | = | engine-driven generator |
| ENERGY STAR® | = | a government-backed program/rating system that helps consumers achieve superior energy efficiency |
| enthalpy | = | the thermodynamic property of a system resulting from the combination of observable properties (per unit mass) thereof: namely, the sum of internal energy and flow work; flow work is the product of volume and specific mass (i.e., energy transmitted into or out of a system or transmitted across a system boundary |
| entropy | = | a measure of the molecular disorder of a system, such that the more mixed a system is, the greater its entropy, and the more orderly or unmixed a system is, the lower its entropy |
| energy source | = | on-site energy in the form in which it arrives at or occurs on a site (e.g., electricty, gas, oil, or coal) |

| | | |
|---|---|---|
| energy resource | = | raw energy that (1) is extracted from the earth (wellhead or mine mouth), (2) is used in the generation of the energy source delivered to a building site (e.g., coal used to generate electricty), or (3) occurs naturally and is available at a site (e.g., solar, wind, or geothermal energy) |
| EPA | = | Environmental Protection Agency |
| EPBD | = | Directive on the Energy Performance of Buildings |
| EPC | = | energy performance certificate |
| ESTIF | = | European Solar Thermal Industry |
| EU | = | European Union |
| F | = | fahrenheit (temperature scale) |
| f-chart | = | method of calculating solar fraction |
| fenestration | = | window treatment |
| GBC | = | Green Building Challenge |
| gpf | = | gallons per fixture |
| gpm | = | gallons per minute |
| Green | = | *see page 3* |
| greenfield | = | real estate property that is pristine (is not contaminated, either potentially or in fact) |
| GreenTip | = | *see page xviii* |
| Guideline | = | within ASHRAE, a document similiar to a Standard but less strict on consensus |
| HID | = | high intensity discharge |
| HVAC&R | = | heating, ventilating, air-conditioning, and refrigerating |
| hybrid ventilation | = | combination of natural and mechanical outdoor air ventilation |
| hydronic | = | pertaining to liquid flow |
| IAQ | = | indoor air quality |
| IDP | = | integrated design process |
| IEA | = | International Energy Agency |
| IEQ | = | indoor environmental quality |
| IESNA | = | Illuminating Engineering Society of North America |
| insolation | = | entry into a building of solar energy |
| IPMVP | = | International Performance Measurement and Verification Protocol |
| K | = | Kelvin or absolute (temperature scale) |
| kW | = | kilowatt |
| kWh | = | kilowatt-hour |
| kWR | = | refrigeration cooling capacity in kW |
| latent load | = | thermal load due strictly to effects of moisture |
| LEED | = | Leadership in Energy and Environmental Design |
| leeward | = | the downwind side—or side the wind blows away from |

| | | |
|---|---|---|
| low-E | = | low emissivity |
| LPD | = | lighting power density |
| L/f | = | liters per fixture |
| L/s | = | liters per second (airflow and water flow) |
| MAT | = | mixed air temperature |
| media | = | energy forms distributed within a building, usually air, water, or electricity |
| MEP | = | minimum energy performance |
| MNEBC | = | Model National Energy Code for Buildings |
| MRT | = | mean radiant temperature |
| M&V | = | measurement and verification |
| NFPA | = | National Fire Protection Association |
| nonrenewables | = | energy resources that can generally be freely used without net depletion or that have the potential to renew in a reasonable period of time |
| $NO_x$ | = | oxides of nitrogen |
| NPDWR | = | national primary drinking water regulations |
| NRC | = | Natural Resources Canada |
| NREL | = | National Renewable Energy Laboratory |
| OC | = | on centers |
| OPRs | = | Owner's Project Requirements |
| PV | = | photovoltaic |
| parametric analysis | = | in situations where multiple parameters affect an outcome, an analysis that determines the magnitude of one or more parameter's impact alone on that outcome |
| plug loads | = | loads (electrical or thermal) from equipment plugged into electrical outlets |
| POTW | = | publicly owned treatment works |
| precooling | = | cooling done prior to the time major cooling loads are anticipated |
| R (as in R-100) | = | resistivity to thermal heat transfer |
| RAIC | = | Royal Architectural Institute of Canada |
| renewables | = | energy resources that have definite, although sometimes unknown, quantity limitations |
| RFP | = | request-for-proposal |
| RMI | = | Rocky Mountain Institute |
| RAIC | = | Royal Architectural Institute of Canada |
| SCADA | = | supervisory control and data acquisition |
| sensible load | = | thermal load due to temperature but not moisture effects |
| skin | = | building envelope |
| stack effect | = | tendency of relatively warm air in a tall column (such as a building) to rise |

| | | |
|---|---|---|
| Standard | = | within ASHRAE, a document that defines properties, processes, dimensions, materials, relationships, concepts, nomenclature, or test methods for rating purposes |
| sustainability | = | providing for the needs of the present without detracting from the ability to fulfill the needs of the future |
| TC | = | technical committee (an ASHRAE group with a common interest in a particular technical subject) |
| TCLP | = | toxicity characteristic leaching procedure |
| TES | = | thermal energy storage |
| Title 24 | = | slang for California's Building Energy Efficiency Standards (Title 24, Part 6 of the California State Building Code) |
| ton | = | cooling capacity, equal to 12,000 Btu/h |
| T&P | = | temperature and pressure |
| TTF | = | thermal test facility |
| TVOC | = | total volatile organic compound |
| USGBC | = | United States Green Building Council |
| VAV | = | variable-air-volume |
| VE | − | value engineering |
| VFD | = | variable-frequency drive |
| VOC | = | volatile organic compound |
| windward | = | the upwind side, or side the wind blows toward |
| zonation | = | how areas in a building are zoned (for A/C purposes) |

# INDEX